SLEEP – PHYSIOLOGY, FUNCTIONS, DREAMING AND DISORDERS

PARASOMNIA DREAMING

EXPLORING OTHER FORMS OF SLEEP CONSCIOUSNESS

SLEEP – PHYSIOLOGY, FUNCTIONS, DREAMING AND DISORDERS

Additional books and e-books in this series can be found on Nova's website under the Series tab.

SLEEP – PHYSIOLOGY, FUNCTIONS, DREAMING AND DISORDERS

PARASOMNIA DREAMING

EXPLORING OTHER FORMS OF SLEEP CONSCIOUSNESS

JAMES FREDERIC PAGEL
EDITOR

nova
Medicine & Health
New York

Library of Congress Cataloging-in-Publication Data

ISBN: 978-1-53617-815-9
Identifiers: LCCN 2020016319 (print) | LCCN 2020016320 (ebook) | ISBN 9781536178159 (hardcover) | ISBN 9781536178180 (adobe pdf)
Subjects: LCSH: Dreams. | Sleep--Stages. | Altered states of consciousness.
Classification: LCC QP426 .P33 2020 (print) | LCC QP426 (ebook) | DDC 612.8/21--dc23
LC record available at https://lccn.loc.gov/2020016319
LC ebook record available at https://lccn.loc.gov/2020016320

Published by Nova Science Publishers, Inc. † New York

Parasomnia Dreaming is a fascinating and important book about the world of dreams emerging with abnormal sleep related behaviors and experiences - the parasomnias. Top experts contribute their research findings and valued clinical and neuroscience perspectives, along with perspectives from literature and the artistic creative process. The parasomnia dream reports will open many eyes and pique curiosity while setting the stage for this engrossing book.

Carlos H. Schenck, M.D.
Professor, University of Minnesota Medical School
Author, Sleep: The Mysteries, The Problems and The Solutions

The wide continuum of dreaming, visionary, and dissociative phenomena associated with the multiple states of sleep -- so well reflected here in Pagel's volume -- offers an important corrective to any simple reduction of these dream-like states to ordinary fantasy and daydreaming. It is only this wider continuum that helps to explain the importance of such states as sources of creativity and insight throughout traditional cultures."

Harry T. Hunt,
Professor Emeritus in Psychology, Brock University,
Author of The Multiplicity of Dreams.

CONTENTS

PREFACE

J. F. Pagel, MS, MD

Dreams are not all the same. There are routine and comforting dreams. There are dreams of the waking day, dreams of the news, of films, of past experience. There are dreams that reflect psychodynamics; those of family and relationships. And then there are the other dreams - the extreme, the transformative, the frightening, and the bizarre. These other dream events can be clinically described as parasomnias - undesirable physical events or experiences that occur during entry into sleep, within sleep, or during arousals from sleep. Most parasomnias include the component of sleep-associated mentation (dreaming).

DREAMS - SHARED CHARACTERISTICS

Sleep is considered a state of consciousness because of its association with dreaming. Dreaming as a state of consciousness is altered from waking, yet all of the forms of dream consciousness have significant similarities with waking consciousness. Dreams utilize waking associated memory processes, address waking concerns and preoccupations, incorporate the point of view of the individual having the dream. Dreams utilize the same processes of thought, emotion and imagery that are used in waking (Pagel 2014). Yet all of the various forms of dream consciousness differ in major ways from waking consciousness. Dreams are constructed in the perceptual isolation of sleep from stored waking memories, emotions, and thoughts; experiencing them primarily as images independent of perceptual input (States 2007). These visual representations are formed without the continuous, explicit, and accessible perceptual input that we use to form images in waking, so that the thought space of dream imagery is very different from our waking visual perception of sensory space (Kosslyn, Thompson & Ganis 2006). Waking, perceptual based images have a vitality and vivacity lacking in non-perceptual imagery such as in dreaming (Scarry 1996). Dreams rarely reflect the incorporation of waking experience into long term memory stores, a function that sleep states play in the processes of learning and memory (Smith 2010). Dream memory incorporates the intrinsic representations also utilized in waking, while emphasizing and extending the associative relationship between memories. Dreams emotions are often more negative than those experienced in waking (Valli et. al. 2008). Dream thought processing as well as the form of consciousness in each stage of sleep has both similarities and differences to that experienced in waking.

All dreams share characteristics. Dreaming is a manifestation of underlying central nervous system (CNS) activity taking place in a defined state of consciousness in which perceptual input is diminished on a continuum somewhere between wake and sleep. In order to be apparent to others, a dream experience must be reported to another in some format (ex. oral story, interview, written report etc.). Dreams include some form of content (visual, sound, thought, or other sense) that can vary from the simple awareness of having been dreaming (the white dream) to the bizarre/ hallucinatory experiences of psychoanalytic transformation (Pagel et. al 2001).

In studies that control for transference and address the similarities between individuals, the most consistent findings as to dream content are gender consistency and continuity with waking experience (Domhoff 2003). Other classic psychoanalytic content constructs such as dream "bizarreness" are more difficult to demonstrate (Colace 2003). Bizarreness as applied to dreaming varies markedly based on definition. The two major approaches for addressing dream bizarreness are the Hobson scale that rates bizarreness from the reported narrative of the dream (addressing incongruities, uncertainties and discontinuous wording in the storyline of the dream report), and the alternative Hunt scale that extends the assessment of bizarreness beyond the written narrative report to address reported hallucinations, mental clouding and confusion, emotion, and altered personal imagery (Hobson et. al. 1987; Hunt 1989). Studies using Hobson's approach consistently rate REM sleep dreams as the most bizarre, while results obtained using Hunt's scale rate many types of non-REM sleep dreams, particularly those associated with parasomnias, as exceedingly bizarre. The question of "bizarreness" is but one of the profound differences between different types of dreaming.

THE DREAMS OF REM SLEEP

The conscious states of sleep have been intensively studied. Just fifty years ago, there were many more studies published on dreaming than there were addressing sleep. Since that time almost all of this work moved from its focus on dream content to a focus on REM sleep, a stage of sleep that independent of evidence was presumed to equate with dreaming. Today, the neuroanatomy, neurochemistry, and electrophysiology of REM sleep has been defined and characterized almost as well as waking forms of consciousness. The forms of conscious mentation, the dreaming, associated with other stages of sleep has received far less study. These states despite their characteristic phenomenology and separate neurophysiology are sometimes thrown into an overall blender and described as non-REM sleep. Dreams associated with non-REM sleep are often studied only in counter-point comparative studies designed to further address and delineate REM sleep dreaming.

Different Dreams - The Parasomnias

Dreaming takes place throughout sleep, not just in REM sleep. The dreams from different sleep stages differ markedly - in content, in length, in narrative structure, in visual presentation, in recall frequency, in the likelyhood of long-term incorporation into memory, their effects on waking, their intensity and type of emotion, their significance, and in their effects on behavior. These phenomenological differences between particular dreams are based, at least in part, on the sleep state in which they occur. Much of what we currently know of these dreams has been derived from research into the remarkable and puzzling set of sleep associated diagnoses called parasomnias. Many of the most common parasomnias are sleep state specific - associated with a particular state of sleep.

The varied parasomnias associated with REM sleep provide insight into the range of varied phenomenology of this high-recall state. Nightmares, occurring in REM sleep except for individuals with the diagnosis of PTSD, are the most commonly reported parasomnia, reported at least once a month by > 70% of sleep laboratory patients (Pagel 2010). The behaviorally active dreams of REM Behavior Disorder take place in a dream realm independent of REM sleep's typical motor paralysis, a motor block that defines the dreams of sleep paralysis. Parasomnia dreams are also commonly reported from the other high recall sleep stage, sleep onset - stage 1, particularly among individuals diagnosed with narcolepsy. In general population samples approximately 40% of individuals report having experienced hypnogogic hallucinations and sleep paralysis (Ohayon et. al. 1996). The parasomnia associated dreams of deep sleep (stage 4) are our primary information source as to the nature of sleep consciousness in this sleep stage neurophysiologically and cognitively most distant from waking. Dreaming is also reported during parasomnias associated with light sleep (stage 2), as well as with other parasomnias that occur independently of sleep stage association. These other parasomnias noted to at least occasionally occur with dreaming include enuresis, exploding head syndrome, sleep groaning, and sleep related eating disorder (AASM 2014).

This book gathers together remarkable input from major researchers working in the field of dream science. These scientists and humanists have reached beyond the accepted paradigms to extend their work into the different forms of sleep consciousness. Chapter 1 addresses what parasomnia dreams can tell us about the mental processing and cognition occurring throughout sleep in different sleep states. Geoff Carre assists in compiling and comparing the nature of dreaming in sleep-state specific parasomnias in order to develop an evidence-based construct that can be use to determine, as based on typical phenomenology, in what sleep stage a reported dream is likely to have occurred.

Chapters 2-7 address the dreams associated with the best studied of the parasomnias in extended detail. In Chapter 2, Katja Valli uses the sleep-onset hypnogogic hallucinations, sleep paralysis, and cataplexy of narcolepsy to describe the dream-reality instability characterizing that diagnosis. In Chapter 3, Michael Schredl presents new research on the differences and similarities between REM sleep and non-REM sleep (primarily stage 2) dream recall and reports. Chapters 4-6 extend our understanding of classic REM sleep dreaming into the dreams occurring in association with the parasomnias that typify that state. Brigitte Holzinger and Lucille Mayer review the typical dreams associated with sleep paralysis in Chapter 4. In Chapter 5, Robert Hoss takes on the important topic of typical PTSD nightmare content, how that content changes with time, and how that content may reflect the underlying pathophysiology of the response to irreconcilable trauma. Isabelle Arnulf addresses REM sleep behavior disorder (RBD) in Chapter 6, an important parasomnia diagnosis in which dreams are sometimes acted out, a diagnosis sometimes associated with underlying neurologic disorders. In Chapter 7 Antonio Zadra addresses the dreamlike mentation reported from parasomnias of Stage 3 - deep sleep - the strange form of sleep consciousness most distant from waking.

In Chapters 8 & 9, authors use and compare our understandings of these various forms of sleep consciousness in the attempt to further describe CNS functioning and the process of mind. Miloslava Kozmova extends from waking into sleep her studies of types of thought and cognition. Bill

Domhoff proposes ways in which sleep-stage associations with different forms of consciousness can be incorporated into sleep laboratory research.

Chapters 10 & 11 are presentations in which the sleep-stage associations of parasomnia dreaming are concretely applied. Kelly Buckeley purviews major works of English Literature for evidence of parasomnia based dreams and addresses their role in narrative and plot. The final chapter uses the construct of sleep state parasomnia phenomenology developed in Chapter 1 as a tool for characterizing the origin of frightening dreams used by successful artists in their work.

OTHER FORMS OF SLEEP CONSCIOUSNESS

Dreams occur throughout sleep. Yet over the last fifty years almost all research and literature has focused on "normal" dreams - the dreams of REM sleep. The dreaming associated with parasomnias, the extreme sleep events producing sometimes strange and unusual behaviors, has rarely been addressed. This book is among the first to focus on such dreaming, not just in REM sleep where parasomnias describe the extremes of that state, but also in the sleep states outside REM sleep in which dream experiences are often even more strange and extreme. Parasomnia dreams include extreme behaviors, thinking and thought sometimes very different from waking, intense visual hallucinations, extreme emotions, an apparently real dream world, as well as confusion, autonomic discharge, and strange automatic behaviors on awakening. Parasomnia dreams describe forms of consciousness in sleep that are very different from the states of waking consciousness. These alternative states are rarely reflective of underlying medical or psychiatric disease, and are quite different from those states induced by drugs in waking. These other dreams, these parasomnia dreams, are quite special. They provide the individual experiencing them with an insight to the other levels of sleep and other forms of consciousness. These are other forms of consciousness that are only recalled by some on awakening, yet are present in the CNS in all of us who enter sleep.

REFERENCES

[1] American Academy of Sleep Medicine. *The international classification of sleep disorders: diagnostic and coding manual*. 3rd ed. Westchester: American Academy of Sleep Medicine; 2014

[2] Colace, C. (2003). *Dream bizarreness reconsidered. Sleep and Hypnosis*, 5(3), 105–128.

[3] Domhoff, G. William. 2003. *The Scientific Study of Dreams: Neural Networks, Cognitive Development, and Content Analysis*. The Scientific Study of Dreams: Neural Networks, Cognitive Development, and Content Analysis. Washington, DC, US: American Psychological Association. https://doi.org/10.1037/10463-000.

[4] Hobson, J. A., S. A. Hoffman, R. Helfand, and D. Kostner. 1987. "Dream Bizarreness and the Activation-Synthesis Hypothesis." *Human Neurobiology* 6 (3): 157–64.

[5] Hunt, Harry T. 1989. *The Multiplicity of Dreams: Memory, Imagination, and Consciousness*. New Haven, CT, US: Yale University Press.

[6] Kosslyn S. Thompson W. Ganis G. (2006) *The Case for Mental Imagery*. Oxford, UK. Oxford University Press.

[7] Ohayon MM. Priest RG. Caulet M. Cuilleminault C. (1996) Hypnagogic and hypnopompic hallucinations: pathological phenomena? *Br J Psychiatry* 169(4):459-67.

[8] Pagel JF (Chair), Blagrove M. Levin R. States B. Stickgold B. White S. (2001) Defining Dreaming – A paradigm for comparing disciplinary specific definitions of dream, *Dreaming*, 11(4);195-202.

[9] Pagel JF (2010) The Nightmares of Sleep Apnea: Nightmare Frequency Declines with Increasing Apnea Hypopnea Index (AHI), *Journal of Clinical Sleep Medicine*, 6(1);69-74.

[10] Pagel J. F. (2014) *Dream Science: Exploring the Forms of Consciousness*. Academic Press (Elsiever) Oxford.

[11] Scarry E. (1995) On vivacity: the difference between daydreaming and imagining-under-authorial-instruction. *Representations* 52:1-26.

[12] Smith C. (2010) Sleep States, Memory Processing, and Dreams, in Dreaming and Nightmares, *Sleep Medicine Clinics* 5 (2), Saunders/Elsevier. Philadelphia, pp. 217-228.

[13] States BO. (2007) *Seeing in the Dark: Reflection on Dreams and Dreaming*. New Haven, Ct. Yale University Press.

[14] Valli K. Strandholm T. Sillanmaki L. Revonsuo A. (2008) Dreams are more negative than real life: Implications for the function of dreaming. *Cognition and Emotion* 22(5);833-861.

In: Parasomnia Dreaming
Editor: James Frederic Pagel
ISBN: 978-1-53617-815-9

Chapter 1

SLEEP-STATE ASSOCIATED PARASOMNIA DREAM REPORTS

J. F. Pagel**, MD and Geoff Carre, PhD

[1]Department of Family Medicine,
Southern Colorado Residency Program Pueblo,
University of Colorado School of Medicine,
Director Rocky Mt. Sleep disorders Center, Pueblo, Colorado
[2]Department of Psychology, Cape Breton University,
Sydney, Nova Scotia

ABSTRACT

Mentation reported from sleep (dreaming) is associated with many described parasomnias (unwanted or disruptive sleep behaviours producing arousal). Parasomnias are classified as to sleep stage in which they occur. Many of the parasomnias are sleep state specific with the parasomnias associated with any particular sleep stage having shared characteristics. Sleep starts, paralysis, and hypnagogic hallucinations occur in sleep onset (stage 1); sleep talking and panic attacks occur in light sleep

* Corresponding Author's Email: pueo34@earthlink.net.

(stage 2); nightmares, sleep paralysis, and REM behaviour disorder dreaming takes place during REM sleep; and somnambulism (sleep walking), night terrors and confusional arousals occur out of deep sleep (stage 3). The dreaming reported from each defined sleep state reflects the underlying electrophysiology, neuroanatomy, neurochemistry, and neurophysiology that defines each sleep stage. The sleep state specific dream reports have consistent formal characteristics, bizarreness, and types of thought processing that can be used to develop and describe the typical phenomenology of dream mentation reported from each state.

The discovery of that dreaming often took place during Rapid-eye Movement (REM)sleep in the 1960's was grafted into extant Freudian psychodyanmic constructs. The result was a wide-spread belief that dreaming occurred only during REM sleep (REMS). REMS = dreaming became a basic tenet of philosophic, psychological, and general scientific belief systems, and a basic foundational supposition for varied theories of neuroconsciousness. Yet even during that early era of dream science, research clearly demonstrated that many dreams occurred in sleep outside the REMS state. (Foulks 1985)). Mentation reported from sleep (dreaming) was associated with many of the described parasomnias (unwanted or disruptive sleep behaviours producing arousal). This was particularly true for those parasomnias occurring in the sleep stages with the highest levels of dream recall: the hypnogogic hallucinations of Sleep onset (Stage 1); and the nightmares and sleep paralysis of REMS. As based on the field of psychoanalysis, dreams were viewed as analyzable, composed of thought contents formed into a story. The day reflective mental activity reported from non-REM Stage 2 sleep, and the cognition reported from deep sleep (Stage 3), most often extreme emotion, fright, and simply structured narratives, was sometimes considered to not actually be dreaming (Nielsen 2000). Some theorists even decided that deep sleep was dreamless sleep (Tonori & Koch 2015). Today it has become generally accepted that dreaming occurs in all stages of sleep, and perhaps though out sleep as dependent on the applied definition for dreaming. (Pagel 2014).

Over the last fifty years, almost all dream research has focused on REM sleep. A side light of this work has been insight into the comparative physiology of the non-REM sleep states, so that today we have achieved considerable insight into the neuroanatomy, neurochemistry and electrophysiology of these other sleep associated states of consciousness. The cognitive aspects of sleep, dreaming and sleep-state associated mentation, has received less study, particularly for those dream states not associated with REMS. While the rare sleep laboratory study has focused on analysis of dreams from non-REM sleep awakenings, much of what we now understand about dreaming outside REMS comes from the study of sleep-state specific parasomnias. This short chapter addresses our current knowledge of the mentation associated with sleep state associated parasomnias, integrating and collating formal characteristics and associated dream phenomenology with a short overview of the described physiology of each sleep state.

SLEEP ONSET

Sleep is entered through a transitional phase, stage 1, characterized by a low amplitude mixed frequency EEG dominated by alpha waves in more than 1/2 of each epoch sampled. Sleep onset is not an on-off phenomena. A thirty second episode of polysomnographic recording in which less than 1/2 of the epoch is alpha is drowsy awake rather than sleep, so that like drowsiness, sleep onset stage 1 is a transitory state from wake into a profoundly different state of consciousness - sleep. For the 10% of individuals without a coherent alpha rhythm, sleep onset is defined by a loss of amplitude, slow rolling eye movements, and transition on to deeper and electrophysiologically better described states of sleep (Iber et al. 2007). Neuroimaging studies indicate that different sites in the CNS maintain activity during this state than those activated during other sleep stages, including those associated with REMS. These activated sites include the bilateral precuneus, cuneus, parietal lobules, and prefrontal and occipto-temporal cortices – sites associated with waking visual perceptual control and not usually associated with dreaming. (Dresler et al. 2012).

If subjects are woken up after sleep onset, 80-90% will report dreaming. These reports tend to be shorter than those reported after awakening from REMS and include intense visual imagery with limited content. Most of us can describe our own sleep onset dreams. They are commonly reality based, relatively bizarre, and sometimes associated with intense anxiety. Eighteen percent of the examples of lucid dreaming reported by Stephen LeBerge, the developer of the concept of conscious dream control (lucid dreaming), were collected at sleep onset (LaBerge 1985). Stage 1 dreams, also show lower levels of motor imagery than dreams associated with REMS (Speth, Frenzel, and Voss 2013). The poets Edgar Allan Poe, John Keats and Charles Baudelaire, and the painter Salvador Dali induced sleep onset dreams as an artist's tool. For example, they would attempt to fall asleep while sitting in a chair and holding a coin over a dish beside the chair. When they entered sleep, the coin would fall in the dish and hopefully awaken them with a dream in memory. The formal characteristics of sleep onset dreams are summarized in Table 1. Comparative bizarreness as well as characteristic dream state associated thought processes are shown as well (from Wolman & Kozmová, 2007).

Sleep paralysis, during which voluntary muscle movements are inhibited, is a parasomnia commonly associated with REMS but can also occur at sleep onset. Sleep paralysis episodes are accompanied by a wide range of bizarre and often terrifying dreams. It is believed that an episode of sleep paralysis served as an inspiration for Fusilli's famous painting "The Nightmare." Poor sleep quality and insomnia symptoms are associated with increased odds of sleep paralysis occurrence (Denis 2018). The lifetime prevalence is estimated to be 7.6% (Sharpless and Barber 2011), and is much more common in those who suffer from narcolepsy.

Sleep starts, a parasomnia that includes generalized myoclonic jerks occurring at sleep onset often induce re-awakening. The experience of a sleep start is sometimes associated with the feeling of "falling," and for some a full blown experience of a hypnogogic hallucination (Oswald 1959).

Table 1. Sleep onset consciousness [formal characteristics, assessment of bizarreness, and thought]

Formal Characteristics	• primarily visual hallucinations perceived as potentially real • coherent dream stories • high recall • potential lucidity • impression of falling (Sleep Starts) • Intense anxiety (PTSD, Sleep Paralysis) • recurrent (reality based) PTSD
Bizarreness	[Ho Scale] * • Discontinuities – high • Incongruities – low • Uncertainties – low [Hu Scale] * • Hallucinations – High • Clouding/confusion – high • General - high (emotion, bizarre personification)
Rational Thought Processing	• Analytical – low • Perceptual – high • Memory and time awareness– high • Affective - high • Executive - low • Subjective - high • Intuitive/projective – low • Operational - low

* The two major scales addressing dream bizarreness include the Hobson scale that rates bizarreness on reported narrative addressing incongruities, uncertainties and discontinuous wording in the reported storyline. Studies using this approach consistently rate REMS dreams as the most bizarre (Hobson et al. 1987). The alternative Hunt scale addresses the reported hallucinations, mental clouding and confusion, emotion, and altered personal imagery not considered bizarre or addressed in narrative dream reports (Hunt 1989).

The term 'hypnagogic hallucination' was created by the French psychologist Alfred Maury in 1848. In 1951, Ernest Jones description of the incubus, a nightmare in which a demon sucks the sleepers soul is a sleep onset dream, its content likely from a hypnogogic hallucination in that it is seemingly real, intensely visual, horrifying, and associated with the experience of an outside presence (Jones 1951). Hypnogogic hallucinations are most often visual, often intense and complex images, but sometimes random speckles, lines or flashes called phosphenes. Rene Descartes described hearing a sharp explosive noise, then seeing flashes so bright that they illuminated his room in the second of his 1619 dreams that led to his discovery of the scientific method (Browne 1977). The experienced hallucinations can include smells, tactile sensations, sounds, and even out of body experiences. While both hypnogogic hallucinations and sleep paralysis are reportedly experienced on rare occasions by up to 45% of the general population, each of these parasomnias is part of the tetrad of narcolepsy, and commonly experienced by individuals with that diagnosis (American Academy of Sleep Medicine 2005). Narcolepsy associated parasomnias are discussed in detail in Chapter Two.

Stage Two

Stage 2 receives the least attention of the defined sleep stages even though, ironically, it is the stage where we spend most (approximately 70%) of our time in sleep. It is characterized electrophysiologically by the appearance of K-complexes, which can be triggered by external stimuli and are made up of a high-amplitude negative sharp wave followed by a positive slow wave, and sleep spindles. Sleep spindles are episodic oscillations of about 12-15 Hz and have been implicated in multiple brain functions, including sleep quality, sensory gating, and memory (Astori, Wimmer, and Lüthi 2013). While dream recall frequency is relatively low during this stage, (dreams reported from approximately 40% of awakenings) the predominance of this stage over the course of sleep means a significant portion of our total dream count occurs during stage 2. It has been found that dreaming is more likely to occur in the presence of fast spindles in a central and posterior cortical region (Siclari et al. 2018).

Stage 2 dreams have rarely been the specific topic of study, but there are many studies of non-REM sleep dreams that methodologically exclude sleep onset - stage 1 by requiring a period of defined sleep before awakening the individual to collect dream content. Since the other non-REM sleep state, Stage 3 - deep sleep, is in general less than 10% of total sleep (even less in individuals over fifty years of age), the dreams collected from non-REM sleep in these studies are overwhelmingly from stage 2. Non-REM sleep dream recall (40%) is lower than in REMS (I. Strauch and Meyer 1996). The associated dreams are described as waking-associated, momentary, and brief with thought-like content and disconnected visual imagery in which the dreamer is emotionally uninvolved (Nielsen, T 1999). Some authors have proposed that this "degraded" non-REM sleep associated mental activity should not be described as dreaming, but rather as sleep-associated cognitive activity (Nielsen 2000).

Of the parasomnias, sleep talking, and sleep panic attacks are associated with this stage. More than half of all people have had the experience of speaking out loud while being asleep, occurring more frequently in childhood and attenuating in the teen years, and sometimes persisting into adulthood (Alfonsi et al. 2019). Since sleep talking reflects the ongoing dream content, it may represent the unique possibility to access the dreamlike mental experience directly. The repeating of topical information, such as the calling out of the name of an upsetting conjugal associate, can lead to a clinical presentation. Panic attacks during this stage may suggest an anxiety disorder. They begin towards the end of stage 2 and result in awaking to a panic attack. They can present with extreme emotion - sometimes described as brief moments of terror - that may are may not be associated specific or developed dream/nightmare content (Craske and Rowe 1997). The formal characteristics, bizarreness and thought processing of Stage 2 dreams and parasomnias are summarized in Table 2.

Table 2. Stage 2 consciousness [formal characteristics, assessment of bizarreness, and thought]

Formal Characteristics	• day reflective content • associated anxiety and panic • variable recall across night
Bizarreness	[Ho Scale] • Discontinuities – Low • Incongruities – Low • Uncertainties - Low [Hu Scale] • Hallucinations – low • Clouding/confusion – low • General - low
Rational Thought Processing	• Analytical – Low • Perceptual – Mod (perceptual integration associated with K-complexes) • Memory and time awareness– Low • Affective – Low • Executive - Low • Subjective - Low • Intuitive/projective – Low • Operational – Mod (potential for arousal)

Stage 3: Deep Sleep

Deep sleep (Stage 3) is EEG defined by prominent large slow waves in the delta range (1 Hz), a rhythm rarely seen in other stages of sleep and almost never observed during waking. Beyond middle age, deep sleep is rarely seen given an age-related decrease in slow wave activity (Pótári et al. 2017). Slow waves occur when neurons are said to be bistable, having a depolarized upstate, which then transitions to a hyperpolarized down-state or off-period characterized by neuronal silence. The occurrence of slow waves in stage 3 sleep can occur locally and asynchronously across brain regions. Deep sleep is sometimes described as dreamless sleep (Siclari and Tononi 2017). It is suggested that the predominance of slow-waves in deep sleep may reflect coordinated bistability across brain regions thereby preventing adequate integration important for consciousness (Kung et al. 2019; Lee et al. 2019).

Awakenings from deep sleep yield accounts of dreaming 50–70% of the time (Nielsen 2000). The reduced dream recall of this state, compared to REMS and sleep onset, may be due to its distance and electrophysiological difference from the consciousness of waking (Koukkou and Lehmann 1983). These dreams are typically fragmentary with a limited story-line (Mutz and Javadi 2017). Interactions with others during the dream may be friendlier than those interactions found in REM dreams (McNamara et al. 2010). The emotions experienced can sometimes be extreme and disturbing.

This stage is also differentiated by a series of remarkable parasomnias categorized as disorders of arousal (American Academy of Sleep Medicine 2014). These include sleep terrors, somnambulism (sleepwalking), and confusional arousals. These parasomnias are reported most often during the ages of 5-25 years among individuals who have a family history of similar parasomnias, although they may persist in some into later adult years (Baldini et al. 2019). These parasomnias rarely occur spontaneously in the sleep laboratory making them difficult to study. In one study, when the patients recalled the incident, violent behaviors were reported more than half the time, and the majority testified they felt confused and disorientated (Baldini et al. 2019). They are more likely to occur after sedative-hypnotic or alcohol use, after sleep deprivation, and during the first deep sleep episode 2-4 hours after sleep initiation.

Confusional arousals, have also been called sleep drunkenness. These are partial arousals from deep sleep. Typically, these events are brief, but can last several hours (American Academy of Sleep Medicine 2014). The individual is confused and disorientated, and events typically include slow speech or mumbling, diminished mentation, and inappropriate behaviors that can be vigorous, highly resistive and sometimes violent. They also show diminished response to external stimuli during the episode.

Table 3. Deep sleep consciousness [formal characteristics, assessment of bizarreness, and thought]

Specific Parasomnias	• SOMAMBULISM (sleepwalking) – autonomic & inappropriate behaviors, frantic attempts to escape a perceived threat, fragmentary recall • CONFUSIONAL AROUSALS – mental confusion & disorientation on arousal from sleep, inappropriate and violent behaviors • NIGHT TERRORS – incoherent vocalizations, extreme fear, intense autonomic discharge, confusion & disorientation, fragmentary recall
Formal Characteristics	• mental confusion & disorientation on arousal • fragmentary recall • autonomic & inappropriate behaviors • frantic attempts to escape a perceived threat
Bizarreness	[Ho scale] • Discontinuities – low • Incongruities – low • Uncertainties – high (explicit vagueness) [Hu Scale] • Hallucinations – low • Clouding/confusion – high (low recall) • General - high (emotion, bizarre personification)
Rational Thought Processing	• Analytical – low • Perceptual – low • Memory and time awareness– low • Affective - high • Executive - low • Subjective - low • Intuitive/projective – low • Operational - low

Sleepwalking (somnambulism) is distinguished by prominent ambulatory behavior. Typically the individual will sit up in bed in a confused manner before displaying a range of behaviors from aimless wandering to remarkably complex behaviors. They behave as though they are wide-awake but with limited awareness. Some individuals have opened doors, prepared food, driven an automobile (with serious consequences), or used the washroom. Usually the individual's eyes are open. Somnambulism associated dreaming is discussed in detail in Chapter Seven.

Sleep terrors often commence with sitting up, but then the individual will emit a loud scream, often demonstrating fear-typical autonomic nervous system activity. Subjects may demonstrate behaviors that suggest attempts to avoid harm.

Dreams associated with the deep sleep parasomnias include minimal conscious control (no lucidity has been reported from this state). Dream reports are short, sometimes intense, and may include only the awareness that dreaming was present before awakening (so-called white dreams). Dream reports are associated with a wide variety of behaviours including automatic 'zombie-like' activity, disorientation, screaming, fright and terror.

Rapid Eye Movement (REM) Sleep

This stage is characterized by a small amplitude, higher frequency EEG pattern that is reminiscent of an alert waking state, conjugate eye movements, and suppression of skeletal muscle activity (EMG). Sporadic muscle twitches, autonomic nervous system activation, genital arousal, and oscillations in body temperature also distinguish this stage from other sleep stages. All mammals, reptiles, and many birds have the electrophysiological state of REMS. Most humans, even those with extensive neurological damage, have episodes of REM sleep that continue to occur into extreme old age. Because of researchers' fascination with REMS and its presumed correlate of dreaming, the neuroanatomy and neurochemistry of this stage has garnered more research than that of any of the other states of sleeping consciousness.

Dream reports collected after REMS awakenings are characterized by vivid imagery, a bizarre or implausible storyline, increased and often negative emotionality, and behaviors with high motivational significance (Schredl 2010). REMS dreaming thoughts are likely to have greater social focus than non-REM and waking thoughts (Perogamvros et al. 2017). In studies that strictly control for transference effects and bias, it has been difficult to demonstrate that dream content reported from REMS varies in any significant way, beyond greater report length, from non-REM sleep reports (Domhoff 2003). The REMS parasomnias include unwanted behaviors or actions during REMS, often occurring during the early hours of the morning just before awakening.

Nightmares are the most commonly reported parasomnia. Except in individuals with posttraumatic stress disorder (PTSD), nightmares almost always occur during REMS. The classic nightmare is accompanied by utter and incomprehensible dread, beginning as a seemingly real and coherent dream sequence that becomes increasingly more disturbing as it unfolds. Nightmares often include the negative emotions of anxiety and fear, as well as, anger, rage, embarrassment, and disgust. Content may focus on imminent physical danger (e.g., threat of attack, falling, injury, death), but may also include aggression toward others, potential personal failures and other distressing themes such as suffocation (Pagel and Nielsen 2005). Most nightmares induce awakening, sometimes as an escape from the storyline of the dream. Some "bad dreams," however, can occur without awakening, and are recalled the next morning (Zadra and Donderi 2000).

In the sleep laboratory population, >70% report having more than one nightmare per month (Pagel and Shocknesse 2007). The lifetime prevalence of frequent nightmares (at least 3/week) is estimated to be from 1% to 8% of the population (American Psychiatric Association 2013). Nightmares have been linked to poor sleep quality and shown to be comorbid with several sleep disorders, including insomnia (Ohayon, Morselli, and Guilleminault 1997). Bizarre negative dreams can occur during all stages of sleep, but rarely do they have the length and detail of classic nightmares, which are more likely to be perceived as terrifyingly real. Higher levels of worry and longer sleep duration have been found to be significantly associated with whether or not participants experienced nightmares, even after controlling for negative affect, PTSD, and stressful life events (Rek,

Sheaves, and Freeman 2017). Emotional arousal may be a necessary condition. When comparing the amplitude of heartbeat-evoked potentials (HEP), as an index of emotional arousal, between nightmare patients and healthy controls during awake, non-REM sleep, and REMS periods, researchers found higher HEP amplitude in nightmare patients compared to healthy controls over a cluster of frontal regions only during REMS (Perogamvros et al. 2019). The high occurrence of nightmares in the "normal" population underlies the possibility that they serve a function, perhaps through emotional processing (Levin and Nielsen 2009).

Individuals with Nightmare Disorder (ND) report frequent nightmares, but do not have a history of significant emotional and/or physical trauma such as that reported by individuals whose nightmares are a symptom of Posttraumatic Stress Disorder (PTSD). The recurring frightening dreams of ND can induce insomnia, waking anxiety and dysphoria. The personality trait of "thin borders," associated with creative personalities is more common among individuals with ND (Hartmann E. 1984).

Nightmares can arise for a number of reasons—stress, anxiety, irregular sleep, and medications; however, the most common and most studied exogenous cause is significant unreconciled trauma. PTSD nightmares can occur outside REM sleep, especially during sleep onset when they can interfere with the initiation of sleep, and contribute to hyperarousal, and intense anxiety (Ross et al. 1989). Those suffering from PTSD can have nightmares that involve reliving the traumatic event and include strong emotions such as fear and rage. PTSD nightmares and their typical content are addressed in detail in Chapter Five.

REM sleep behavior disorder (RBD) is a diagnosis most common in older males and in some neurologic illnesses (i.e., Parkinson's). Those affected report physically acting out their dreams. This may occur because the mechanisms that disable motor activity during sleep are not functioning normally. Individuals with RBD may also report frequent nightmares and disorders of arousal. RBD and its associated dream content is addressed in detail in Chapter Six.

Another typically REMS associated parasomnia is sleep paralysis, where your mind is conscious, but you are unable to move your body. This parasomnia often includes negative and frightening content, developed in great detail, and associated with distress that extends into awakening. Sleep paralysis is discussed in detail in Chapter Four.

Table 4. Rem stage consciousness [formal characteristics, assessment of bizarreness, and thought]

Formal Characteristics	• coherent dream sequences • detailed plot, character or actions • high recall • Increased length of report • disturbing, intense emotions (nightmares & sleep paralysis) • Potential lucidity • recurrent (reality based) nightmares • Comparative reality based memory
Bizarreness	[Ho Scale] • Discontinuities – High, • Incongruities – High • Uncertainties - High [Hu Scale] • Hallucinations – low • Clouding/confusion – low • General - high – (uncanny emotion)
Rational Thought Processing	• Analytical – High • Perceptual – High • Memory and time awareness– High • Affective - High • Executive - High • Subjective - High • Intuitive/projective – High • Operational - Mod

CONCLUSION

The electrophysiology used to define each state, the neurochemistry, and the described neuroanatomy differs markedly between the sleep states. The sleep stage-associated dreams linked to the sleep behaviours that we call parasomnias are waking descriptions of the cognitive activity occurring in these different forms of sleep-associated consciousness. The typical dreams of each state differ markedly, with the dreaming reported from each sleep state having a characteristic phenomenology that can be correlated with what is known of that sleep state physiology. Because of this wide variation in phenomenology, studies that ascribe general characteristics to all dreaming are likely to be conflating and definitionally diffuse. Later chapters in this book address the better studied parasomnias in greater detail: nightmares, sleep paralysis, the dream associated parasomnias of narcolepsy, somambulisn, and the dreams of RBD. An understanding of sleep state associated dream phenomenology can be used (as in Chapter Ten & Eleven) in the attempt to determine whether a reported dream was most likely to have originated in REM sleep, sleep onset (Stage 1), Stage 2 or in deep sleep (Stage 3). Further study of parasomnia associated cognition offers the possibility for extending our understanding of the different forms of sleep consciousness that we call dreams.

REFERENCES

Alfonsi, Valentina, Aurora D'Atri, Serena Scarpelli, Anastasia Mangiaruga, and Luigi De Gennaro. 2019. "Sleep Talking: A Viable Access to Mental Processes during Sleep." *Sleep Medicine Reviews* 44: 12–22. https://doi.org/10.1016/j.smrv.2018.12.001.

American Academy of Sleep Medicine. 2005. *International Classification of Sleep Disorders*. 2nd ed. ICD-10.

———. 2014. *International Classification of Sleep Disorders*.

American Psychiatric Association. 2013. *Diagnostic and Statistical Manual of Mental Disorders.* Fifth Edition. American Psychiatric Association. https://doi.org/10.1176/appi.books.9780890425596.

Astori, Simone, Ralf D. Wimmer, and Anita Lüthi. 2013. "Manipulating Sleep Spindles--Expanding Views on Sleep, Memory, and Disease." *Trends in Neurosciences* 36 (12): 738–48. https://doi.org/10.1016/j.tins.2013.10.001.

Baldini, Tommaso, Giuseppe Loddo, Elisa Sessagesimi, Francesco Mignani, Fabio Cirignotta, Susanna Mondini, Laura Licchetta, Francesca Bisulli, Paolo Tinuper, and Federica Provini. 2019. "Clinical Features and Pathophysiology of Disorders of Arousal in Adults: A Window into the Sleeping Brain." *Frontiers in Neurology* 10 (May). https://doi.org/10.3389/fneur.2019.00526.

Browne, Alice. 1977. "Descartes's Dreams." *Journal of the Warburg and Courtauld Institutes* 40: 256–73. https://doi.org/10.2307/750999.

Craske, Michelle G., and Melissa K. Rowe. 1997. "Nocturnal Panic." *Clinical Psychology: Science and Practice* 4 (2): 153–74. https://doi.org/10.1111/j.1468-2850.1997.tb00107.x.

Denis, Dan. 2018. "Relationships between Sleep Paralysis and Sleep Quality: Current Insights." *Nature and Science of Sleep* 10 (November): 355–67. https://doi.org/10.2147/NSS.S158600.

Domhoff, G. William. 2003.. *The Scientific Study of Dreams: Neural Networks, Cognitive Development, and Content Analysis.* Washington, DC, US: American Psychological Association. https://doi.org/10.1037/10463-000.

Dresler, Martin, Renate Wehrle, Victor I. Spoormaker, Stefan P. Koch, Florian Holsboer, Axel Steiger, Hellmuth Obrig, Philipp G. Sämann, and Michael Czisch. 2012. "Neural Correlates of Dream Lucidity Obtained from Contrasting Lucid versus Non-Lucid REM Sleep: A Combined EEG/FMRI Case Study." *Sleep* 35 (7): 1017–20. https://doi.org/10.5665/sleep.1974.

Foulks D. (1985) *Dreaming: A Cognitive-Psychological Analysis*. Hillsdale, NJ. Lawrence-Erlbaum Associates.

Hartmann E. 1984. *The Nightmare*. New York: Basic Books.

Hobson, J. A., S. A. Hoffman, R. Helfand, and D. Kostner. 1987. "Dream Bizarreness and the Activation-Synthesis Hypothesis." *Human Neurobiology* 6 (3): 157–64.

Hunt, Harry T. 1989. *The Multiplicity of Dreams: Memory, Imagination, and Consciousness*. New Haven, CT, US: Yale University Press.

Iber, C., S. Ancoli-Israel, A. L. Chesson, and S. F. Quan. 2007. "*The AASM Manual for the Scoring of Sleep and Associated Events American Academy of Sleep Medicine*." Westchester: American Academy of Sleep Medicine.

Jones, E. 1951. *On the Nightmare*. New York, NY: Liveright.

Koukkou, M., and D. Lehmann. 1983. "Dreaming: The Functional State-Shift Hypothesis." *The British Journal of Psychiatry* 142 (3): 221–31. https://doi.org/10.1192/bjp.142.3.221.

Kung, Yi-Chia, Chia-Wei Li, Shuo Chen, Sharon Chia-Ju Chen, Chun-Yi Z. Lo, Timothy J. Lane, Bharat Biswal, Changwei W. Wu, and Ching-Po Lin. 2019. "Instability of Brain Connectivity during Nonrapid Eye Movement Sleep Reflects Altered Properties of Information Integration." *Human Brain Mapping* 40 (11): 3192–3202. https://doi.org/10.1002/hbm.24590.

LaBerge, Stephen. 1985. *Lucid Dreaming*. 1st ed.. Los Angeles: JPTarcher.

Lee, Minji, Benjamin Baird, Olivia Gosseries, Jaakko O. Nieminen, Melanie Boly, Bradley R. Postle, Giulio Tononi, and Seong-Whan Lee. 2019. "Connectivity Differences between Consciousness and Unconsciousness in Non-Rapid Eye Movement Sleep: A TMS–EEG Study." *Scientific Reports* 9 (1): 1–9. https://doi.org/10.1038/s41598-019-41274-2.

Levin, Ross, and Tore Nielsen. 2009. "Nightmares, Bad Dreams, and Emotion Dysregulation: A Review and New Neurocognitive Model of Dreaming." *Current Directions in Psychological Science* 18 (2): 84–88. https://doi.org/10.1111/j.1467-8721.2009.01614.x.

McNamara, Patrick, Patricia Johnson, Deirdre McLaren, Erica Harris, Catherine Beauharnais, and Sanford Auerbach. 2010. "REM and NREM Sleep Mentation." *International Review of Neurobiology* 92: 69–86. https://doi.org/10.1016/S0074-7742(10)92004-7.

Mutz, Julian, and Amir-Homayoun Javadi. 2017. "Exploring the Neural Correlates of Dream Phenomenology and Altered States of Consciousness during Sleep." *Neuroscience of Consciousness* 2017 (1). https://doi.org/10.1093/nc/nix009.

Nielsen, T. 1999. "Mentation During Sleep. The NREM/REM Distinction." In *Handbook of Behavioral State Control: Cellular and Molecular Mechanisms*, edited by Ralph Lydic and Helen A. Baghdoyan, 101–28. CRC Press.

Nielsen, Tore A. 2000. "A Review of Mentation in REM and NREM Sleep: 'Covert' REM Sleep as a Possible Reconciliation of Two Opposing Models." *Behavioral and Brain Sciences* 23 (6): 851–66. https://doi.org/10.1017/S0140525X0000399X.

Ohayon, M. M., P. L. Morselli, and C. Guilleminault. 1997. "Prevalence of Nightmares and Their Relationship to Psychopathology and Daytime Functioning in Insomnia Subjects." *Sleep* 20 (5): 340–48. https://doi.org/10.1093/sleep/20.5.340.

Oswald, I. 1959. "Sudden Bodily Jerks on Falling Asleep." Brain: A *Journal of Neurology* 82 (1): 92–103. https://doi.org/10.1093/brain/ 82.1.92.

Pagel, J. F., and T. Nielsen. 2005. "Parasomnias: Recurrent Nightmares." In *The International Classification of Sleep Disorders – Diagnostic and Coding Manual* (ICD-11). American Academy of Sleep Medicine.

Pagel, J. F., and Shannon Shocknesse. 2007. "Dreaming and Insomnia: Polysomnographic Correlates of Reported Dream Recall Frequency." *Dreaming* 17 (3): 140–51. https://doi.org/10.1037/1053-0797.17.3.140.

Pagel J. F. (2014) *Dream Science: Exploring the Forms of Consciousness.* Academic Press (Elsiever) Oxford.

Perogamvros, Lampros, Benjamin Baird, Mitja Seibold, Brady Riedner, Melanie Boly, and Giulio Tononi. 2017. "The Phenomenal Contents and Neural Correlates of Spontaneous Thoughts across Wakefulness, NREM Sleep, and REM Sleep." *Journal of Cognitive Neuroscience* 29 (10): 1766–77. https://doi.org/10.1162/jocn_a_01155.

Perogamvros, Lampros, Hyeong-Dong Park, Laurence Bayer, Aurore A. Perrault, Olaf Blanke, and Sophie Schwartz. 2019. "Increased Heartbeat-Evoked Potential during REM Sleep in Nightmare Disorder."

NeuroImage: Clinical. January 29, 2019. https://doi.org/10.1016/j.nicl.2019.101701.

Pótári, Adrián, Péter P. Ujma, Boris N. Konrad, Lisa Genzel, Péter Simor, János Körmendi, Ferenc Gombos, Axel Steiger, Martin Dresler, and Róbert Bódizs. 2017. "Age-Related Changes in Sleep EEG Are Attenuated in Highly Intelligent Individuals." *NeuroImage* 146: 554–60. https://doi.org/10.1016/j.neuroimage.2016.09.039.

Rek, Stephanie, Bryony Sheaves, and Daniel Freeman. 2017. "Nightmares in the General Population: Identifying Potential Causal Factors." *Social Psychiatry and Psychiatric Epidemiology* 52 (9): 1123–33. https://doi.org/10.1007/s00127-017-1408-7.

Ross, R. J., W. A. Ball, K. A. Sullivan, and S. N. Caroff. 1989. "Sleep Disturbance as the Hallmark of Posttraumatic Stress Disorder." *The American Journal of Psychiatry* 146 (6): 697–707. https://doi.org/10.1176/ajp.146.6.697.

Schredl, Michael. 2010. "Characteristics and Contents of Dreams." *International Review of Neurobiology* 92: 135–54. https://doi.org/10.1016/S0074-7742(10)92007-2.

Sharpless, Brian A., and Jacques P. Barber. 2011. "Lifetime Prevalence Rates of Sleep Paralysis: A Systematic Review." *Sleep Medicine Reviews* 15 (5): 311–15. https://doi.org/10.1016/j.smrv.2011.01.007.

Siclari, Francesca, Giulio Bernardi, Jacinthe Cataldi, and Giulio Tononi. 2018. "Dreaming in NREM Sleep: A High-Density EEG Study of Slow Waves and Spindles." *The Journal of Neuroscience* 38 (43): 9175–85. https://doi.org/10.1523/JNEUROSCI.0855-18.2018.

Siclari, Francesca, and Giulio Tononi. 2017. "Local Aspects of Sleep and Wakefulness." Current Opinion in Neurobiology, *Neurobiology of Sleep*, 44 (June): 222–27. https://doi.org/10.1016/j.conb.2017.05.008.

Speth, Jana, Clemens Frenzel, and Ursula Voss. 2013. "A Differentiating Empirical Linguistic Analysis of Dreamer Activity in Reports of EEG-Controlled REM-Dreams and Hypnagogic Hallucinations." *Consciousness and Cognition* 22 (3): 1013–21. https://doi.org/10.1016/j.concog.2013.07.003.

Strauch, I., and B. Meyer. 1996. “Dreams in Different Sleep Stages.” In *In Search of Dreams: Results of Experimental Dream Research*, edited by Inge Strauch and Barbara Meier, 135–36. SUNY Press.

Tononi G. Koch C. (2015) Consciousness: here, there, everywhere? *Philosophical Transactions of the Royal Society.* https://doi.org/10.1098/rstb.2014.0167.

Wolman, Richard N., and Miloslava Kozmová. 2007. “Last Night I Had the Strangest Dream: Varieties of Rational Thought Processes in Dream Reports.” *Consciousness and Cognition* 16 (4): 838–49. https://doi.org/10.1016/j.concog.2006.09.009.

Zadra, Antonio, and D. Donderi. 2000. “Nightmares and Bad Dreams: Their Prevalence and Relationship to Well-Being.” *Journal of Abnormal Psychology* 109 (2): 273–81.

In: Parasomnia Dreaming
Editor: James Frederic Pagel
ISBN: 978-1-53617-815-9

Chapter 2

Dreaming and Parasomnias in Narcolepsy

Katja Valli[1,2,3,*], PhD, Juha Markkula[2,4], MD, PhD and Salla Lamusuo[2,5], MD, PhD

[1]Department of Psychology and Speech-Language Pathology, University of Turku, Turku, Finland
[2]Sleep and Breathing Center, Turku University Hospital, Turku, Finland
[3]Department of Cognitive Neuroscience and Philosophy, University of Skövde, Skövde, Sweden
[4]Department of Psychiatry, University of Turku, Turku, Finland
[5]Department of Neurology, University of Turku, Turku, Finland

Abstract

Narcolepsy affects less than 0.05% of the population, and is most commonly diagnosed during adolescence and early adulthood. Diagnostically, narcolepsy is divided into two subtypes. Narcolepsy type 1 (NT1) is characterized by the presence of orexin deficiency and cataplexy, and the pathophysiology in NT1 involves the partial or complete cessation

* Corresponding Author's Email: katval@utu.fi.

of orexin production by neurons in the lateral hypothalamus. Narcolepsy type 2 (NT2) presents without cataplexy and orexin levels are normal or only slightly reduced. The pathophysiological mechanism is thus less clear in NT2. The main symptoms of narcolepsy are excessive daytime sleepiness (EDS) which culminates in involuntary lapses into sleep during daytime, and disrupted nocturnal sleep. These symptoms are often more severe in NT1 than NT2 patients. In addition to sleep-wake regulation problems, unpleasant phenomena during nocturnal sleep and daytime naps and sleep attacks are common in narcolepsy patients. Dreams of narcolepsy patients are typically rated to be more vivid, bizarre and emotional, and they report recalling dreams and nightmares more often than healthy controls. Other unpleasant experiences, such as Non-Rapid Eye Movement (NREM) and Rapid Eye Movement (REM) sleep parasomnia symptoms, including sleep walking, sleep terrors and REM Sleep Behaviour Disorder (RBD), are also more prevalent in narcolepsy patients than in general population. Undesirable phenomena during sleep are often accompanied by equally distressing sleep-wake transition experiences. Sleep paralysis episodes are more common in narcolepsy patients than in healthy individuals and frequently accompanied by frightening hypnagogic or hypnopompic hallucinations, occurring at the onset or offset of sleep, respectively. The waking lives of narcolepsy patients are affected in many ways by their sleep-related experiences. Many patients with narcolepsy experience confusion of memory sources, and may hold delusional beliefs, sometimes for extended periods of time, that a memory created by a dream is a memory of a real experience. Nightmares, parasomnias, sleep paralysis and hypnagocic and hypnopompic hallucinations, as well as the inability to distinguish dream memories from waking memories, can be highly distressing for the patients. Narcolepsy patients also have a higher prevalence of psychiatric symptoms, especially depression and anxiety. The illness usually has significant effects on the patients' daily lives, functional abilities and life courses. Therefore, awareness of narcolepsy and its associated symptoms, and efficient treatment of not only daytime sleepiness but sleep-related distressing symptoms, might increase the psychological well-being and quality of life of the patients.

INTRODUCTION

The origin of the term narcolepsy is Greek, 'narke' signifying numbness or stupor and 'lepsis' meaning attack or to seize. The earliest account of narcolepsy was likely reported in 1600s by Willis who described patients "with a sleepy disposition who suddenly fall fast asleep" (Lennox 1939).

The first case reports of narcolepsy were described separately by the German physicians Westphal and Fischer in late 1870s, and a couple of years later by the French physician Gélineau who also introduced the term 'narcolepsy' in 1880 (Todman 2007).

The first diagnostic criteria for the narcoleptic syndrome, including excessive daytime sleepiness, cataplexy, sleep paralysis and hypnagogic hallucinations, were outlined in 1957 by Yoss and Daly (1957). Just few years later, Vogel recorded for a first time a sleep onset Rapid Eye Movement (REM) sleep episode in a patient with narcolepsy when investigating dreaming in this patient (Vogel 1960). Later sleep laboratory investigations by Kleitman and Dement verified the findings of dissociated REM sleep in narcolepsy (Kleitman 1963).

Another great leap forward in understanding narcolepsy took place in 1990s when the relationship between narcolepsy and immune system was discovered. Matsuki and colleagues (1992) identified a specific gene variant, HLA DQB1*0602 on chromosome 6, in majority of narcolepsy patients presenting with cataplexy. This led to a suggestion that narcolepsy may be an autoimmune disorder. Combined with the discovery of neuropeptide orexin/hypocretin by two independent research groups in late 1990s (Sakurai and colleagues (1998) who named the peptide orexin, and Lecea and colleagues (1998) who named the peptide hypocretin), this has resulted in the proposition of a disease mechanism that orexin producing cells in the hypothalamus may be affected by an autoimmune process in HLA-associated narcolepsy.

In the past few decades, substantial advances have been made in understanding the main clinical manifestations, aetiology, and pathophysiology of narcolepsy. Literature on the adjacent phenomena related to narcolepsy, such as intensified dreaming, nightmares, lucid dreaming, parasomnia symptoms, sleep paralysis, hypnagogic and hypnopompic hallucinations and confusion of memory sources, have also been accumulating, but the significance of these symptoms to patients' overall psychological well-being and functional abilities remains unclear. This chapter focuses on describing what is currently known about the altered sleep-related experiences that closely associate with the core symptoms of narcolepsy.

EPIDEMIOLOGY, SYMPTOMS, DIAGNOSTIC CRITERIA, PATHOPHYSIOLOGY, AND COMORBIDITY

Narcolepsy is a chronic neurological disorder which usually manifests in adolescence, around the age of 15, with a second peak around the age of 35 years (Antelmi et al. 2019). Narcolepsy affects approximately 0.026-0.050% of the population (Hublin et al. 1994, Ohayon et al. 2002, Silber et al. 2002). The exact prevalence remains somewhat unclear, due to the fact that some people present borderline symptoms and may not be correctly diagnosed or diagnosis may be delayed. There are also indications that the incidence may vary between populations, being highest in Japanese (Juji et al. 1988) and lowest in Jewish (Wilner et al. 1988) and Arabic (al Rajeh et al. 1993) populations. However, these estimates are based on studies with uncertain methodology, and should be treated with caution. If the incidence would vary between populations, the differences would likely be explained by both genetic and environmental factors (Bassetti et al. 2019).

The first and foremost symptom of narcolepsy is excessive daytime sleepiness (EDS) which typically culminates in involuntary sleep attacks and daytime napping. Narcolepsy diagnosis requires complaints of EDS for more than three months (AASM, 2014). The patients exhibit an inability to remain in sleep or wake states for the normal length of time, and have short sleep latencies, often enter Rapid Eye Movement (REM) sleep after sleep onset, have disrupted nighttime sleep and sleep-wake rhythm abnormalities. In Multiple Sleep Latency Test (MSLT), the presence of short sleep latency (mean < 8 minutes) and two or more sleep-onset REM periods (SOREMP) are typical findings. Drugs that stimulate the central nervous system, such as modafinil or methylphenidate, are the primary treatment to combat excessive daytime sleepiness. Scheduled brief naps during the day and regular bedtime routine are recommended. The daytime naps are usually brief (20-30 min) and refreshing.

Sleep and wake states also dissociate in narcolepsy patients in the sense that sleep-related phenomena can inappropriately occur during wakefulness (such as REM sleep atonia extending to wakefulness in the form of cataplexy), or during transitions (such as REM sleep atonia extending to wakefulness during sleep paralysis, or dream-like hypnagogic or hypnopompic hallucinations while falling asleep or waking up). Automatic behaviors, i.e., purposeful or semi-purposeful behaviors with amnesia of the activity afterwards, are also common in narcolepsy patients when sleep inertia persists in the waking state. Selective serotonin reuptake inhibitors (SSRIs) or serotonin and norepinephrine reuptake inhibitors (SNRIs), which suppress REM sleep, may be prescribed to alleviate the symptoms of cataplexy, hypnagogic hallucinations and sleep paralysis. Sodium oxybate is effective for cataplexy and also helps to improve the often poor nighttime sleep. In high doses it may also reduce daytime sleepiness.

In the International Classification of Sleep Disorders (AASM 2014), two subtypes of narcolepsy are distinguished. Narcolepsy type 1 (NT1) is characterized by the presence of orexin deficiency (< 110 pg/mL in cerebrospinal fluid) and cataplexy, i.e., brief episodes of loss of muscle tone or muscle weakness while consciousness is preserved. Cataplexy is typically triggered by emotional reactions, such as laughter, and is often partial. Cataplexy symptoms may appear years after excessive daytime sleepiness manifests and even after narcolepsy has been diagnosed. Narcolepsy type 2 (NT2) presents without cataplexy and in NT2 patients orexin levels are normal or only slightly reduced. The symptoms of NT2 patients are usually less severe (Reading 2019). Although NT2 has been reported to be less common than NT1, recent studies suggest that NT2 may be more frequent in the general population than previously estimated (Cohen et al. 2018).

The underlying pathophysiology of NT1 involves the partial or complete loss or dysfunction of neurons in the lateral hypothalamus that produce the neuropeptide orexin. Orexin system functions as a regulator and stabilizer of vigilance states by modulating the arousal threshold which in turn allows the organism to maintain appropriate and adequate wakefulness (Sutcliffe and de Lecea 2002). When orexin neurons are activated, they contribute to maintain an aroused state, and their activity is inhibited during NREM and REM sleep. Autopsy findings of patients with NT1 have revealed a selective loss of majority of the 50,000–70,000 orexin neurons in the lateral hypothalamus (Peyron et al. 2000). It is, however, possible that some orexin-producing neurons may not be irreversibly lost, but the production of orexin is merely silenced. Deficient secretion of orexin leads to instability of sleep-wake regulation, which manifests as transition abnormalities between vigilance states: excessive daytime sleepiness, sudden sleep attacks and abnormal sleep structure with sleep onset REM periods. Because in NT2 patients orexin levels are normal or only slightly reduced, the pathophysiological mechanism of NT2 is suspected to differ to some extent from the pathophysiology of NT1, although the core symptoms (excluding cataplexy) are similar.

Narcolepsy likely results from the combined effects of genetic predisposition, environmental factors and triggering events which together lead to the selective destruction or dysfunction of orexin-producing neurons in NT1. Genetically, narcolepsy is strongly associated with a specific gene variant in the Human Leucocyte Antigen (HLA) class II, namely HLA-DQB1*06:02. This variant is expressed in 86–98% of patients who have NT1 and in approximately half of patients with NT2, but notably, it is also expressed in 5–38% of the general population (Mignot et al. 2001, Tafti et al. 2014). Much weaker associations between narcolepsy and HLA class I genes have also been reported (Ollila et al. 2015). These observations suggest that the development of narcolepsy involves cytotoxic immune mechanisms, i.e., that the loss or dysfunction of orexin-producing neurons is mediated by the immune system (Bassetti et al. 2019).

The low concordance of narcolepsy in identical twins (Dauvilliers et al. 2004) indicates that also environmental factors have a large role in the development of narcolepsy. It has been suggested that childhood exposure to agents that compromise the immune system, such as viruses or bacteria, might alter the development of the immune system and predispose individuals to narcolepsy. Subsequent exposure might then later trigger an immune response that leads to the destruction of orexin neurons (e.g.,

Bassetti et al. 2019). After the influenza H1N1 pandemic 2009-2010, significant increase in narcolepsy diagnoses was reported, both in conjunction with the Pandemrix vaccination in northern European countries as well as in China in the absence of vaccination. Thus, it is suspected that influenza H1N1 infection itself, as well as some influenza vaccines, might have provoked the development of narcolepsy.

Narcolepsy is associated with a substantial psychiatric and medical comorbidity (Ohayon 2013, Cohen et al. 2018). Based on a community-based cohort study, Cohen and colleagues (2018) found the prevalence of narcolepsy to be 32.4 per 100.000 of which 41% were of type 1 narcolepsy. The prevalence of psychiatric comorbidities at the time of narcolepsy diagnosis was substantial: psychiatric disorders in general were present in 46% of patients. In particular, depression (40%) and anxiety disorders (21%) were common. The prevalence of psychiatric disorders among narcolepsy patients increased during the observation period of 9.9 yrs to 57%, depression to 50%, and anxiety disorder to 25%. Obstructive sleep apnea, with a prevalence of 29%, was among the three most prevalent comorbidities and its prevalence increased to 34% during the follow-up period. As compared to controls, at the time of diagnosis the Odds Ratio (OR) for patients was significantly elevated for anxiety disorder, and the associations between narcolepsy and depression, psychiatric disorders in general, as well as obstructive sleep apnea, were significantly increased both at the time of diagnosis and in the follow-up. Narcolepsy also has a major impact on the Health Related Quality of Life (Raggi, Plazzi, and Ferri 2019)

Dream Recall

In retrospective questionnaires and interviews, patients with narcolepsy report significantly higher overall, monthly and weekly dream recall frequency than healthy controls (Dodet et al. 2015, Rak et al. 2015, Schredl 1998). However, the reported rates of dream recall vary considerably from study to study for both patients and controls. Schredl (1998) asked 23 patients diagnosed with narcolepsy to estimate their average dream recall by using a five-point Likert scale ranging from never (1) to very often (5) and compared the rates to those reported by a representative sample of more than 700 control participants. The average recall was 3.0 for patients, which on the Likert sale signifies reporting recalling dreams sometimes, and 2.3 for controls, which is closest to the Likert value of recalling dreams rarely. In an interview study, using monthly recall frequency as measure, Dodet and colleagues (2015) reported an average retrospectively estimated dream recall rate of 49.0 dreams per month for narcolepsy patients while the corresponding recall rate in matched healthy controls was 15.5 dreams per month. Rak and colleagues (2015) also investigated monthly dream recall using a retrospective interview, but then calculated weekly recall rates, finding that patients with narcolepsy reported an average recall of 4.3 dreams per week (17 per month) compared to 0.8 dreams per week (3.2 per month) reported by members of the general population. In the most recent questionnaire study, Schredl and Olbrich (2019) noted that narcolepsy patients have significantly higher self-estimated dream recall frequency compared to other sleep disorder patients (with various diagnoses such as insomnia, Insufficient Sleep Syndrome, Restless legs Syndrome (RLS), primary hypersomnia, parasomnia, sleep apnea). In this study, narcolepsy patients estimated recalling dreams on average approximately once per week while the other patients approximately 2-3 times per month.

Although the average recall rate assessed with retrospective measures varies considerably between these studies, the results systematically imply that patients diagnosed with narcolepsy have higher self-estimated dream recall rates than healthy matched controls, members of the general population or other sleep disorder patients. Notably, however, the estimations reported above are based on self-reports and spontaneous dream recall. The story becomes more complicated when dream recall frequency is measured using spontaneous or forced awakenings in a setting where sleep is monitored with polysomnogram (PSG).

Schredl (1998) collected dream reports from patients and controls in a sleep laboratory. The participants wrote down their dreams after morning awakenings. The 23 patients with narcolepsy recorded a total of 14 dreams and the 30 control participants reported altogether 56 dreams. Although

Schredl does not report whether the difference is statistically significant, this study shows that spontaneous morning recall is not necessarily higher in patients with narcolepsy. Further, in REM awakenings conducted at home or in sleep laboratory, dream recall frequency of narcolepsy patients has been found similar to that of controls. Fosse, Stickgold and Hobson (2002) monitored the sleep of 15 narcolepsy patients and nine controls at home with professional polysomnogram equipment, and the experimenter who spent the night in a separate room, woke up the sleepers via two-way intercom. Awakenings were performed between 5 and 15 minutes into REM periods. Nighttime REM awakenings yielded almost equal dream recall rates for patients (87% of awakenings) and for healthy controls (88% of awakenings). Cipolli and colleagues (2008) studied 17 patients with narcolepsy as well as 17 healthy controls in a sleep laboratory, waking them up from first and third REM period, eight minutes after stage onset. They found no significant differences in dream recall rates between patients and controls, neither in awakenings from the first (94% vs 82%, respectively) nor third REM period (77% vs 88%, respectively).

Dream recall rates in narcolepsy patients have also been studied utilizing daytime awakenings during or after naps. In possibly the very first systematic investigation on dreaming in patients with narcolepsy, Vogel (1976) conducted 25 awakenings in 18 narcolepsy patients from daytime naps. Patients were awakened two to four minutes into SOREM periods, yielding a dream recall rate of 85%, i.e., very close to the recall rate obtained from healthy individuals from nighttime REM awakenings (Nielsen 2000). In a later study by Roth and Bruhova (1969a) utilizing afternoon nap awakenings, they acquired 97% dream recall rate from REM sleep and 34% from NREM sleep in narcolepsy and hypersomnia patients. In a study by Fosse and colleagues (2002) conducted at home setting, the patients were awakened from SOREM periods during daytime naps as well as during the experimental nights. Recall rates were higher in SOREM awakenings (98% of daytime and 100% of nighttime SOREM awakenings, respectively) than from nighttime REM periods (87%).

Waihrich and colleagues (2006) utilized MSLT to investigate the occurrence of dreaming in narcolepsy patients and patients with daytime sleepiness of non-narcoleptic origin. Overall, narcolepsy patients had a significantly higher dream recall rate than non-narcoleptic patients (57% vs 28%), but surprisingly, a much lower recall rate from naps including only Non-Rapid Eye Movement (NREM) sleep (28% vs 75%). In a similar study, Schredl and Olbrich (2019) compared dream recall rates in MSLT naps between patients with narcolepsy diagnosis and patients with various other sleep disorder diagnoses (insomnia, Insufficient Sleep Syndrome, RLS, primary hypersomnia, parasomnia, sleep apnea). They also found dream recall rates to be significantly higher in narcolepsy patients (58%) than in other sleep disorder patients (30%) when questioned after a nap which included REM sleep. Naps including REM sleep vs NREM sleep yielded higher recall rates for all patients but the NREM dream recall rate did not differ between the narcolepsy patients (26%) and other sleep disorder patients (22%). Thus, the narcolepsy patients' NREM recall rates were very similar to those of other sleep disorder patients, but remarkably higher when naps included REM sleep. However, one methodological limitation here is that the patients were not questioned directly after awakening from a nap but completed the questionnaire later.

The above reviewed studies on dream recall frequency point to a slight discrepancy between narcolepsy patients' own judgments about their dream recall, and how often they report dreams when questioned during experimental awakenings, compared to controls. Why is it that narcolepsy patients systematically report higher dream recall frequency while there is less systematic evidence that the actual ability to recall dreams from especially nighttime REM awakenings or NREM sleep naps would differ between patients and comparison groups? The overactive REM sleep system in narcolepsy (Cao and Guilleminault 2017) might obviously be one factor related to higher recall. It is plausible that narcolepsy may actually increase the amount of subjective experiences in this patient group. Although in systematic nocturnal awakenings narcolepsy patients do not recall dreaming more frequently than healthy controls, they did report longer dreams in early night REM awakenings (1st REM period), even when SOREM periods were not present (Cipolli et al. 2008, Mazzetti et al. 2010). Further, they do have a very high dream recall rate when they wake up from naps or involuntary sleep attacks that include SOREM periods. These could affect the patients' perception of having higher dream recall and thus their retrospective estimates. Interestingly, Schredl and Olbrich (2019) recently found a significant correlation between self-estimated dream recall rate and dream recall from MSLT naps, suggesting that dream recall from naps and sleep attacks may be a factor affecting retrospective estimates. This may imply that the patients' self-estimates could provide tentatively reliable, even if not completely accurate, information about their dream recall.

Further, there may be other factors that also affect the recall rates. Fragmented sleep and frequent nocturnal awakenings in narcolepsy could prompt patients to remember dreams more frequently than controls. Perhaps also differences in the content of dreams could partially explain why retrospective estimates are so much higher in patients while systematic nocturnal awakening yield less pronounced differences. If the dreams of narcolepsy patients are richer in content, more vivid, intense, realistic or emotional, these dream characteristics might affect recall processes. Emotional and personally significant experiences leave a stronger and more easily retrievable trace in the long-term memory, and thus, vivid and intense emotional dreams are easier to recall than mundane dreams. This possible memory bias for emotional content might affect self-estimated recall rates when retrospectively assessed. Next, we will therefore discuss whether the content of dreams is different in patients with narcolepsy than in control groups.

Dream Content

The dream content of narcolepsy patients seems to be altered to some degree compared to healthy controls, and to the direction which may at least partially explain the higher self-estimated dream recall frequency. The studied qualitative dream features include emotionality (presence, valence and intensity of emotions), bizarreness and vividness.

In studies using retrospective questionnaires and interviews, narcolepsy patients have reported experiencing more emotions and more negative emotions (such as fear, anxiety, anger, sorrow, guilt, disgust) in their dreams than controls (Sturzenegger and Bassetti 2004, Wamsley et al. 2014). These self-report findings have not, however, been fully corroborated by studies utilizing polysomnographic monitoring and/or systematic awakenings.

In the very first systematic study on the dream content of narcolepsy patients, Vogel (1976) noted that dream mentation during daytime SOREM naps included pleasant experiences in 19% and unpleasant experiences in 19% of those reports where content was explicitly described. He compared these to figures obtained from healthy controls in another sample at that time. Although no bias towards more negative emotions in dreams of narcolepsy patients was observed, Vogel noticed that narcolepsy patients had pleasant experiences less frequently (19% vs 38%), i.e., there was a bias towards less positive content. Some later studies, however, have reported more emotions, and more negative emotions in dreams of narcolepsy patients.

In a sleep laboratory study by Schredl (1998) where dream reports were collected after morning awakenings, the ratio of positive to negative emotions indicated that the patients' dreams tended to be slightly more negative in emotional tone, although the difference was not statistically significant. There were no differences in the intensity of emotions in patients' dreams compared to dreams of controls. Partially similar results were observed in the study by Fosse and colleagues (2002), conducted at homes of the participants with polysomnographic monitoring and systematic awakenings. Overall, emotions were approximately equally often present in the dreams of narcolepsy patients than healthy controls (85% vs 76%), but statistically significantly more often present in SOREM periods (92%) as opposed to nighttime REM periods (81%) in patients compared to controls. This higher prevalence of emotions in SOREM dreams included higher levels of both positive and negative emotion. In fact, in dreams of both patients and controls, positive and negative emotions appeared in a balanced ratio. Joy/elation and anxiety/fear were the two most common emotions in dreams, but the prevalence of anxiety/fear in SOREM dreams was enhanced compared to both control groups' and narcolepsy patients' nighttime REM dreams. Further, anxiety/fear was found to be more intense in narcoleptic

SOREM dreams than in the nighttime REM dreams of the control group (but not nighttime REM dreams of patients).

Methodological differences seem to well explain the minor differences in results of the above mentioned studies. When dream reports were elicited only in the morning (Schredl 1998) as opposed to also including early-night and daytime nap SOREM dreams (Fosse, Stickgold, and Hobson 2002) no differences in emotions between patients and controls were found. In contrast, it appears that both nightime and daytime SOREM dreams are more negative and more intensively negative compared to nighttime REM dreams. This may well contribute to the higher retrospective dream recall estimates in the patient group as well as estimates of having more negative dreams.

Other possible alterations in dream content of narcolepsy patients have also been addressed, mainly the bizarreness and vividness of dreams. However, empirical results on these features have produced somewhat inconsistent results which may partially depend on how the concepts have been defined and measured in different studies. Schredl (1998) measured report-level bizarreness with a scale ranging from 1 (like waking reality) to 4 (several bizarre elements) in a study utilizing morning awakenings in a laboratory, and found narcolepsy patients' dreams to be significantly more bizarre than those of healthy controls. Fosse (2000) scored each occurrence of bizarreness in dream reports collected both during nighttime and nap awakenings, categorizing bizarreness as discontinuity, improbability or vagueness. In contrast with Schredl's findings, he found improbabilities and discontinuities to be overall more prevalent in the dreams of healthy controls than narcolepsy patients. In both groups, improbabilities increased from early to late night REM awakenings, but in narcolepsy patients, daytime nap SOREM dreams included the highest levels of improbabilities and discontinuities.

Vividness has also been defined in various ways, and has, for example, been used to refer to visual qualities as well as emotional intensity of the dreams. Self-estimates have suggested dreaming to be more vivid in narcolepsy patients (Lee et al. 1993, Wamsley et al. 2014), but the concept of vividness was probably not defined for the participants in these studies. Early sleep laboratory studies using afternoon naps also noted the vivid and bizarre quality of narcolepsy patients' dreams (Roth and Bruhova 1969a, b). Roth and Bruhova (1969a) remark that 80% of the dreams in narcolepsy and hypersomnia patients were vivid "with a strong affective component and visual and acoustic perceptions". They thus included emotional intensity and visual qualities in their conception of vividness. In contrast, in the study conducted at home with PSG monitoring and systematic awakenings, Fosse (2000) conceptualized vividness to be composed of visual brightness and clarity (not affective intensity). He measured these features by altering the visual features of a photograph, producing 16 versions of the same photograph with varying degrees of clarity and brightness. After each awakening, the participants chose the photograph that best matched their most bright and clear (vivid) visual imagery. Fosse found no evidence of more vivid dreams in narcolepsy patients. Vividness increased across the night in both patients and controls, and tended to be higher in late night REM sleep in controls than in patients. In narcolepsy patients, daytime SOREM dreams exhibited the overall highest levels of vividness, but were not statistically significantly more vivid than nighttime SOREM and nocturnal late REM dreams. In fact, vividness in daytime SOREM dreams was similar to the level of late night REM dreams in controls.

In conclusion, narcolepsy patients estimate their dreams to be more emotional, negative, vivid and bizarre than controls, but systematic awakenings show that these dream features are typically more prevalent in SOREM dreams, and especially daytime SOREM, rather than nighttime REM dreams. This may possibly partially explain why narcolepsy patients rate their dream experiences differently from controls.

Thus far, there are no studies on the content of NREM dreams in narcolepsy patients, which might shed further light on dream alterations in this patient group. It is plausible that sleep stages are more diffuse and blended in narcolepsy patients, as shown in rodent model utilizing orexin knockout mice (Behn et al. 2010). More global but subtle alterations in sleep architecture might affect dream features, both during REM and NREM sleep. Also, given that narcolepsy patients report more nightmares and recurrent dreams, this may affect their estimates of the general emotional tone, vividness and bizarreness of their dreams.

Nightmares

Nightmares are defined as vivid and dysphoric well-remembered dreams that often involve a threat to the dreamer and lead to an awakening. Nightmares are often reported by narcolepsy patients, which was noted already in the early studies on dreaming in narcolepsy. For example, Roth and Bruhova (1969a, b) mention that dreams of narcoleptic patients are often described as vivid and terrifying, i.e., as nightmares. Roth and Bruhova (1969b) also noted that nightmares are more commonly reported by narcolepsy patients presenting with cataplexy.

Because nightmare prevalence and frequency cannot be studied in laboratory conditions, the prevalence and frequency estimates are based on self-reports collected with questionnaires and interviews. The percentage of narcolepsy patients reporting having nightmares, as well as the frequency of their nightmares, varies significantly between studies, but is systematically higher than in comparison populations (Dodet et al. 2015, Lee et al. 1993, Mayer et al. 2002, Rak et al. 2015, Sturzenegger and Bassetti 2004). In the general population, nightmare prevalence estimates have centered around 3% to 5% (Sandman et al. 2013). In contrast, 33-70% of narcolepsy patients report having nightmares (Dodet et al. 2015, Lee et al. 1993, Mayer et al. 2002, Pisko et al. 2014, Rak et al. 2015, Sturzenegger and Bassetti 2004). Nightmare frequency is also elevated, and patients estimate having 4.5 to 7.3 nightmares per month (compared to less than 1 per month in controls) (Dodet et al. 2015, Rak et al. 2015). Further, Dodet and colleagues (2015) report that a higher number of patients with type 1 narcolepsy (83%) have nightmares than of patients with type 2 narcolepsy (41%), and that NT1 patients also have significantly higher nightmare frequency than NT2 patients (10.2 vs 1.4 nightmares per month). These corroborate the early findings by Roth and Bruhova (1969b) that dreaming is potentially more negative in patients presenting with cataplexy. Yet, it is unknown how many narcolepsy patients may suffer from actual nightmare disorder.

Frequent nightmares can be expected to have a negative effect on sleep, by causing awakenings and difficulty falling asleep again, and the distress caused by nightmares may significantly decrease well-being and quality of life in patients. In clinical practise, anecdotal observations indicate that especially in early stages of the disorder the terrifying and frequent nightmares may cause significant distress for the patients, but as the patients gradually adapt to the illness, they also learn to cope better with the nightmares. There are, however, no systematic studies that would have investigated whether the prevalence, frequency or nature of nightmares changes with the duration of the illness, or whether and how the patients adapt to having nightmares. There are also thus far no systematic content analytic studies on the nightmare content of narcolepsy patients, only anecdotal reports of the typical themes (such as murder, being pursued) (Roth and Bruhova 1969b).

Disturbing nightmares may be a major overlooked problem in narcolepsy patients, and most patients suffering from nightmares are likely not offered any treatment to cope with their nightmares. Cognitive psychotherapeutic approaches, such as imagery rehearsal therapy (IRT) which alleviates the frequency and intensity of nightmares in chronic nightmare sufferers (Krakow and Zadra 2006), may be used by some informed clinicians to treat narcolepsy patients. However, there are no studies that would have investigated the efficacy of IRT in narcolepsy patients and whether the method should be modified for this patient group. Further, sodium oxybate may be efficient in reducing nightmares (Rak et al. 2015). Some patients also report finding relief for their nightmares from the ability to have lucid dreams.

LUCID DREAMING

Lucid dreaming is the phenomenon of becoming aware that one is dreaming while having the dream, i.e., lucid dreaming entails the activation of self-reflective ability during sleep that enables the person to understand that the ongoing experience is a dream. Some lucid dreamers are able to guide and control the content of their dreams, perform pre-determined tasks in the dream (such as count to ten, hold their breath) and remember to execute eye movement signals in the dream to communicate with the researchers via eye movement monitoring.

Vogel noted already in 1976 that in 73% of awakenings from SOREM periods with recalled content the narcoleptic subjects reported that they were to some extent aware that the experience was not real, and in 45% of awakenings the participants reported that they had some control over the content of their mentation. Further, in approximately 20% of the SOREM awakenings, the patients experienced a combination of being aware of dreaming, having control over the content of the experience, and awareness of being in a sleep laboratory, indicating preserved self-awareness during a significant number of SOREM periods. Later studies have confirmed that narcolepsy patients experience lucid dreaming more commonly than the general population.

In a telephone interview study, Rak and colleagues (2015) asked the patients to estimate their lucid dreaming frequency on a seven-point scale ranging from never to several times a week, and calculated a monthly lucid dreaming frequency based on averages within each category. Of the 60 patients interviewed, 47 (78%) reported having lucid dreams, with a monthly average of 6.9 lucid dreams (compared to 0.7 in healthy controls). This difference between patients and controls remained significant after adjusting for self-reported dream recall frequency, which is higher in narcolepsy patients than controls. Similarly, in a face-to-face interview study Dodet and colleagues (2015) found the self-estimated prevalence of lucid dreaming to be 77% in 53 narcolepsy patients compared to 49% in controls, with the average number of lucid dreams being 7.6 per month in patients and only 0.3 in controls. When studied in sleep laboratory, 33 of 57 (58%) naps including REM sleep led to a report of having a lucid dream, and in 14 instances, the patients succeeded in communicating the lucid dream occurrence to the observers with eye movement signal.

Dodet and colleagues (2015) also asked how the patients utilize their ability to have lucid dreams, and 78% reported that lucid dreams were useful in reducing the impact of nightmares. Thus, lucid dreams could offer one tool to manage the nightmares the patients experience. However, 58% of patients also reported that lucid dreaming promotes awakenings, and 32%

of patients that lucid dreaming promotes tiredness. Thus, lucid dreams may also further disrupt the already fragmented sleep in these patients.

Notably, however, not all narcolepsy patients report experiencing lucid dreams, and the possible correlates of lucid dreaming in narcolepsy remain largely unexplored. Rak and colleagues (2015) found lucid dreaming frequency to be higher in women, and to decrease with age, and this applied to narcolepsy patients as well as controls. Similar findings have also been observed in other studies on healthy lucid dreamers. Rak and colleagues (2015) also compared drug naïve patients to those on medication, as well as the currently medicated patients to their previous drug-free periods, and noted that being on medication was not associated with lucid dreaming frequency. They, however, found that patients taking sodium oxybate had lower dream recall and nightmare recall frequency, and thus sodium oxybate may reduce nightmares without affecting lucid dreams.

Regardless, in narcolepsy patients self-reflective awareness is, on average, more often activated during sleep than in healthy subjects. Given that lucid dreaming is almost purely a REM sleep phenomenon (Stumbrys and Erlacher 2012), this suggest that the high incidence and frequency of lucid dreaming is related to altered REM sleep processes. In healthy lucid dreamers, lucidity seems to be associated with increased prefrontal cortex activation (Dresler et al. 2012, Voss et al. 2014), and it could be suggested that in narcoleptics those brain areas that are typically less active during REM sleep in healthy individuals, such as prefrontal areas, have abnormally high activation levels, which would provide an optimal background state for lucid dreaming to occur.

Sleep Paralysis and Hallucinations

Sleep paralysis is the inability to move during the transition from sleep to wakefulness or from wakefulness to sleep. It results when REM sleep atonia extends into wakefulness, and skeletal muscles become or remain paralyzed while the person is awake. Sleep paralysis typically lasts few seconds, but can go on even up to few minutes. Shortness of breath and the feeling of difficulty breathing may accompany the episodes which is likely due to the paralysis of accessory respiratory muscles while the diaphragm remains fully functional. Eyes can typically be opened, but not always, and eyelids may feel unusually heavy.

Although sleep paralysis is well-known to often accompany narcolepsy, there are surprisingly few studies as to its prevalence and frequency in narcolepsy patients compared to controls. In an interview study by Dodet and colleagues (2015), sleep paralysis was found to be significantly more common in patients with narcolepsy (59%) than in healthy controls (15%). In the same study, sleep paralysis was also found to be more often present in NT1 than NT2 (64% vs 47%) patients. Frauscher and colleagues (2013) have reported a similar sleep paralysis prevalence (50%) in the Innsbruck narcolepsy cohort, and Luca and colleagues (2013), in a multicenter stydy involving more than one thousand patients, have reported a comparable prevalence of 53%.

Sleep paralysis episodes often involve hypnagogic or hypnopompic hallucinations. Hypnagogic hallucinations are abnormal perceptions that occur while falling asleep, whereas hypnopompic hallucinations occur upon awakening. Assessed retrospectively in an interview study by Dodet and colleagues (2015), hypnagogic (51%) and hypnopompic hallucinations (28%) are also often experienced by narcolepsy patients, while the prevalence of both is approximately 8% in controls. Frauscher and colleagues (2013) also found 56% prevalence of hypnagogic hallucinations in their patient sample. In the multicentre study by Luca and colleagues (2013), a very similar prevalence of 63% was reported. In Dodet et al.'s (2015) study both hypnagogic (66%) and hypnopompic (33%) hallucinations were more often experienced by patients with type 1 than type 2 narcolepsy (29%; 18%, respectively), and similar findings were reported by Frauscher and colleagues (2013), with 59% of NT1 and 38% of NT2 patients reporting hypnagogic hallucinations. These hallucinations are most often visual or auditory, but may also be tactile or proprioceptive (D'Agostino and Limosani 2016). The content of these hallucinations has not been systematically assessed in any study, but D'Agostino and Limosani

(2016) report that the most common themes seem to be related to attack and aggression. Out-of-body experiences may also occur during sleep paralysis.

Leu-Semenescu and colleagues (2011) studied hallucinations in narcolepsy patients, and noted that most of the hallucinations were typically experienced at sleep onset (55%), or both onset/offset (42%), but rarely only at sleep offset (3%). The hallucinations occurred with sleep paralysis in 40% of patients with hallucinations. However, in 28% of patients with cataplexy, and in 6% without cataplexy, hallucinations also occurred during daytime while wide awake. Overall, hallucinations (whether they were sleep-related or occurred during full wakefulness) were more often experienced by patients with (59%) than without (28%) cataplexy (Leu-Semenescu et al. 2011). Of the patients who had hallucinations, 58% also experienced sleep paralysis while only 16% of those without hallucinations reported sleep paralysis. Most notably, the hallucinations are often experienced as terrifying or anxiety-provoking, and especially when they co-occur with sleep paralysis.

Sleep paralysis and hypnagogic and hypnopompic hallucinations in narcolepsy patients are considered to be caused by the intrusion of REM-like dream mentation and REM sleep atonia into the waking state. While these phenomena also occur in healthy individuals, the dissociation between REM sleep and wakefulness is much more pronounced in narcolepsy patients, and may create optimal conditions for such experiences during transitional states. Regardless, very little research has been conducted on the phenomenology of these states in narcolepsy patients, and although the prevalence and frequency of sleep paralysis and hypnagogic and hypnopompic hallucinations differ between narcolepsy patients and healthy controls, it remains unexplored whether the experiences are phenomenologically different.

Notably, hallucinatory and delusional phenomena of narcolepsy, such as hypnagogic and hypnopompic hallucinations, can be mistaken for signs of psychosis. Talih (2011) found in a literature review on psychotic symptoms in schizophrenia and narcolepsy that the most useful distinguishing features are visual hallucinations in narcolepsy versus mainly auditory hallucinations in schizophrenia. Plazzi and colleagues (2015), on the other, hand examined

28 NT1 and 21 schizophrenia patients and found that while almost all forms of hallucinatory phenomena are present in both groups, in schizophrenia patients unimodal verbal auditory hallucinations during wakefulness are more typical, whereas hypnagogic and hypnopompic hallucinations, nonverbal auditory hallucinations and shadow hallucinations point more towards narcolepsy.

PARASOMNIA

Parasomnias are disorders characterized by abnormal or unusual movements, behaviours or emotional or perceptual experiences during sleep. They arise from dissociated states when features of wakefulness, REM sleep or NREM sleep mix (AASM 2014). Parasomnias are classified based on whether they originate from NREM or REM sleep. NREM parasomnias are considered to be disorders of arousal that mainly arise from slow-wave sleep, and include, for example, confusional arousals, sleepwalking and sleep terrors. REM parasomnias arise from REM sleep, and include REM sleep behaviour disorder (RBD), isolated recurrent sleep paralysis and nightmare disorder. RBD is characterized by loss of muscle atonia and repeated episodes of movements during REM sleep, culminating in dream-enacting behaviors.

In narcolepsy, which is more typically considered to involve REM sleep rather than NREM sleep abnormalities, both NREM and REM parasomnia symptoms are more prevalent than in healthy controls. Sleepwalking has been reported occur in 8% to 14% of narcolepsy patients (Mayer et al. 2002, Strurzenegger and Bassetti, 2004), while the comparable number in the general adult population is between 2% and 4%. Sleep terrors are also experienced by 9% of narcolepsy patients (Mayer et al. 2002). Frauscher et al. (2013) and Luca et al. (2013) have also noted NREM parasomnias to affect approximately 8%-10% of narcolepsy patients.

While the prevalence estimates for NREM parasomnias in narcolepsy patients (in those few studies that have investigated these symptoms) are highly similar, estimates for the presence of REM parasomnias, mainly

RBD, vary significantly between studies. These discrepancies are likely due to small sample sizes, differences in the diagnostic criteria of RBD and whether narcolepsy type 1 and type 2 patients have been studied separately. Notably, RBD has been reported in both NT1 and NT2 patients and not all the patients with NT1 suffer from RBD.

Addressed by questionnaires and interviews, RBD has been estimated to be evident in 36%–61% of narcolepsy patients (Dauvilliers, Jennum, and Plazzi 2013, Nightingale et al. 2005), but studies utilizing polysomnographic criteria have found the incidence to vary between less than 10% and more than 70%. In one of the earliest polysomnography studies, Schenck and Mahowald (1992) reported that of 17 suspected RBD cases (which comprised 12% of all narcolepsy patients studied at the clinic at that time) ten fulfilled the diagnostic criteria for RBD, i.e., the prevalence of RBD in the total sample was below 10%. Slightly higher estimates were obtained by Mayer and colleagues (1993, 2002) who noted the prevalence of polysomnographically verified RBD to be 15%-19% in narcolepsy patients. Frauscher and colleagues (2013) found an even higher prevalence of 24% in the Innsbruck narcolepsy cohort, and Mattarozzi and colleagues (2008) report that 43% of their patient sample evidenced RBD in polysomnography. The European Narcolepsy Network multicenter study, with a substantial number of participants evaluated for RBD (N = 295), revealed the presence of RBD in 50% of patients (Luca et al. 2013). The highest prevalence of polysomnographically verified RBD (73%) in narcolepsy patients with cataplexy has been reported by Cipolli and colleagues (2011). RBD also seems to be more common in those narcolepsy patients with (38%) than without (11%) hallucinations (Leu-Semenescu et al. 2011), and narcolepsy with RBD compared to narcolepsy without RBD is related to increased prevalence of vivid dreams (33% vs. 26%) and nightmares (52% vs. 20%) (Pisko et al. 2014).

RBD episodes in narcolepsy patients are usually milder and less pronounced than in isolated RBD or RBD related to neurodegenerative diseases, although rare cases of violent RBD in narcolepsy have been also described (Ingravallo, Schenck, and Plazzi 2010). The pathophysiology differs between isolated RBD and RBD concomitant with narcolepsy which implies that narcolepsy patients have low risk of conversion to neurodegenerative disorders (Antelmi et al. 2019). Given that RBD can be present in both NT1 and NT2 patients, and is thus not directly related to orexin deficiency, the exact mechanism of RBD in narcolepsy remains unclear. However, anti-cataplectic drugs and selective serotonin reuptake inhibitors, used in many patients to treat cataplexy or psychiatric symptoms, may provoke or cause of RBD, and thus RBD may not always be directly induced by narcolepsy itself.

Misattribution of Memory Sources in Narcolepsy

The instability of sleep-wake transitions in narcolepsy patients is reflected in their memory processes. Notably, memories that originate from dream experiences or hypnagogic and hypnopompic hallucinations are easily confused with being memories of waking experiences in these patients. Although perfectly healthy people may also occasionally confuse dream-generated memories as recollections of real waking events, this is much more pronounced in narcolepsy patients. Case reports (Hays 1992, Szucs et al. 2003) have illustrated accusations of sexual misconduct or assault based on false memories created by dreams, leading even to legal procedures. These case reports highlight the sincerity of the patients' beliefs, and the possible severe consequences that the confusion of memory sources for experiences may have for the patients and the accused.

Wamsley and colleagues (2014) have conducted the first systematic investigation of the dream-generated delusional memories in 46 narcolepsy patients compared to healthy age-matched controls. The participants were asked in a telephone interview whether they "have ever had the experience of being unsure whether something was real, or if it was from a dream". While only 15% of healthy controls reported that they had confused dream-generated memories with waking memories, 83% of narcolepsy patients reported having had this kind of an experience at least once a week. Furthermore, in addition to being more prevalent, delusional memories were also more frequent in the patient group. Of the six control participants who reported delusions, only two had experienced this more than once in their lives, while of the patients who completed the follow-up interview, 65% reported experiencing dream delusions at least once a week, and 95% at least once a month.

Wamsley and colleagues (2014) also confirmed the previous case-study findings that the delusional memories are highly realistic and credible for the patients. The patients with delusional memories reported that although some delusions could be resolved within minutes after awakening, many often persisted for hours, days, or even weeks. For example, several patients had dreamed that their parents, children, or pets had died, believing that this was true until they met the presumed deceased, and were surprised to find them alive. The delusions can also be so credible that patient may act accordingly, for example, start a legal process (Szűcs et al., 2003), call the police about an intruder, or make funeral arrangements for presumed dead relative (Wamsley et al., 2014). Wamsley and colleagues (2014) further hypothesized that highly abnormal dreaming, such as high dream recall rate or self-reported vividness of dreams, could be related to higher rate of delusional memories. However, they found that self-reported dream recall frequency and vividness of dreaming were unrelated to dream-generated delusions. Dream-generated delusional memories were also not explained by differences in memory impairment, measured with Prospective-Retrospective Memory Questionnaire (PRMQ), between narcolepsy patients with and without delusions. However, the patients scored higher for both retrospective and prospective memory problems than healthy controls. Wamsley and colleagues (2015) have therefore suggested that the delusions could result from an abnormal memory encoding of dream memories. Given monoaminergic and cholinergic neurons, which are involved in encoding memories, are also major targets of the orexin neurons, the lack of activity of orexin neurons could alter the monoaminergic and cholinergic transmission and encoding of dream content into long-term memory.

In summary, delusional confusion of memory sources seems to characterize the lives of narcolepsy patients as they are much more likely than healthy controls to misperceive that something that happened in a dream is a waking life memory. These delusions are not only common but highly distressing for the patients. Although the mechanism of dream delusions remains undetermined, narcolepsy is associated with high likelihood of misattribution of memory sources, which complicates the everyday lives of the patients.

CONCLUSION

Excessive daytime sleepiness, disrupted nighttime sleep and cataplexy in NT1 are the major symptom complaints in narcolepsy patients. Predominantly sleepiness, but also psychiatric comorbidities and the symptoms typically associated with narcolepsy, such as nightmares, parasomnia, sleep paralysis, hypnagogic and hypnopompic hallucinations, and delusional memories, may significantly contribute to the patients' daily living, socialization, education, and work, and may have a deleterious effect on quality of life and life satisfaction.

The adjacent symptoms associated with narcolepsy, i.e., nightmares, parasomnia, sleep paralysis, hypnagogic and hypnopompic hallucinations, and delusional memories, vary in occurrence and severity, but affect a substantial number of patients and may cause significant distress. Although such associated symptoms may be less disrupting for the patients than daytime sleepiness and sleep-wake instability, in clinical practise these symptoms should be explored and potentially treated more systematically. Many patients may, unless directly questioned, even avoid mentioning these symptoms to their physicians, in the fear that the symptoms may be labelled as psychiatric problems.

Because the psychiatric comorbidity is high and symptoms overlapping, narcolepsy can be mistaken for a psychiatric disorder, or missed as a comorbidity behind the psychiatric disorder. In contrast, psychiatric comorbidity may be easily overlooked in purely somatic treatment settings. Since there is a remarkable comorbidity and several overlapping symptoms with psychiatric disorders, there is a risk of diagnostic confusion and misinterpretations in clinical settings. It is not unusual that patients with narcolepsy are first suspected to have a psychiatric disorder or chronic fatigue syndrome.

Given the psychiatric comorbidities, disturbing associated symptoms, and the major impact on the quality of life, narcolepsy patients are often in need of psychosocial support in addition to proper adjustment of medication. The patients would therefore likely benefit from additional attention to the adjacent symptoms, the acknowledgement and treatment of which might significantly improve life satisfaction and quality of life in narcolepsy patients.

REFERENCES

American Academy of Sleep Medicine. (2014). *The International Classification of Sleep Disorders*. (ICSD-3). Darien, IL: AASM.

Antelmi, Elena, Fabio Pizza, Christian Franceschini, Raffaele Ferri & Giuseppe Plazzi. (2019). "REM Sleep Behavior Disorder in Narcolepsy: A Secondary Form or an Intrinsic Feature?" *Sleep Medicine Reviews*, *50*, 101254. https://doi.org/10.1016/ j.smrv.2019.101254.

Bassetti, Claudio L. A., Antoine Adamantidis, Denis Burdakov, Fang Han, Steffen Gay, Ulf Kallweit, Ramin Khatami, et al. (2019). "Narcolepsy — Clinical Spectrum, Aetiopathophysiology, Diagnosis and Treatment." *Nature Reviews Neurology*, *15* (9), 519–39. https://doi. org/10.1038/s41582-019-0226-9.

Behn Diniz, Cecilia, G., Elizabeth B. Klerman, Takatoshi Mochizuki, Shih Chieh Lin & Thomas E. Scammell. (2010). "Abnormal Sleep/Wake Dynamics in Orexin Knockout Mice." *Sleep*, *33* (3), 297–306. https:// doi.org/10.1093/sleep/33.3.297.

Cao, Michelle T; Christian, Guilleminault. (2017). "Narcolepsy: Diagnosis and Management." In *Principles and Practice of Sleep Medicine* (6th Edition), edited by Meir Kryger, Thomas Roth, and William C Dement, 873–82. Philidelphia: Elsevier.

Cipolli, Carlo, Claudia Bellucci, Katia Mattarozzi, Michela Mazzetti, Giovanni Tuozzi & Giuseppe Plazzi. (2008). "Story-like Organization of REM-Dreams in Patients with Narcolepsy–Cataplexy." *Brain Research Bulletin*, *77* (4), 206–13. https://doi.org/10.1016/j.brainresbull.2008.07.012.

Cohen, Alexander, Jay Mandrekar, Erik K. St. Louis, Michael H. Silber & Suresh Kotagal. (2018). "Comorbidities in a Community Sample of Narcolepsy." *Sleep Medicine*, *43*, 14–18. https://doi.org/10.1016/j.sleep.2017.11.1125.

D'Agostino, Armando & Ivan Limosani. (2016). "Hypnagogic Hallucinations and Sleep Paralysis." In *Narcolepsy*, edited by Meeta Goswami, Michael J. Thorpy, and S.R. Pandi-Perumal, 81–93. Springer International Publishing. https://doi.org/10.1007/978-3-319-23739-8_8.

Dauvilliers, Yves, Stephanie Maret, Claudio Bassetti, Bertrand Carlander, Michel Billiard, Jacques Touchon & Mehdi Tafti. (2004). "A Monozygotic Twin Pair Discordant for Narcolepsy and CSF Hypocretin-1." *Neurology*, *62* (11), 2137–38. https://doi.org/10.1212/wnl.62.11.2137.

Dauvilliers, Yves, Poul Jennum & Giuseppe Plazzi. (2013). "Rapid Eye Movement Sleep Behavior Disorder and Rapid Eye Movement Sleep without Atonia in Narcolepsy." *Sleep Medicine*, *14* (8), 775–81. https://doi.org/10.1016/j.sleep.2012.10.006.

Dodet, Pauline, Mario Chavez, Smaranda Leu-semenescu, Jean-louis Golmard & Isabelle Arnulf. (2015). "Lucid Dreaming in Narcolepsy." *Sleep*, *38* (3), 487–97. https://doi.org/10.5665/sleep.4516.

Dresler, Martin, Renate Wehrle, Victor I Spoormaker, Stefan P Koch, Florian Holsboer, Axel Steiger, Hellmuth Obrig, Philipp G Samann & Michael Czisch. (2012). "Neural Correlates of Dream Lucidity Obtained from Contrasting Lucid versus Non-Lucid REM Sleep: A Combined EEG/FMRI Case Study." *Sleep*, *35*, 1017–20. https://doi.org/10.5665/sleep.1974.

Fosse, Roar. (2000). "REM Mentation in Narcoleptics and Normals : An Empirical Test of Two Neurocognitive Theories." *Consciousness and Cognition*, *509* (488), 488–509. https://doi.org/10.1006/ccog.2000.0466.

Fosse, Roar, Robert Stickgold & Allan Hobson, J. (2002). "Emotional Experience during Rapid-Eye-Movement Sleep in Narcolepsy." *Sleep*, *25*, 724–32. https://doi.org/10.1093/sleep/25.7.724.

Frauscher, Birgit, Laura Ehrmann, Thomas Mitterling, David Gabelia, Viola Gschliesser, Elisabeth Brandauer, Werner Poewe & Birgit Hogl. (2013). "Delayed Diagnosis, Range of Severity, and Multiple Sleep Comorbidities: A Clinical and Polysomnographic Analysis of 100 Patients of the Innsbruck Narcolepsy Cohort." *Journal of Clinical Sleep Medicine*, *9* (8), 805–12. https://doi.org/10.5664/jcsm.2926.

Hays, Peter. (1992). "False but Sincere Accusations of Sexual Assault Made by Narcoleptic Patients." *Medico-Legal Journal*, *60*, 265–71.

Hublin, Christer, Markku Partinen, Jaakko Kaprio, Markku Koskenvuo & Christian Guilleminault. (1994). "Epidemiology of Narcolepsy". *Sleep*, *17*, (8 (Suppl)), S7-12. https://doi.org/10.1093/sleep/17.suppl_8.s7.

Ingravallo, Francesca, Carlos H Schenck & Giuseppe Plazzi. (2010). "Injurious REM Sleep Behaviour Disorder in Narcolepsy with Cataplexy Contributing to Criminal Proceedings and Divorce." *Sleep Medicine*, *11* (9), 950–52. https://doi.org/10.1016/j.sleep.2010.07.001.

Juji, Takeo, Kazumasa Matsuki, Katsushi Tokunaga, Tohru Naohara & Yutaka Honda. (1988). "Narcolepsy and HLA in the Japanese." *Annals of the New York Academy of Sciences*, *540* (1), 106–14. https://doi.org/10.1111/j.1749-6632.1988.tb27056.x.

Kleitman, Nathaniel. (1963). *Sleep and Wakefulness*. Chicago, London: The University of Chicago Press.

Krakow, Barry & Antonio Zadra. (2006). "Clinical Management of Chronic Nightmares: Imagery Rehearsal Therapy." *Behavioral Sleep Medicine*, *4* (1), 45–70.

Lecea, Luis de, Thomas S Kilduff, Christelle Peyron, Xiao-Bing Gao, Pamela E Foye, Patria E Danielson, Chiaki Fukuhara, et al. (1998). "The Hypocretins: Hypothalamus-Specific Peptides with Neuroexcitatory Activity." *Proceedings of the National Academy of Sciences of the United States of America*, *95* (1), 322–27. https://doi.org/10.1073/pnas.95.1.322.

Lee, Jung H; Donald L Bliwise, Emmanuelle Labret-Bories, Christian Guilleminault & William C Dement. (1993). "Dream-Disturbed Sleep in Insomnia and Narcolepsy." *Journal of Nervous and Mental Disease*, *181*, 320–24. https://doi.org/10.1097/00005053-199305000-00008.

Lennox, William G. (1939). "Thomas Willis on Narcolepsy." *Archives of Neurology And Psychiatry*, *41* (2), 348–51. https://doi.org/10.1001/archneurpsyc.1939.02270140134010.

Leu-Semenescu, Smaranda, Valerie Cochen De Cock, Valerie Dauriac Le Masson, Rachel Debs, Sophie Lavault, Emmanuel Roze, Marie Vidailhet & Isabelle Arnulf. (2011). "Hallucinations in Narcolepsy with and without Cataplexy: Contrasts with Parkinson's Disease." *Sleep Medicine*, *12* (5), 497–504. https://doi.org/10.1016/j.sleep.2011.03.006.

Luca, Gianina, José Haba-Rubio, Yves Dauvilliers, Gert Jan Lammers, Sebastiaan Overeem, Claire E. Donjacour, Geert Mayer, et al. (2013). "Clinical, Polysomnographic and Genome-Wide Association Analyses of Narcolepsy with Cataplexy: A European Narcolepsy Network Study." *Journal of Sleep Research*, *22* (5), 482–95. https://doi.org/10.1111/jsr.12044.

Matsuki, Kazumasa, F Carl Grumet, Xiaoyan Lin, Michael Gelb, Christian Guilleminault, Willam C Dement & Emmanuel Mignot. (1992). "DQ (Rather than DR) Gene Marks Susceptibility to Narcolepsy." *The Lancet*, *339* (8800), 1052. https://doi.org/10.1016/0140-6736(92)90571-j.

Mattarozzi, Katia, Claudia Bellucci, Claudio Campi, Carlo Cipolli, Raffaele Ferri, Christian Franceschini, Michela Mazzetti, et al. (2008). "Clinical, Behavioural and Polysomnographic Correlates of Cataplexy in Patients with Narcolepsy/Cataplexy." *Sleep Medicine*, *9* (4), 425–33. https://doi.org/10.1016/j.sleep.2007.05.006.

Mayer, Geert, Karl Kesper, Helga Peter, Thomas Ploch, Thomas Leinweber & Jorg H Peter. (2002). "Untersuchung Zur Komorbidität Bei Narkolepsiepatienten [Examination of comorbidity in narcolepsy patients]." *Deutsche Medizinische Wochenschrift*, *127*, 1942–46.

Mazzetti, Michela, Claudia Bellucci, Katia Mattarozzi, Giuseppe Plazzi, Giovanni Tuozzi & Carlo Cipolli. (2010). "REM-Dreams Recall in

Patients with Narcolepsy-Cataplexy." *Brain Research Bulletin*, *81*, 133–40. https://doi.org/10.1016/j.brainresbull.2009.10.021.

Mignot, Emmanuel, Ling Lin, William Rogers, Yutaka Honda, Xiaohong Qiu, Xiaoyan Lin, Michele Okun, et al. (2001). "Complex HLA-DR and -DQ Interactions Confer Risk of Narcolepsy-Cataplexy in Three Ethnic Groups." *American Journal of Human Genetics*, *68* (3), 686–99. https://doi.org/10.1086/318799.

Nielsen, Tore A. (2000). "Covert REM Sleep Effects on REM Mentation: Further Methodological Considerations and Supporting Evidence." *Behavioral and Brain Sciences*, *23* (6), 1040–56. https://doi.org/10.1017/S0140525X00974028

Nightingale, Sam, Jane C Orgill, Irshaad O Ebrahim, Simone F de Lacy, Swastik Agrawal & Adrian J Williams. (2005). "The Association between Narcolepsy and REM Behavior Disorder (RBD)." *Sleep Medicine*, *6* (3), 253–58. https://doi.org/10.1016/j.sleep.2004.11.007.

Ohayon, Maurice M; Robert G Priest, Jurgen Zulley, Salvatore Smirne & Teresa Paiva. (2002). "Prevalence of Narcolepsy Symptomatology and Diagnosis in the European General Population." *Neurology*, *58* (12), 1826–33. https://doi.org/10.1212/wnl.58.12.1826.

Ohayon, Maurice M. (2013). "Narcolepsy Is Complicated by High Medical and Psychiatric Comorbidities: A Comparison with the General Population." *Sleep Medicine*, *14* (6), 488–92. https://doi.org/10.1016/j.sleep.2013.03.002.

Ollila, Hanna M; Jean Marie Ravel, Fang Han, Juliette Faraco, Ling Lin, Xiuwen Zheng, Giuseppe Plazzi, et al. (2015). "HLA-DPB1 and HLA Class i Confer Risk of and Protection from Narcolepsy." *American Journal of Human Genetics*, *96* (1), 136–46. https://doi.org/10.1016/j.ajhg.2014.12.010.

Peyron, Christelle, Juliette Faraco, William Rogers, Beth Ripley, Sebastiaan Overeem, Yves Charnay, Sona Nevsimalova, et al. (2000). "A Mutation in a Case of Early Onset Narcolepsy and a Generalized Absence of Hypocretin Peptides in Human Narcoleptic Brains." *Nature Medicine*, *6* (9), 991–97. https://doi.org/10.1038/79690.

Pisko, Juraj, Lukas Pastorek, Jitka Buskova, Karel Sonka, Sona Nevsimalova. (2014). "Nightmares in Narcolepsy: Underinvestigated Symptom?" *Sleep Medicine*, *15* (8), 967–72. https://doi.org/http://dx.doi.org/10.1016/j.sleep.2014.03.006.

Plazzi, Giuseppe, Chiara Fabbri, Fabio Pizza & Alessandro Serretti. (2015). "Schizophrenia-Like Symptoms in Narcolepsy Type 1: Shared and Distinctive Clinical Characteristics." *Neuropsychobiology*, *71* (4), 218–24. https://doi.org/10.1159/000432400.

Raggi, Alberto, Giuseppe Plazzi & Raffaele Ferri. (2019). "Health-Related Quality of Life in Patients with Narcolepsy: A Review of the Literature." *Journal of Nervous and Mental Disease*, *207* (2), 84–99. https://doi.org/10.1097/NMD.0000000000000918.

al Rajeh, Saad, Olajide Bademosi, Hassan Ismail, Adnan Awada, Adekunle Dawodu, Hussein Al-Freihi, Sulaiman Assuhaimi, et al. (1993). "A Community Survey of Neurological Disorders in Saudi Arabia: The Thugbah Study." *Neuroepidemiology*, *12* (3), 164–78. https://doi.org/10.1159/000110316.

Rak, Michael, Pierre Beitinger, Axel Steiger, Michael Schredl & Martin Dresler. (2015). "Increased Lucid Dreaming Frequency in Narcolepsy." *Sleep*, *38* (5), 787–92. https://doi.org/10.5665/sleep.4676.

Reading, P. J. (2019). "Update on Narcolepsy." *Journal of Neurology*, *266* (7), 1809–15. https://doi.org/10.1007/s00415-019-09310-3.

Roth, Bedrich & Sona Bruhova. (1969a). "A Clinical and Polygraphic Study of Dreams in Narcolepsy and Hypersomnia." *Activitas Nervosa Superior*, *11* (3), 223–28.

Roth, Bedrich & Sona Bruhova. (1969b). "Dreams in Narcolepsy, Hypersomnia and Dissociated Sleep Disorders." *Experimental Medicine and Surgery*, *27*, 187–209.

Sakurai, Takeshi, Akira Amemiya, Makoto Ishii, Ichiyo Matsuzaki, Richard M Chemelli, Hirokazu Tanaka, S Clay Williams, et al. (1998). "Orexins and Orexin Receptors: A Family of Hypothalamic Neuropeptides and G Protein-Coupled Receptors That Regulate Feeding Behavior." *Cell*, *92* (4), 573–85. https://doi.org/10.1016/ s0092-8674(00)80949-6.

Sandman, Nils, Katja Valli, Erkki Kronholm, Hanna M Ollila, Antti Revonsuo, Tiina Laatikainen & Tiina Paunio. (2013). "Nightmares: Prevalence among the Finnish General Adult Population and War Veterans during 1972-2007." *Sleep*, *36* (7). https://doi.org/10.5665/sleep.2806.

Schenck, Carlos H. & Mark W. Mahowald. (1992). "Motor Dyscontrol in Narcolepsy: Rapid-eye-movement (REM) Sleep without Atonia and REM Sleep Behavior Disorder." *Annals of Neurology*, *32* (1), 3–10. https://doi.org/10.1002/ana.410320103.

Schredl, Michael. (1998). "Dream Content in Patients with Narcolepsy: Preliminary Findings." *Dreaming*, *8* (2), 103–7. https://doi.org/10.1023/B:DREM.0000005900.63728.a7.

Schredl, Michael & Kate Isobel Olbrich. (2019). "Dream Recall after Multiple Sleep Latency Test Naps with and without REM Sleep." *International Journal of Dream Research*, *12* (2), 81–84. https://doi.org/10.11588/ijodr.2019.2.64730.

Silber, Michael H; Lois E Krahn, Eric J Olson & Shane, Pankratz V. (2002). "The Epidemiology of Narcolepsy in Olmsted County, Minnesota : A Population-Based Study". *Sleep*, *25* (2), 197–202. https://doi.org/10.1093/sleep/25.2.197.

Stumbrys, Tadas & Daniel Erlacher. (2012). "Lucid Dreaming during NREM Sleep: Two Case Reports." *International Journal of Dream Research*, *5*, 151–55. https://doi.org/http://dx.doi.org/10.11588/ijodr.2012.2.9483.

Sturzenegger, Christian & Claudio L. Bassetti. (2004). "The Clinical Spectrum of Narcolepsy with Cataplexy: A Reappraisal." *Journal of Sleep Research*, *13* (4), 395–406. https://doi.org/10.1111/j.1365-2869.2004.00422.x.

Sutcliffe, J. Gregor & Luis de Lecea. (2002). "The Hypocretins: Setting the Arousal Threshold." *Nature Reviews Neuroscience*, *3*(5), 339-349. https://doi.org/10.1038/nrn808.

Szucs, A., Janszky, J., Hollo, A., Migleczi, G. & Halasz, P. (2003). "Misleading Hallucinations in Unrecognized Narcolepsy." *Acta*

Psychiatrica Scandinavica, *108* (4), 314. https://doi.org/10.1034/j.1600-0447.2003.00114.x.

Tafti, Mehdi, Hyun Hor, Yves Dauvilliers, Gert J. Lammers, Sebastiaan Overeem, Geert Mayer, Sirous Javidi, et al. (2014). "DQB1 Locus Alone Explains Most of the Risk and Protection in Narcolepsy with Cataplexy in Europe." *Sleep*, *37* (1), 19–25. https://doi.org/10.5665/sleep.3300.

Talih, Farid Ramzi. "Narcolepsy Presenting as Schizophrenia: A Literature Review and Two Case Reports." *Innovations in Clinical Neuroscience*, *8* (4), 30.

Todman, D. (2007). "Narcolepsy: A Historical Review." *The Internet Journal of Neurology*, *9* (2). http://ispub.com/IJN/9/2/7361.

Vogel, Gerald. (1960). "Studies in Psychophysiology of Dreams." *Archives of General Psychiatry*, *3* (4), 421-428. https://doi.org/10.1001/archpsyc.1960.01710040091011.

Vogel, Gerald W. (1976). "Mentation Reported from Naps of Narcoleptics." *Advances in Sleep Research*, *3*, 161–68.

Voss, Ursula, Romain Holzmann, Allan Hobson, J., Walter Paulus, Judith Koppehele-Gossel, Ansgar Klimke & Michael A Nitsche. (2014). "Induction of Self Awareness in Dreams through Frontal Low Current Stimulation of Gamma Activity." *Nature Neuroscience*, *17* (6), 810–12. https://doi.org/10.1038/nn.3719.

Waihrich, Eduardo Siqueira, Raimundo Nonato Delgado Rodrigues, Henrique Aragao Silveira, Fernando da Fonseca Melo Froes & Guilherme Henrique da Silva Rocha. (2006). "Comparative Analysis of Multiple Sleep Latency Tests (MSLT) Parameters and Occurrence of Dreaming in Patients with Daytime Sleepiness of Narcoleptic and Non-Narcoleptic Origin." *Arquivos de Neuro-Psiquiatria*, *64* (4), 958–62. http://dx.doi.org/10.1590/S0004-282X2006000600014.

Wamsley, Erin, Claire E H M Donjacour, Thomas E Scammell, Gert Jan Lammers & Robert Stickgold. (2014). "Delusional Confusion of Dreaming and Reality in Narcolepsy." *Sleep*, *37* (2), 419–22. https://doi.org/10.5665/sleep.3428.

Wilner, A., Lawrence Steinman, Perez Lavie, Ron Peled, Adam Friedmann & Chaim Brautbar. (1988). "Narcolepsy-Cataplexy in Israeli Jews Is Associated Exclusively with the HLA DR2 Haplotype A Study at the Serological and Genomic Level." *Human Immunology*, *21* (1), 15–22. https://doi.org/10.1016/0198-8859(88)90077-8.

Yoss, Robert E; David D Daly. (1957). "Criteria for the Diagnosis of the Narcoleptic Syndrome." *Proceedings of the Staff Meetings. Mayo Clinic*, *32* (12), 320–28.

In: Parasomnia Dreaming
Editor: James Frederic Pagel
ISBN: 978-1-53617-815-9

Chapter 3

DREAMING IN NREM PARASOMNIAS: AN EMPIRICAL STUDY

Michael Schredl*
Central Institute of Mental Health, Medical Faculty
Mannheim/Heidelberg University, Germany

ABSTRACT

NREM parasomnias – also termed disorders of arousal – are characterized by recurrent episodes of incomplete awakening from sleep (including screams, walking, etc.) often with complete amnesia for the episode. Although research had focused on mentation reports related to these episodes, dream studies in this patient group has not been carried out in a systematic way. Based on questionnaire data of 83 adult NREM parasomnia patients, diagnosed in the sleep lab, the present findings of impaired sleep quality, increased tiredness during the day, and heightened daytime stress levels are in line with previous research. Dream reports provided by 42 patients indicate that their dreams are more negatively toned compared to those of controls, which is compatible with the finding of elevated nightmare frequency in the patient group. In order to corroborate the findings of the present study, sleep laboratory

* Corresponding Author's Email: Michael.Schredl@zi-mannheim.de.

investigations with REM and NREM awakenings to elicit dream reports have to be carried out.

Keywords: NREM parasomnia, sleepwalking, sleep terror/*Pavor nocturnus*, dreaming, nightmares

Introduction

NREM parasomnias are defined as recurrent episodes of incomplete awakening from sleep with inappropriate or absent responsiveness, limited cognition, and partial or complete amnesia for the episode (American Academy of Sleep Medicine, 2014). Due to this incomplete awakening, these parasomnias are also called disorders of arousal (Broughton, 1968). Three types can be differentiated: Confusional arousals (mental confusion and/or confused behavior without leaving the bed), sleepwalking (complex behavior outside the bed), and sleep terrors or pavor nocturnus (abrupt panic, often accompanied by a scream). NREM parasomnias should be differentiated from nocturnal frontal lobe epilepsy which is typically characterized by stereotypical behavior and frequent occurrences per night (Derry, Harvey, Walker, Duncan, & Berkovic, 2009). The prevalence rates (occurrence during the last 12 months) for sleepwalking is higher in children (5.0%) than in adults (1.5%) (Stallman & Kohler, 2016), the prevalence of the two NREM parasomnias are within a similar range (American Academy of Sleep Medicine, 2014). The frequency of NREM parasomnia episodes is often increased by physiological stressors (e.g., fever, sleep deprivation, alcohol consumption) and psychological stressors (e.g., occupational stress) (Avidan, 2017). Although the occurrence of NREM parasomnias are associated with psychiatric disorders like major depression or anxiety disorders (Ohayon, Guilleminault, & Priest, 1999; Waters, Moretto, & Dang-Vu, 2017), depression and anxiety scores elicited via questionnaires are not heightened in comparison to the general population (Labelle, Desautels, Montplaisir, & Zadra, 2013). Interestingly, other sleep disorders like sleep-related breathing disorders or periodic limb movements during sleep can trigger NREM parasomnia events (Bargiotas et al., 2017; Lundetræ et al., 2018) and patients with NREM parasomnias experience increased levels of daytime sleepiness measured with the Epworth Sleepiness Scale (Carrillo-Solano, Leu-Semenescu, Golmard, Groos, & Arnulf, 2016). Also several types of medication, e.g., antidepressants, benzodiazepine receptor agonists, and antipsychotics have been associated with NREM parasomnias (Stallman, Kohler, & White, 2018). The etiological model includes an interaction between slow-wave sleep instability (with possible genetic disposition) and precipitating factors like physiological or psychological stressors (Castelnovo, Lopez, Proserpio, Nobili, & Dauvilliers, 2018).

Despite the often complete amnesia, about 90% of adult persons with the NREM parasomnias of sleepwalking and/or sleep terrors can report at least one episode with recalling dream-like mentation. Systematic studies evaluating the percentage of sleepwalking and/or sleep terror episodes with recall of some mentation in larger patient samples have not been carried out. With the strict definition of dreaming as subjective experiences occurring during sleep (Schredl, 2018), one might argue that the subjective experiences or mentation during the hybrid state of the brain between sleep and waking should be called dreams. These dream-like mentation reports are typically very short, a single image, often negatively toned like collapsing walls, spiders, apartment being on fire, or dangerous intruders (Arnulf, 2019; Oudiette et al., 2009). Although Oudiette et al. (2009) mentioned that the NREM parasomnia patients also report "classical" dreams that were remembered in the morning and can differentiate those from NREM parasomnia-related mentation; the content of these dreams were not analyzed. The occurrence of nightmares seems to be associated with NREM parasomnias (Ohayon et al., 1999) and especially NREM parasomnia patients with marked psychopathology suffer more often from nightmare disorders (Labelle et al., 2013). Interestingly, within this study (Labelle et al., 2013) about 50% of the patients report that the occurrence of bad dreams can increase the frequency of sleepwalking episodes. So far, research has not systematically investigated dreaming that is not related to NREM parasomnia episodes in this patient group.

The aim of this pilot study is to study dream recall, nightmare frequency, and dream content in patients with sleepwalking and/or sleep terrors in comparison to healthy controls. Due to the problems with fully awakening it was hypothesized that dream recall and dream length is reduced in these patients. As nightmares and NREM parasomnias are related to stress in waking life (Avidan, 2017; Schredl, 2018), an increased nightmare frequency in the patient group was expected. Moreover, the nightmare frequency would be related to waking-stress levels.

METHODS

Participants

In the present study, 83 patients (48 women, 35 men) with the diagnosis of Non-REM parasomnia (sleepwalking and/or sleep terrors) and 208 healthy participants (136 women, 35 men) were included. The gender distribution was not significantly different (χ^2 = 1.5, p = .2276). Thirty-one of the patients had night terrors and fifty-two sleepwalked. The patients had an average age of 28.49 ± 8.30 years (range: 13 – 53) and the healthy participants of 30.08 ± 12.03 years (range: 18 – 76). Age also did not differ between the two groups (t = 1.3, p = .2002). Dream content has been reported by 42 of the patients (17 women, 15 men) with an average age of 28.40 ± 9.57 years (range: 16 – 64) and 99 healthy participants (62 women, 37 men) who were on average 30.71 ± 11.42 years old (range: 18 – 76). Gender (χ^2 = 0.0, p = .8515) and the ages of the two groups (t = 1.2, p = .2538) have not been significantly different.

Measurements

Dream Questions

Dream recall frequency in the last few months was measured with a seven-point scale: 0 = never, 1 = less than once a month, 2 = about once a month, 3 = two to three times a month, 4 = about once a week, 5 = several times a week, 6 = nearly every day (Schredl, Berres, Klingauf, Schellhaas, & Göritz, 2014a). This item showed a high retest reliability of r = .83 (Schredl, 2004b). For eliciting nightmare frequency an eight-point scale (0 = never, 1 = less than once a year, 2 = about once a year, 3 = about two to four times a year, 4 = about once a month, 5 = two to three times a month, 6 = about once per week, 7 = several times a week). This scale also showed a high retest reliability of r = .75 (Stumbrys, Erlacher, & Schredl, 2013).

Morning Dream Questionnaire

In a dream questionnaire participants had to rate if they remembered a dream (0 – no, 1 – I dreamed but I cannot remember the content, 2 – yes, I remember one or several dreams with following content) and they had to write down the dream content and rate the intensity of positive and negative dream emotions on two four-point scales (0 = none, 1 = somewhat, 2 = moderate, 3 = strong). The emotional tone has been calculated as the difference of both values. Results for dream recall frequency of the retrospective and the laboratory dream questionnaire showed high correlations of r = .683 and r = .710 (Schredl, 1991; Schredl, Kraft-Schneider, Kröger, & Heuser, 1999).

Sleep Questionnaires

The sleep questionnaire B (Görtelmeyer, 1986), the short SFB-B questionnaire, contains 28 items which ask for sleep habits and subjective sleep parameters during the previous two weeks. In the present analyses the 11 items concerning sleep quality (retest reliability after 4 weeks: r = .68), 8 items about their feeling of being refreshed in the morning (retest reliability after 4 weeks: r = .78) and one about nocturnal awakenings (retest reliability after 4 weeks: r = .61; Görtelmeyer, 1986) were included. These items were transformed – if necessary – to five-point scales and averaged. The item "How did you feel in the morning after awakening? – well rested" with the answer categories: 1 = never, 2 = rare, 3 = sometimes, 4 = often, 5 = very often, would be one example for measuring the feeling of being refreshed in the morning. The scale and item values vary between 1 and 5. Validity of the sleep quality item has been measured by comparing it to medical diagnosis which correlated highly r = -.63 (Görtelmeyer, 1986).

The Landecker Inventory for sleep disorders questionnaire (LISST) contains 75 items (Weeß, Schürmann, & Steinberg, 2002); the following items or indices were analyzed in the present study: nightmare frequency, problematic sleep-wake rhythm with 7 items, complaints of insomnia with 6 items, sleep quality which was coded as higher scores mean worse quality on 8 items, tiredness during the day with 5 items and complaints of parasomnia with 6 items. High reliability could be measured for sleep quality (Cronbach's alpha: 0.898) and tiredness (Cronbach's alpha: 0.830) (Weeß et al., 2002). The sleep quality item showed a high correlation (r = .74) with the sum score of the Pittsburg Sleep Quality Index (Buysse, Reynolds III, Monk, Berman, & Kupfer, 1989). An example which was part of the scale for sleep quality would be "During the night, I get up because I can no longer sleep." All the used items have been measured on a six-point scale (1 = never, 2 = rare, 3 = occasional, 4 = often, 5 = mostly, 6 = always). That is the range of the nightmare item whereas the range for the indices varied as they were computed as sums of the single items.

Stress Questionnaire

The recreation-distress questionnaire EBF (Kallus, 1995) consists of 72 items which have been divided into 12 subtests consisting each of 6 items which all were rated on seven-point scales (0 = never, 1 = rarely, 2 = sometimes, 3 = several times, 4 = often, 5 = very often, 6 = always). Participants had to answer items like "In the last 3 days and nights, I have been aggressive." Of every subtest an average has been calculated and then averaged with the results of other subtests, seven subtests (42 items) were used to compute the distress index and five subtests measured recreation aspects. Internal consistency is sufficient for all the subtests (r = .80 to r = .97; Kallus, 1995). A 24-hour retest of a similar questionnaire showed correlations between r = .79 and r = .91 (Kallus, 1995).

Dream Content Analysis

To evaluate the dream reports, a dream manual has been used (Schredl, Burchert, & Grabatin, 2004). The rater had to judge the level of bizarreness on a four-point scale (1 = in the waking world possible and can happen in the daily activities, 2 = a lot of usual items but some unusual connections but still possible, 3 = one or two fantasy objects, unusual connections and impossible in waking life, 4 = a lot of fantasy objects, bizarre connections and actions and impossible in waking life). Positive and negative feelings have been coded in the following on a four point scale (0 = none, 1 = somewhat, 2 = moderate, 3 = strong). The emotional tone has been calculated as the difference of both values. If the laboratory setting occurred in a dream, this was rated as: (1 = present, 0 = absent. A three-point scale measured the occurrence of problems within the dream: No problems (= 0), minor problems (= 1), e.g., arguing, and major problems (= 2), e.g., of being chased. Interrater reliability showed sufficient results: bizarreness scale with r = .765 and problems scale with r = .816 (Schredl et al., 2004).

Procedure

All patients underwent a diagnostic interview and two nights of polysomnography; these data served as basis for the diagnosis of a NREM parasomnia by a sleep specialist and also ruled out any other sleep disorders like sleep-related breathing disorder or periodic limb movement during sleep. The healthy participants underwent a similar procedure and, if a previously unknown sleep-disorder was detected, the participant was excluded. Before the first polysomnographic recording, participants were asked to complete the sleep, dream, and stress questionnaires. After each of the two consecutive nights in the sleep laboratory with standard sleep measurements (EEG, horizontal and vertical eye movements, electrocardiogram and the submental electromyogram, leg movements and respiration measurements), participants were presented with the dream questionnaire in order to record any dreams – if recalled. In addition, they rated positive and negative emotions. One participant slept three times 2 consecutive nights for diagnostic reasons in the laboratory, questionnaires of the first stay have been used together with the dream reports of all six nights. All dream reports have been transcribed and rated in a randomized order according to the dream manual (see above).

Statistical analyses have been carried out with SAS 9.4 software package for Windows. Dream content of the patients has been compared to dream content of the healthy participants by ordinal (cumulative logit analyses) and parametric regressions. Regressions have been used to analyze the effects of distress, age, gender and word count on dream recall frequency, nightmare frequency and other sleep questionnaire items. Correlations were used to show the relationship between dream recall, nightmare frequency and distress and were controlled for age and gender.

RESULTS

As shown in Table 1, there was no significant difference between NREM parasomnia patients and healthy controls regarding dream recall frequency. However, the patient group showed a higher nightmare frequency (both measures) compared to healthy controls. Increasing age was related to lower dream recall frequency and nightmare frequency, whereas women tended to report higher dream recall and more nightmares. The two nightmare scales correlated with $r = .743$ ($p < .0001$, $n = 259$).

When distress (EBF index) has been added to the regression analyses of nightmare frequency, patients with NREM parasomnia still showed a significantly higher nightmare frequency than healthy controls (Standardized estimate = .4946, $\chi^2 = 47.4$, p = <.0001) with significant effects of age (Standardized estimate = -.2741, $\chi^2 = 15.9$, p = <.0001, gender (Standardized estimate = .2063, $\chi^2 = 10.6$, p = .0011), and distress (Standardized estimate = .2452, $\chi^2 = 12.8$, p = .0003). The LISST item for nightmare frequency showed a similar pattern: group effect (Standardized estimate = .5486, $\chi^2 = 51.9$, p = <.0001), gender (Standardized estimate = .1759, $\chi^2 = 7.1$, p = .0076) and distress (Standardized estimate = .2640, χ^2 =

13.9, $p = .0002$). Only the age effect was no longer significant (Standardized estimate = -.0901, $\chi^2 = 1.7$, $p = .1897$).

The results of the sleep and stress questionnaires are presented in Table 2. Overall, the patients with NREM parasomnia showed higher scores for distress, rhythm problems, complaints of insomnia, sleep quality, parasomnia, and tiredness during the day. The SF-B indices indicate also that the sleep quality and the feeling of being refreshed in the morning were reduced whereas they reported more awakenings during night.

As shown in Table 3, the number of dreams reported by the NREM parasomnia group did not differ significantly from the number of dreams reported by the control group. However, parasomnia patients reported significantly shorter dreams. Whereas dream bizarreness and the occurrence of laboratory references did not differ between the two groups, the patients' dreams more often included some kind of problem the dreamer had to deal with. Whereas there were no differences for positive dream emotions, the dreams were rated as more negatively toned by the patient group compared to the healthy controls.

Table 4 shows that there are no significant effects of age, gender or group membership on positive feelings. Negative feelings occurred slightly more often in the patients than in the healthy controls. Age and gender had no influence on negative feelings. The difference of both did show slightly higher scores for healthy participants but this effect did not reach significance. Age and gender had no effect on this variable.

As the dream reports that were collected in the morning do not allow any conclusions as to whether these reports might be actually referring to a NREM parasomnia episode (although sleepwalking and sleep terror did not occur in the sleep lab), the reports were inspected retrospectively.

Table 1. Dream recall frequency and nightmare frequency including ordinal regression analyses

Variable	NREM Parasomnia N = 83	Healthy Controls N = 208	Analysis								
			Group effect			Age Effect			Gender Effect		
	M ± SD	M ± SD	SE	χ^2	p	SE	χ^2	p	SE	χ^2	p
Dream recall frequency	3.78 ± 1.84	3.58 ± 1.60	.0896	2.3	.1277	-.2089	12.4	.0004	.1377	5.5	.0188
Nightmare frequency	4.73 ± 2.08	2.37 ± 1.89	.5609	66.2	<.0001	-.3054	21.1	<.0001	.2157	12.1	.0005
Nightmare frequency item of LISST	3.30 ± 1.44	1.82 ± 0.78	.5890	70.2	<.0001	-.1486	5.3	.0215	.1658	7.0	.0081

SE = Standardized estimates.

Table 2. Means and standard deviations and parametric regressions for sleep and stress-related variables

Variable		NREM Parasomnia	Healthy Controls	Analysis								
				Group Effect			Age Effect			Gender Effect		
		M ± SD	M ± SD	SE	t	p	SE	t	p	SE	t	p
EBF	Recovery	2.65 ± 0.84	3.48 ± 0.77	-.4275	-7.6	<.0001	.0777	1.4	.1715	.0293	0.5	.6037
	Distress	1.82 ± 0.85	1.24 ± 0.67	.3307	5.7	<.0001	-.1661	-2.9	.0048	.0329	0.6	.5799
LISST	Rhythm problems	19.73 ± 7.16	16.56 ± 6.33	.1871	3.1	.0015	-.1847	-3.2	.0017	-.0885	-1.5	.1285
	Insomnia	16.13 ± 7.23	11.50 ± 4.51	.3465	6.2	<.0001	-.1325	-2.4	.0190	.0158	0.3	.7782
	Sleep Quality	20.91 ± 8.43	14.85 ± 4.76	.4088	7.4	<.0001	-.0640	-1.2	.2490	.0380	0.7	.4921
	Tiredness	14.89 ± 6.00	10.86 ± 3.59	.3759	6.7	<.0001	-.0605	-1.1	.2825	.0079	0.1	.8884
	Parasomnia	19.39 ± 5.82	9.36 ± 2.62	.7714	20.1	<.0001	-.0074	-0.2	.8480	.1099	2.9	.0045
SFB-B	Sleep Quality	3.33 ± 0.73	4.15 ± 0.46	-.5469	-10.9	<.0001	.0889	1.8	.0779	-.0090	-0.2	.8586
	Feeling of being refreshed in the morning	2.66 ±0.81	3.38 ± 0.74	-.3671	-6.8	<.0001	.2222	4.1	<.0001	-.0262	-0.5	.6288
	Nocturnal Awakenings	2.99 ± 1.29	2.27 ± 0.89	.3103	5.4	<.0001	.0438	0.8	.4420	.0990	1.7	.0823

SE = Standardized estimates.

Table 3. Dream content analysis

Variable	NREM Parasomnia N = 42	Healthy Controls N = 99	Analysis											
			Group effect			Age Effect			Gender Effect			Word count		
	M ± SD	M ± SD	SE	t	p	SE	t	p	SE	t	p	SE	t	p
Number of dreams	1.33 ± 0.57	1.25 ± 0.44	.0733	0.9	.3924	-0.346	-0.4	.6866	.0411	0.5	.6301			
Word count	42.92 ± 44.23	76.21 ± 72.32	-.2384	-2.9	.0046	-.0844	-1.0	.3096	.1020	1.2	.2182			
Bizarreness	1.71 ± 0.85	1.82 ± 0.84	.0629	0.8	.4117	-.0176	-0.3	.8132	-.1068	-1.4	.1514	.5227	6.9	<.0001
Problems	0.45 ± 0.60	0.29 ± 0.47	.1976	2.3	.0213	-.0367	-0.4	.6574	-.0235	-0.3	.7757	.2603	3.1	.0026
Laboratory	35.71%	37.37%	-0.558	0.3	.5845	-.2330	4.3	.0384	-0.231	0.1	.8148	-.0737	0.1	.4821

SE = Standardized estimates.

Table 4. Self-rated positive and negative feelings

Variable	NREM Parasomnia N = 42	Healthy Controls N = 99	Analysis								
			Group effect			Age Effect			Gender Effect		
	M ± SD	M ± SD	SE	χ^2	p	SE	χ^2	p	SE	χ^2	p
Positive emotions	1.17 ± 1.04	1.29 ± 0.95	-0.507	-0.5	.6046	.0403	0.4	.6816	.0669	0.7	.4925
Negative emotions	1.55 ± 1.09	0.99 ± 0.92	.2352	2.5	.0143[1]	-.0844	-0.9	.3753	.0967	1.0	.3061
Emotional tone	-0.38	0.30 ± 1.49	-.1820	-1.9	.0615[1]	.0785	0.8	.4186	-.0219	-0.2	.8194

SE = Standardized estimates, [1]one-tailed.

Based on the findings reported in the literature (Arnulf, 2019), eight reports might be associated with a NREM parasomnia episode, e.g., “The dream was related to the sleep investigation. I was with people who also underwent polysomnography. I remember everything only faintly, but I felt small and scared of the others.” or “If someone does something in the electrode, I would have to die, and then it finally happened”. One patient reported recalling a parasomnia episode and a normal dream – differentiating between the two: “1. Typical incident early in the night (feeling half sleep/half awake): I am looking for my daughter. I feel like I'm looking for her in the bed. I think she's behind the closet door, but emotions are going downhill. I cannot get to her. Panic, I sit up and hit against the closet. Intense panic, palpitations. When I realized that I'm in bed in the sleep lab, I calmed down. 2. "Normal Dream": I am somehow traveling with several people (cannot remember with whom) and looking for a place to stay.”

In order to validate the findings reported in Tables 3 and 4, we repeated the analysis but excluded all patients who reported NREM parasomnia-like mentation. The following regression analyses were similar to those presented in Tables 3 and 4, even though only the group effect is reported here. The mean word count was still different between the patient group (47.13 ± 47.26, N = 35) and the healthy controls (Standardized estimate = -.1980, t = -2.3, p = .0225). Similar to the analysis in Table 3, the patients' dreams included more problematic themes (0.44 ± 0.62) than the dreams of the control group (Standardized estimate = .1743, t = 2.2, p = .0450). In addition, the self-rated negative emotions were also more intense in the patient group (1.42 ± 1.06) compared to the control group (Standardized estimate = .1767, t = 1.8, p = .0363, one-tailed).

No relationship between dream recall frequency and distress was shown for both groups (see Table 5). But distress was positively related to both nightmare frequency in NREM parasomnia patients and – to a smaller extent – in the healthy control group.

DISCUSSION

Overall, the findings indicate that patients with NREM parasomnias also report more often impaired sleep quality, tiredness during the day, and higher daytime stress levels. Moreover, the dreams not associated with NREM parasomnia episodes are more negatively toned, reflected also in the finding of an increased nightmare frequency.

Prior to discussing the findings in detail, several methodological issues have to be addressed. The major issue is the method of dream collection. The dreams were reported in the morning not immediately after awakening (detaching the electrodes was the first thing). So, there is no possibility of matching reports with sleep stages or NREM parasomnia episodes. Since Oudiette et al. (2009) reported that none of their 43 patients had any recollections regarding NREM parasomnia episodes, one might argue that most of the reports are "classic" dreams. Another aspect is that in our sleep lab no *pavor nocturnus* or sleepwalking episodes (limiting length of electrode wires) occurred while mild and brief episodes of confusional arousal (sitting up, looking around) did occur rarely but were not recorded and analyzed systematically. Moreover, the findings didn't change by removing reports that might be recollections of sleepwalking or sleep terrors. The next step would be to carry out REM (and NREM) awakenings and elicit dream reports directly via intercom (see: Schredl, 2018). The findings regarding the gender difference in nightmare frequency (women tend to report more nightmares than men) and the relationship between stress and nightmare frequency support the validity of the present findings as this has been reported in previous studies (Levin & Nielsen, 2007; Schredl & Reinhard, 2011). Another issue is addressed by Labelle et al. (2013): can the heightened psychopathology, increased daytime sleepiness/tiredness, or presence of other sleep disorders in adult NREM parasomnia patients that has been reported previously (Avidan, 2017; Bargiotas et al., 2017; Carrillo-Solano et al., 2016) be explained by the fact that most samples consisted of help-seeking patients and that patients with no psychopathology or impaired sleep quality might not consult a sleep center. Our clinical experience would indicate that most NREM parasomnia patients seek help because of previously experienced dangerous situations and/or injuries, like cutting bread with a big, sharp knife or jumping out of the window thinking the apartment is on fire. However, empirical data regarding the reasons as to why these patients seek help have not been collected in a systematic way. One of the strengths of the study is that all patients underwent at least two nights of polysomnography thereby ruling out the presence of any other sleep disorder. Furthermore, the presence of a mental disorder like depression, anxiety disorder can also be ruled out since the sleep specialists are also psychiatrists who carefully evaluated the patients for possible psychopathology.

Home dream recall did not differ between the adult patients with sleepwalking/sleep terrors and the healthy controls. In the habitual sleep-wake rhythm at home, there are no possible negative effects of prolonged sleep inertia that can be seen in these patients (Avidan, 2017). It would be very interesting to carry out a REM and NREM awakening study in order to test whether dream recall is especially reduced after awakenings in the first part of the night but not in the second part. However, patients reported shorter dreams, and as dream length is also an indicator for the ability to recall dreams (Schredl, 2004a) this might indicate that the basic mechanisms present in the arousal disorders also affect dream recall.

Regarding nightmares, the hypothesis of heightened nightmare frequency in patients with NREM parasomnias was confirmed. Although daytime stress correlated with nightmare frequency in the patient sample, the stress measure did not explain the differences in nightmare frequency between NREM parasomnia patients and controls (controlled in a second set of regression analyses). Although some patients completed a nightmare frequency scale that included the following definition: "Nightmares are dreams with strong negative emotions that result in awakening from the dreams. The dream plot can be recalled very vividly upon awakening." (Schredl, Berres, Klingauf, Schellhaas, & Göritz, 2014b) most of the patients and controls did not have this specific definition, so one might argue that some patients did not differentiate between sleep terrors and nightmares (Oudiette et al., 2009). On the other hand, one has to keep in mind that NREM parasomnias in adults occur quite rarely, thus, the over-estimation of nightmare frequency due to including sleep terrors would not be that pronounced. It would be desirable to study this methodological issue in a more detailed way.

The finding of elevated nightmare frequency raises the question of how NREM parasomnias might be linked to the REM parasomnia nightmares. One study (Haridi et al., 2017) reported higher phasic activity in NREM parasomnia patients during REM sleep compared to controls and they thus might also be experiencing more intense REM dreaming. This hypothesis must be tested with REM awakening studies. The more intense negative dream emotions found in the patient group would support this line of thinking. The dream contents of the patients included problems more often than healthy controls' dreams but the number of dream reports is relatively small. Large-scaled home-based studies using dream diaries or voice-recorders (Schredl et al., 2019) are needed.

To summarize, this pilot study provided a first glimpse into dreaming in NREM parasomnia patients that is not related to sleepwalking/sleep terror episodes. The findings indicate that dreams are more negatively toned but systematic sleep lab studies carrying out REM and NREM awakenings to elicit dream reports have to be carried out to corroborate the present findings.

REFERENCES

American Academy of Sleep Medicine. (2014). *The international classification of sleep disorders.* (ICSD-3). Darien, IL: AASM.

Arnulf, I. (2019). Dreaming in parasomnias. In K. Valli & R. J. Hoss (Eds.), *Dreams: Understanding biology, psychology, and* culture - Volume 1 (pp. 238-249). Santa Barbara: Greenwood.

Avidan, A. Y. (2017). Non-rapid eye movement parasomnias: Clinical spectrum, diagnostic features, and management. In M. Kryger, T. Roth, & W. C. Dement (Eds.), *Principles and practice of sleep medicine* (6th edition) (pp. 981-992). Philidelphia: Elsevier.

Bargiotas, P., Arnet, I., Frei, M., Baumann, C. R., Schindler, K., & Bassetti, C. L. (2017). Demographic, clinical and polysomnographic characteristics of childhood- and adult-onset sleepwalking in adults. *European Neurology,* 78(5-6), 307-311.

Broughton, R. J. (1968). Sleep disorders: Disorders of arousal? *Science,* 159, 1070-1078.

Buysse, D., Reynolds III, C., Monk, T., Berman, S., & Kupfer, D. (1989). The Pittsburgh Sleep Quality Index: A new instrument for psychiatric practice and research. *Psychiatry Research,* 28, 193-213.

Carrillo-Solano, M., Leu-Semenescu, S., Golmard, J.-L., Groos, E., & Arnulf, I. (2016). Sleepiness in sleepwalking and sleep terrors: a higher sleep pressure? *Sleep Medicine,* 26, 54-59. doi: https://doi.org/10.1016/j.sleep.2015.11.020.

Castelnovo, A., Lopez, R., Proserpio, P., Nobili, L., & Dauvilliers, Y. (2018). NREM sleep parasomnias as disorders of sleep-state dissociation. *Nature Reviews Neurology,* 14(8), 470-481. doi: 10.1038/s41582-018-0030-y.

Derry, C. P., Harvey, A. S., Walker, M. C., Duncan, J. S., & Berkovic, S. F. (2009). NREM Arousal Parasomnias and Their Distinction from Nocturnal Frontal Lobe Epilepsy: A Video EEG Analysis. *Sleep,* 32(12), 1637-1644. doi: 10.1093/sleep/32.12.1637.

Görtelmeyer, R. (1986). Schlaf-Fragebogen A und B (Sf-A, Sf-B) (Sleep Questionnaires A and B). In S. Collegium Internationale Psychiatriae (Ed.), *International Skalen für Psychiatrie.* Weinheim: Beltz.

Haridi, M., Weyn Banningh, S., Clé, M., Leu-Semenescu, S., Vidailhet, M., & Arnulf, I. (2017). Is there a common motor dysregulation in sleepwalking and REM sleep behaviour disorder? *Journal of Sleep Research,* 26(5), 614-622. doi:10.1111/jsr.12544

Kallus, K. W. (1995). *Erholungs-Belastungs-Fragebogen (EBF) (Recreation-Distress Questionnaire).* Frankfurt: Swets & Zeitlinger.

Labelle, M.-A., Desautels, A., Montplaisir, J., & Zadra, A. L. (2013). Psychopathologic correlates of adult sleepwalking. *Sleep Medicine,* 14(12), 1348-1355. doi:http://dx.doi.org/10.1016/j.sleep.2013.05.023.

Levin, R., & Nielsen, T. A. (2007). Disturbed dreaming, posttraumatic stress disorder, and affect distress: a review and neurocognitive model. *Psychological Bulletin,* 133, 482-528.

Lundetræ, R. S., Saxvig, I. W., Pallesen, S., Aurlien, H., Lehmann, S., & Bjorvatn, B. (2018). Prevalence of Parasomnias in Patients With

Obstructive Sleep Apnea. A Registry-Based Cross-Sectional Study. *Frontiers in Psychology,* 9, 1140.

Ohayon, M. M., Guilleminault, C., & Priest, R. G. (1999). Night terrors, sleepwalking, and confusional arousals in the general population: their frequency and relationship to other sleep and mental disorders. *Journal of Clinical Psychiatry,* 60(4), 268-276; quiz 277.

Oudiette, D., Leu, S., Pottier, M., Buzare, M.-A., Brion, A., & Arnulf, I. (2009). Dreamlike mentations during sleepwalking and sleep terrors in adults. *Sleep,* 32, 1621-1627.

Schredl, M. (1991). *Traumerinnerungshäufigkeit und Trauminhalt bei Schlafgestörten, psychiatrischen Patienten und Gesunden (Dream recal frequency and dream content in patients with sleep disorders, psychiatric patients and healthy controls).* Universität Mannheim: Unveröffentlichte Diplomarbeit.

Schredl, M. (2004a). Dream report length and dream recall frequency: are they affected by the same factors? In S. P. Shohov (Ed.), *Advances in Psychology Research* (Vol. 28, pp. 15-25). New York: Nova Science Publisher.

Schredl, M. (2004b). Reliability and stability of a dream recall frequency scale. *Perceptual and Motor Skills,* 98, 1422-1426. doi:http://dx.doi.org/10.2466/pms.98.3c.1422-1426.

Schredl, M. (2018). *Researching Dreams: The Fundamentals.* Cham: Palgrave Macmillan.

Schredl, M., Berres, S., Klingauf, A., Schellhaas, S., & Göritz, A. S. (2014a). The Mannheim Dream questionnaire (MADRE): German and English versions. *International Journal of Dream Research,* 7(2), (https://doi.org/10.11588/ijodr.12014.11582.16798).

Schredl, M., Berres, S., Klingauf, A., Schellhaas, S., & Göritz, A. S. (2014b). The Mannheim Dream questionnaire (MADRE): Retest reliability, age and gender effects. *International Journal of Dream Research,* 7, 141-147. doi:https://doi.org/10.11588/ijodr.2014.2.16675.

Schredl, M., Burchert, N., & Grabatin, Y. (2004). The effect of training on interrater reliability in dream content analysis. *Sleep and Hypnosis,* 6, 139-144.

Schredl, M., Dreer, J., Mösle, A., Rall, M., Rauch, L., Rose, S., & Seuffert, S. (2019). Voice-recorded vs. written dream reports: A research note. International *Journal of Dream Research,* 12(1), 138-140.

Schredl, M., Kraft-Schneider, B., Kröger, H., & Heuser, I. (1999). Dream content of patients with sleep apnea. *Somnologie,* 3, 319-323.

Schredl, M., & Reinhard, I. (2011). Gender differences in nightmare frequency: a meta-analysis. *Sleep Medicine Reviews,* 15, 115-121.

Stallman, H. M., & Kohler, M. (2016). Prevalence of Sleepwalking: A Systematic Review and Meta-Analysis. *PLoS ONE,* 11(11), e0164769. doi:10.1371/journal.pone.0164769.

Stallman, H. M., Kohler, M., & White, J. (2018). Medication induced sleepwalking: A systematic review. *Sleep Medicine Reviews,* 37, 105-113. https://doi.org/10.1016/j.smrv.2017.01.005.

Stumbrys, T., Erlacher, D., & Schredl, M. (2013). Reliability and stability of lucid dream and nightmare frequency scales. *International Journal of Dream Research,* 6, 123-126. http://dx.doi.org/10.11588/ijodr.2013.2.11137.

Waters, F., Moretto, U., & Dang-Vu, T. T. (2017). Psychiatric Illness and Parasomnias: a Systematic Review. *Current Psychiatry Reports,* 19(7), 37. doi:10.1007/s11920-017-0789-3.

Weeß, H.-G., Schürmann, T., & Steinberg, R. (2002). Das Landecker Inventar für Schlafstörungen (The Landeck Inventory for Sleep Disorders (III-2.3.2). In H. Schulz (Ed.), *Kompendium Schlafmedizin (Compendium Sleep medicine)* (pp. 1-3). Landsberg: Ecomed.

In: Parasomnia Dreaming
Editor: James Frederic Pagel
ISBN: 978-1-53617-815-9

Chapter 4

SLEEP PARALYSIS

Brigitte Holzinger[1,2], PhD and Lucille Mayer[2]
[1]Sleep Coaching, Medical University Vienna, Vienna, Austria
[2]Institute for Consciousness and Dream Research, Vienna, Austria

ABSTRACT

Sleep paralysis (SP) is defined as a temporary inability to move at sleep onset or awakening which is caused by muscle atonia, a hallmark of rapid eye movement (REM) sleep, persisting into wakefulness. Consciousness during the episode is preserved and subjects fully recall the episodes afterwards. Recurrent isolated SP is recognized as a formal diagnosis (G47.51) in the ICSD-3 but not in the ICD-10 and DSM-V measures. SP episodes are typically accompanied by bizarre and often terrifying hallucinations and many denote a sense of pressure on their body or feeling a scary presence in the room. During SP, the experient is aware of being paralyzed, alone in the dark, and in a defenseless supine position. Therefore, SP has always been fascinating and a topic prone to supernatural beliefs and is mentioned since ancient times. SP is most likely to develop in adolescence and has its peak occurrence in the 20s and 30s, a second phase of development can be seen in the elderly above 60 years. The risk for SP is higher in case of stressful life phases, prolonged sleep deprivation or irregular sleep-wake rhythms like shift work or alcohol consumption. A maternal form of transmission has been suggested. The disorder has been associated with worse sleep quality and insomnia, as well as higher daytime sleepiness, depression, anxiety and PTSD symptoms. For differential diagnosis, similar disorders like false awakenings or nightmare disorders have to be ruled out. Several pharmacological and therapeutical interventions have been proven effective. A lucid dream (LD) is defined as a dream during which one is aware of the dream state and able to volitionally control its content. Lucid Dreaming Therapy may lead to the disappearance of frightening hallucinations and anxiety related to SP episodes and increase ego strength as well as emotional safety.

Introduction

> "One night I woke up to what I thought was the sound of footsteps in my hallway. I was laying on my back, which immediately made me confused and a bit nervous because I never sleep on my back. I tried to turn my head to see where the sound was coming from, but I couldn't move at all. The panic set in and I heard the source of the footsteps come into my room. Now, although I could not see the figure, I knew it was a home invader. I could feel him standing on the opposite side of my bed, silently staring down on me. He had some sort of weapon in his right hand, I believe it was a knife. I swear my heart has never dropped and raced like that in my entire life. I kept trying and trying to move or scream, but I was frozen in place. It felt like hours." – From gingeroblivion, Reddit.

Sleep paralysis (SP) is a recurring temporary inability to move the trunk and all of the limbs at sleep onset (hypnagogic or predormital form) or upon awakening from sleep (hypnopompic or postdormital form) (American Academy of Sleep Medicine 2014; Cheyne, Rueffer, and Newby-Clark 1999; Denis et al. 2018; B. Sharpless and Doghramji 2015). The characteristic clinical feature of SP is the complete inability to move (not only heaviness) in the presence of full wakefulness (Högl and Iranzo 2017) which is, in most cases, accompanied by intense and often frightful hallucinations (Cheyne and Girard 2007). During the episodes, consciousness is preserved and full recall is present afterwards (Sharpless and Kliková 2019). Due to the feelings of helplessness and fear that often accompany the experiences, SP can lead to anxiety of initiating sleep, but is generally regarded a benign condition (American Academy of Sleep

Medicine 2014). The inability to move is caused by muscle atonia, otherwise a hallmark of physiologic rapid eye movement (REM) sleep, and in case of SP persists and extends into wakefulness (Högl and Iranzo 2017). The parasomnia is therefore closely linked to REM-sleep and might easily be confused with other REM-sleep disorders, anxiety dreams or nightmares. In this chapter, we will offer diagnostic criteria for SP and explain neurophysiological correlates responsible for the episodes and concurrent hallucinations. At-risk groups and vulnerable individuals with high probability to suffer from SP will be discussed. Then, psychopharmacological and non-pharmacological treatment options are shortly introduced. Last but not least, sleep-related factors that influence the occurrence of SP, including dreams and lucid dreaming, are discussed. Investigating SP and its relation to lucid dreaming seems holding potential information for a deeper understanding of sleep, especially REM sleep. Differential diagnoses are listed to show what sets SP apart from similar phenomena. First, however, we will summarize the occurrence of SP over the years and across different cultures around the world.

Occurence in Other Cultures and Former Times

Reports of SP-like experiences reach back to hellenistic times and have over the centuries given fuel to fear driven explanations such as demons, visitations and abductions (Wróbel-Knybel et al. 2018). It has always been a fascination and a topic prone to supernatural beliefs and is mentioned since ancient times. SP is a global phenomenon, the experience being interwoven with a culture´s folklore in many places (Law and Kirmayer 2005; Ness 1978). The experience of SP is first found in the writings of Hippocrates (~400 BC), and the Greeks named it *ephialtes*, roughly translated as "to pounce upon someone". In his book "Oneirocritica", Artemidorus of Daldis associated *ephialtes* with the god Pan, who could have sex with the dreamer and this was viewed as a promise of a great future. In the Christian era of the Middle Ages the parasomnia was connected to demons sexually harassing their victims, often in the form of very beautiful women, called succubus or their male counterpart, incubuses. Supernatural lovers feature in the mythology of other cultures too, including the Arabian *jinns*, Greek *satyrs*, Hindu *bhoots* and Celtic *dusii* (de Sá and Mota-Rolim 2016). *Sharpless and Doghramji (2015)* found a strong convergence of SP descriptions across time and place, which speaks for a certain cultural invariance for the phenomena in the sense that many denote a sense of pressure or crushing on the body. The resulting sense of fear or dread seems to be universal as well. Nevertheless, the creature that is the "oppressor" varies across the world. In China for instance, it was or is believed that SP is caused by ghosts tying down those affected. In Newfoundland SP was, and sometimes still is, called the „Old Hag Phenomenon", because the paralysis is said to be caused by old witches who haunt the victim (Wing, Lee, and Chen 1994). The Brazilian folklorist Cascudo (2012) describes the "Pisadeira", a crone with long fingernails who lurks on roofs at night and is the "nightmare personified in an old man or woman". The creature would trample on the chest of the sleeping and go away only when the person woke up. The Inuit believe that the shamans are the main responsible for the *uqumangirniq,* an experience closely linked to the spiritual world which shares the characteristics of a SP experience. In the Japanese and Far East tradition, the *kanashibari* is the cultural equivalent of SP and a popular phenomenon often represented in *mangas* (Cascudo 2012). *Kanashibari* can be caused by possession, ghosts, or even humans who summon a vengeful spirit to suffocate their enemies (de Sá and Mota-Rolim 2016). While some cultures typically see common animals from their environment, others envision shape-shifting humans, spirits, demons or other supernatural entities (Sharpless and Doghramji 2015). These beliefs are not a thing of the past, for instance some people in the United States still take SP as an abduction by aliens (McNally and Clancy 2005).

Diagnostic Criteria

While secondary SP is one of the symptoms of the neurological disorder narcolepsy, isolated SP occurs in the absence of narcolepsy (American

Academy of Sleep Medicine 2014). Narcolepsy is a chronic neurological disease manifesting as difficulty in maintaining continuous wake and sleep. Clinical presentation usually requires excessive daytime sleepiness occurring alone or together with features of rapid-eye movement (REM) sleep dissociation (e.g., cataplexy, hypnagogic/hypnopompic hallucinations, sleep paralysis), and disrupted nighttime sleep (American Academy of Sleep Medicine 2014). If an individual suffers from SP, it should be investigated if the individual also suffers from narcolepsy, which is usually done with the help of polysomnography. Sometimes the term *fearful isolated sleep paralysis* is used to indicate cases where episodes are causing clinically significant fear and/or distress (Sharpless and Doghramji 2015). In case of repeated episodes, the term *recurrent (fearful) isolated sleep paralysis* (RISP) can be used, though there is no agreement on exactly how often episodes need to occur in order to be considered recurrent (Ohaeri et al. 2004). An episode can last seconds to minutes but can be aborted by sensory stimulation or by the patient making intense effort to move (American Academy of Sleep Medicine 2014).

At present, RISP is recognized as a formal diagnosis (G47.51) in the ICSD-3 (table 1), but lacks a diagnostic code in both the tenth edition of the International Statistical Classification of Diseases and Related Health Problems (ICD-10; *World Health Organization, 1992*) and the fifth edition of Diagnostic and Statistical Manual of Mental Disorders (DSM-5; *American Psychiatric Association, 2013*). However, RISP can be coded as an unspecified sleep–wake disorder (DSM-5: 307.40) or other specified sleep–wake disorder (DSM-5: 307.49) (Sharpless, 2016). The diagnosis is not based upon any formal medical tests (eg, polysomnography), but upon structured and semistructured clinical interviews and/or questionnaires. The former are preferred due to the oftentimes-confusing phenomenology of clinical presentations and the necessity for differential diagnosis. A discrepancy between assessment by questionnaire and clinical interview might occasionally be observed (Bassetti 2003). The assessment of RISP should be made within the broader diagnostic context of other psychiatric conditions, for instance by using the Structured Clinical Interview for DSM-5 Disorders by First, Williams, Karg, and Spitzer (2014) (Sharpless 2016).

Table 1. International Classification of Sleep Disorders, Third Edition, Diagnostic Criteria for Recurrent Isolated Sleep Paralysis (American Academy of Sleep Medicine 2014)

Diagnostic Criteria for Recurrent Isolated Sleep Paralysis	
Criteria A-D must be met	
A	A recurrent inability to move the trunk and all of the limbs at sleep onset or upon awakening from sleep.
B	Each episode lasts seconds to a few minutes.
C	The episodes cause clinically significant distress including bedtime anxiety or fear of sleep.
D	The disturbance is not better explained by another sleep disorder (especially narcolepsy), mental disorder, medical condition, medication, or substance use.

SP – A REM-SLEEP DISORDER

During sleep, the brain cycles regularly between wakefulness, nonrapid eye movement (NREM) sleep and rapid eye movement (REM) sleep (Fleetham and Fleming 2014). When transitions between these stages are blurred, parasomnias occur, one of which is SP. More specifically, SP combines elements of REM sleep with elements of wakefulness (Denis 2018). Therefore, SP is characterized as a "mixed state of consciousness". Electroencephalographic (EEG) activity during SP episodes was found to show abundant alpha activity, mixed with more typical REM activity. Alpha activity in the EEG is typically associated with quiet wakefulness and is normally absent during REM sleep (Denis 2018). Due to this activity, individuals are aware of their condition and are able to describe the experience vividly afterwards (Hishikawa et al. 1976). They are also able to perceive environmental events during the episode and report accurately on them after waking (Hishikawa et al. 1976). Denis (2018) states, that the conscious state of SP can be modeled within the activation-input-modulation (AIM; *Hobson and McCarley 1977*) framework which describes three parameters within which different conscious states differ. Waking can be modeled as: 1 – a state of high neural activation; 2 – a dominance of external (environmental) over internal input; and 3 – high monoaminergic modulation. Normal REM sleep consists of: 1 – high activation, 2 – dominance of internal over external input, and 3 – cholinergic neuromodulation thought to account for the bizarre features of dream mentation. SP can be modeled as a mixture of these two states: 1 – high activation, 2 – mixed external and internal input, 3 – mixed neuromodulation.

REM sleep is characterized by rapid eye movements, cortical activation, vivid dreaming, skeletal muscle paralysis (atonia), and muscle twitches (Aserinsky and Kleitman 1953; Dement and Kleitman 1957). Possibly to prevent dangerous movements during dreams from occuring, muscle atonia in all parts of the body (except eyes and respiratory system) is a natural feature of REM sleep (Fraigne et al. 2015). The state of SP is thus believed to arise from an ongoing continuation of the REM-induced muscle paralysis into a waking state (Mahowald and Schenck 2005), which has also been shown in polysomnographic studies. As Shelley R. Adler argues: "It is important to remember that sleep paralysis is a normal physiological correlate of REM sleep; it is the consciousness of the paralysis that is 'abnormal'" (Adler, 2011: 79)

During SP episodes, eyes are usually closed, but the possibility to open them during SP has been proven (Denis 2018).

Brain Physiology behind REM-Sleep and SP

REM sleep is most closely linked to SP. In the transition from non-REM sleep to REM sleep, the brain increasingly focuses inwards and becomes less effected by input from the external environment (Waters et al. 2016). The connectivity between anterior frontal and posterior sensory regions is suspended. Evidence suggests that a brainstem region in the pons known as the sub-laterodorsal nucleus (SLD), also called the subcoeruleus, is important for REM sleep generation (Jouvet 1962, as cited in Torontali, Grace, Horner, & Peever, 2014). While pharmacological and electrical activation of SLD neurons produces a REM-like state, neurodegeneration of the SLD region is associated with REM sleep behaviour disorder in humans and SLD lesions produce REM sleep without atonia in animals (Torontali et al. 2014). REM sleep is marked by intense cortical activation and absence of skeletal muscle tone. REM sleep paralysis is a natural feature of normal REM sleep, triggered by the pontine reticular formation (pons) and ventromedial medulla which suppress skeletal muscle tone via interneurons of the spinal cord, through the inhibitory neurotransmitters GABA and glycine (Brooks and Peever 2012; Jalal and Hinton 2013). Acetylcholine appears to suppress respiratory motoneuron activity during natural REM sleep (Fraigne et al. 2015). Furthermore, the authors suggest that acetylcholine not only acts to directly entrain the core of the REM sleep circuitry, but also modulates the glutamatergic mechanisms that underlie REM sleep muscle control. Grace, Vanstone, and Horner (2014) showed that inactivation of cholinergic receptors increased the latency of NREM-to-REM sleep transitions and caused a greater proportion of these transitions to fail into entering REM sleep. Cholinergic activity and subcoeruleus interact and reinforce each other to let neural systems alternate between states of consciousness (Fraigne et al. 2015). Activation of melanin-concentrating hormone (MCH) neurons of the hypothalamus prolong the duration of REM sleep. Over-expression of REM sleep characteristics or untimely activation of REM sleep circuitry are the pathological causes of several sleep disorders (Fraigne et al. 2015). Accordingly, Hishikawa and Shimizu (1995) suggest that SP may be produced by hyperactivation of cholinoceptive and/or cholinergic sleep-on neurons or hypoactivation of noradrenergic or serotonergic Sleep-off neurons in the pons. A distinctive feature of SP, Cheyne, Rueffer, and Newby-Clark (1999) argue, is the anomalous combination of high levels of exogenous and endogenous sources of cortical activation.

Low CSF hypocretin-1 has been linked with increased frequency of SP and narcolepsy (Denis, 2018).

HALLUCINATIONS IN SP

In addition to the ongoing muscle atonia, around 75% of SP episodes are typically accompanied by a wide range of bizarre and often terrifying hallucinations (Cheyne et al., 1999). The sense of dread and terror is consistent with observed elevated heart rate or tachycardia during SP (Fukuda 1989). Some patients experience predormital or postdormital hallucinations at separate times from episodes of SP (American Academy of Sleep Medicine 2014). SP hallucinations may include sensing the presence of others, the sensation of external pressure on the chest, hearing odd sounds such as footsteps, and feelings of falling or spinning. As previously described, these episodes are often interpreted as supernatural experiences that are elaborated in culturally distinct ways (see for example, Cheyne et al., 1999; Hufford, 2005). Three broad groups of hallucination types could be elaborated: vestibular motor (V-M), intruder, and incubus (Cheyne and Girard 2007). V-M hallucinations often include feelings of floating, falling, or flying, as well as out-of-body experiences (OBEs), autoscopy (seeing oneself from an external station point), and illusory motor movements, which can range from simple arm movements to sitting up and moving around the room. Intruder hallucinations involve the feeling of a presence like reports of paranormal creatures (e.g., aliens, witches), sensory hallucinations (e.g., footsteps, verbal threats), or the sensation that something is pulling at the bed covers. Incubus hallucinations involve difficulty breathing, sensations of pressure (typically on the chest), feelings of strangulation or choking, and the feeling of impending death. For intruder and incubus hallucinations these sensations occur both as isolated sensations and as more elaborate narratives. Intruder and Incubus hallucinations usually co-occur and are accompanied by fear, whereas V-M hallucinations are more positive, involving feelings of bliss (Cheyne, 2003). The vast majority (up to 90%) of sleep paralysis episodes are associated with fear (Cheyne, Rueffer, and Newby-Clark 1999). SP typically occurs when sleeping in supine position (Fukuda et al. 1998; B. A. Sharpless et al. 2010) and with full stomach (de Sá and Mota-Rolim 2016). Hallucinatory episodes may be present but are not essential for the diagnosis (American Academy of Sleep Medicine 2014).

Brain Physiology behind SP hallucinations

Hishikawa et al. (1978) report that SP and hypnagogic hallucinations were associated with sleep-onset REM periods with a latency of less than 2 minutes from physiological indications of REM onset and that transient alpha activity in the EEG is higher during SP.

Jalal (2018) describes that activation of the serotonin 2A receptor (5-HT2A) has been shown to induce visual hallucinations, "mystical" subjective states, OBEs, and can modulate fear circuits. 5-HT2A receptor activation induces visual hallucinations by increasing cortical excitability and altering visual-evoked cortical responses (Kometer et al. 2013). A high number of 5-HT2A receptors can be found in the visual cortex (Moreau, Amar, Le Roux, Morel, & Fossier, 2010), and alterations of their density is associated with experiencing visual hallucinations in schizophrenia and Parkinson´s disease (González-Maeso et al. 2008; Huot et al. 2010). Serotonergic hallucinations ("pseudo-hallucinations"), induced by hallucinogenic drugs, tend to be "dream-like", with the dreamer having meta-awareness or insight. Hallucinatory experiences during SP have the classic features of serotonergic hallucinations (awareness, dream-like & mystical hallucinations, separation from the self, sensory distortions) and are similar to those induced by hallucinogenic drugs (Jalal, 2018). 5-HT2A receptors are distributed in higher-level cortical networks such as the default mode network (DMN) (e.g., Tagliazucchi et al., 2016) which mediates processing of self-related information. Hallucinogenic drugs like psilocybin and LSD (5-HT2AR agonists) can induce OBE-type events like feelings of derealization and ego dissolution (Lebedev et al. 2015; Tagliazucchi et al. 2016). The temporoparietal junction (TPJ) is a region of the DMN and directly implicated in OBEs and hallucinating human-like figures (Arzy et al. 2006). Serotonin 2A receptors are also densely concentrated in the limbic system and individuals under the influence of 5-HT2A agonists may occasionally have extreme fear reactions and experience paranoia (Jalal, 2018). Research suggests that 5-HT2A receptor activation leads to greater attribution of personal relevance to otherwise meaningless cues (Preller et al. 2017), which can be seen in the dysfunctional "personal relevance attribution" in psychotic disorders. In effect, in SP multisensory cues are more likely to be interpreted in a personalized way and incorporated into an elaborate narrative (Jalal, 2018). Whereas the neurology and chemistry described here is in some ways related to the experience of dreaming, the SP experience is quite different than a vivid dream - as will be discussed below.

SP and Fear

Surveys report that over 80% of respondents report fear and two thirds rate their fearfulness during episodes near the top of the scales provided (Cheyne and Girard 2007). During SP the experient is aware of being paralyzed, alone in the dark, and in a defenseless supine position. This awareness will almost certainly increase fear and further reinforce the activation of the threat activated vigilance system (TAVS; Cheyne, 2003) and therefore increase the probability that SP episodes contain more fear and violence than do dreams (Parker and Blackmore 2002). The TAVS includes the amygdala and related structures and helps to detect predators and assess risks under conditions of uncertain, hidden, or ambiguous threat (Blanchard and Blanchard 1972). Cheyne (2001) argues, that the endogenous activation of the TAVS during REM and SP are experienced as an ominous but elusive sense of presence. The experience of SP and the often concurring threatening hallucinations can be terrifying, and people often report marked distress following episodes (Postepisode Distress) (Cheyne and Pennycook 2013).

Prevalence

In the general population, prevalence estimates between 25–30% are common but vary considerably, with reports of 6–40% having had at least one experience of SP (Cheyne, Rueffer, and Newby-Clark 1999; Fukuda et al. 1998; Spanos et al. 1995). In a meta-analysis of 35 studies, *Sharpless & Barber (2011)* found a lifetime prevalence of 7.6% in the general population. Experiences of SP might not easily be reported to a physician due to fear of being declared "insane" or actually developing a mental disorder, therefore, dark numbers might be a lot higher than official ones (Fuhrmann and Mayer 2016).

PREDISPOSING AND PRECIPITATING FACTORS

General Factors

A maternal form of transmission has been postulated and onset is usually in adolescence between 14 and 17 (American Academy of Sleep Medicine 2014; Dahlitz and Parkes 1993). Precipitating factors in the general population are jetlag, irregular sleep-wake patterns (e.g., in shift work; e.g., Kotorii et al., 2001), sleep deprivation (American Academy of Sleep Medicine 2014), or elevated levels of stress (Denis, French, and Gregory 2018). SP is also more likely to develop in higher age due to fragmentation of REM sleep, however, studies that have investigated age differences in SP prevalence have found no significant effect of age (Denis et al. 2015; Hsieh et al. 2010; Ramsawh et al. 2008). Higher rates of SP have also been found in samples of college students, wherein 25–42% of individuals reported at least one episode (Awadalla et al. 2004; Cheyne, Rueffer, and Newby-Clark 1999; Fukuda et al. 1998). Student rates have been found to be slightly lower than psychiatric patients and of those, panic disorder patients evidenced the highest overall rates. Slightly more women (18.9%) experienced lifetime SP than men (15.9%) (Sharpless and Barber 2011). A study in an American sample found that males were more likely to have experienced lifetime isolated SP than females, but there were no differences in lifetime fearful or recurrent fearful isolated SP episodes (Sharpless et al. 2010). Evidence is mixed with regards to ethnicity. One study found African American individuals experienced a higher incidence of SP compared with Caucasians

02/07/2020 14:37:00. Sharpless et al. (2010) found that non-Caucasians had a higher prevalence of fearful and recurrent fearful isolated SP compared to Caucasians. Two other studies however found no significant effect of ethnicity (Otto et al. 2006; Paradis et al. 2009). In a survey of Chinese adolescents, higher rates of SP were shown in rural compared to urban areas (Ma, Wu, and Pi 2014). Many SP experients report that their episodes, particularly first episodes, occur in novel settings and when they are alone (Cheyne, 2016).

Substance Use

It is unclear whether SP is associated with general substance use (Denis, French, & Gregory, 2018). Ma, Wu and Pi (2014) showed that those who reported drinking at least one alcoholic drink per day over the last month were significantly more likely than others to report experiencing SP. However, in a large UK sample, alcohol intake over the past week did not predict SP independently of anxiety symptoms, depressed mood, threatening events, and sleep quality (Denis et al. 2015). Evidence for the association between smoking behaviour and SP is also mixed with no effect found in some studies (e.g., Denis et al., 2015). However, increased odds of experiencing SP in those who smoked (defined as at least one cigarette a day) (15%) compared with those who did not (8%) were found in a Japanese survey (Munezawa et al. 2011). Surprisingly, no evidence of a significant association between caffeine intake and SP has been found (Denis et al. 2015; Ma, Wu, and Pi 2014), given the negative impact caffeine can have on sleep (Drake et al. 2013).

Personality Traits

Personality factors have not been shown to play a major role, certain personality traits may however be risk factors (Sharpless, 2016). For instance, paranoia, dissociation, imaginativeness, and beliefs in the paranormal/supernatural have been linked to SP (McNally and Clancy 2005; Ramsawh et al. 2008; Spanos et al. 1995).

Physical Health

So far, no specific health related factors could be identified. Nevertheless, a connection of general physical health, assessed with general physical health quality of life scales, and the occurrence of SP could be found in multiple studies (Mume and Ikem 2009; Ohayon et al. 1999; Takeuchi et al. 2002). Results regarding the association between BMI and the prevalence of SP are mixed (Ohayon et al. 1999; B. A. Sharpless et al. 2010).

Mental Health

General mental health was shown to significantly influence the frequency of SP. The worse the mental health measures, the higher the frequency of SP (Hsieh et al. 2010; Spanos et al. 1995; Takeuchi et al. 2002). In most studies, patients with higher self-reported measures of depression had higher chances of experiencing SP (Denis and Poerio 2017; Paradis et al. 2009; B. A. Sharpless et al. 2010). *Otto et al. (2006)* found a rate of 20% of isolated SP episodes in a sample of outpatients diagnosed with various anxiety disorders, yet in general, mixed evidence could be obtained in regard to SP and anxiety disorders (Ohayon et al. 1999; Paradis et al. 2009; Sharpless et al. 2010).

PTSD

SP appears to be particularly prevalent in post-traumatic stress disorder. Two studies (Abrams et al. 2008; McNally and Clancy 2005) evaluated the relationships between childhood sexual abuse and (CSA) and SP and found that those reporting CSA reported more frequent and more distressing episodes of SP. Participants reporting SP and CSA also reported higher scores on measures of depression, dissociation, absorption and post-traumatic symptoms. The same participants also reported higher frequencies and intensities of incubus and intruder episodes of SP relative to vestibular motor episodes and relative to the no CSA group. Individuals with PTSD experience more distressing dreams or sudden awakenings during the night, which is a way of reexperiencing the traumatic event and constitutes a hallmark of the disorder (Ohayon & Shapiro, 2000). It is plausible that episodes of SP may constitute a less common mode of reexperiencing (Hinton et al. 2005). There is also evidence to suggest that mechanisms underlying waking hypervigilence in CSA survivors may be responsible for dysfunctions during REM sleep (Hetzel and McCanne 2005). Moreover, dysfunctional sleep patterns common to PTSD (e.g., frequent re-awakenings), along with states of hyperarousal, may be contributing factors to increased frequency of SP episodes through dysregulation of REM sleep patterns (Abrams et al. 2008; Ohayon and Shapiro 2000).

Treatment and Prevention

Fortunately, there are several promising psychopharmacological and non-pharmacological approaches to treatment of SP. However, no randomized controlled trials have yet been conducted (Sharpless, 2016).

Non-Pharmacological Treatments

Psychoeducation and sleep hygiene can help prevent and treat SP. Sleep hygiene includes a number of practices to promote good sleep quality and limit daytime sleepiness. To prevent and deal with SP factors such as getting not enough sleep, an irregular sleep-wake cycle and stress, bad physical health (e.g., being excessively obese) should be addressed in (Cheyne, 2002). Certain foods (e.g., beans), too much caffeine and the use of nicotine or alcohol may also trigger episodes of SP and their avoidance can help prevent SP. Treatment of sleep fragmentation due to sleep apnea and other sleep disorders might be sufficient in some cases. It might be helpful, not to fall asleep in a supine position, since the occurrence of SP is increased when lying on your back. A cognitive behavioural therapy for sleep paralysis now exists and includes a screening for insomnia at initial assessment (Sharpless & Doghramji, 2015). By combining several approved non-pharmacological techniques into a holistic therapy to resolve sleep complaints, Holzinger and Klösch (2013) created an alternative concept of "sleep coaching" (SC) based on the concept of Gestalt therapy which can be applied to the SP as well. Lucid Dreaming (LD) Therapy has been shown to reduce nightmare frequency and nightmare suffering (Spoormaker and van den Bout 2006). Holzinger et al. (2015) showed that in a Gestalt therapy setting, combined with LD techniques, patients suffering from nightmares learned not only to control their dreams but also became aware of the problems behind those nightmares and were able to resolve them. Being able to gain control of SP episodes with the help of LD to control the content may be a useful technique to disrupt episodes and turn them into positive experiences. This approach also had positive effects on sleep complaints and daytime functioning and is further explained in the section about lucid dreaming.

Psychopharmacological Treatments

The same drugs used for cataplexy can be tried from treatment of SP. The most commonly used agents are tricyclic antidepressants and selective serotonin reuptake inhibitors. Their hypothesized mechanism of action is the suppression of REM sleep. They are commonly used in the treatment of other symptoms of narcolepsy as well. Regarding the tricyclics, clomipramine (25–50 mg qd), imipramine (25–150 mg qd), protriptyline (10–40 mg), and desmethylimipramine (25–150 mg qd) have all been reported to reduce SP (Sharpless, 2016).

Sleep Related Factors and SP

In the following, we will focus on SP as considered a dissociative state, its relation with other similar sleep- and wake-related phenomena such as hallucinations, false awakenings and nightmares. We will also discuss how the quality of sleep has an impact on the occurrence of SP.

Similar Sleep-Wake Phenomena

Exploding Head Syndrome

Exploding head syndrome implies hearing loud noises such as explosions at wake-sleep or sleep-wake transitions and is more common in individuals reporting episodes of fearful isolated sleep paralysis (37%) than in those not reporting sleep paralysis (14%) (Sharpless & Doghramji, 2015). Hallucinations are usually briefer (often 1 second or less) than SP hallucinations, always loud, and undifferentiated, not recognizable speech or part of a developed narrative. No paralysis occurs and conscious awareness of surroundings is absent (B. Sharpless and Doghramji 2015).

False Awakenings

The phenomenon of false awakenings describes a very similar state as SP. A false awakening is an experience in which individuals falsely believe that they have woken up, only to discover that the perceived awakening was part of a dream. Like sleep paralysis, false awakenings tend to be highly realistic and can lead to anxiety (Denis, 2018). False awakenings often occur in relation to the mixed conscious state of LD, as soon as the individual realizes they are dreaming and tries to wake themselves (Brigitte Holzinger, LaBerge, and Levitan 2006).

Pseudo Sleep Paralysis

Pseudo-SP is associated with hysterical states and depression. In these conditions, muscle weakness is only partial and patients can usually raise a finger or roll over. Patients have excessive difficulties to get up in the morning (Bassetti, 1997).

Paradoxical Insomnia

In paradoxical insomnia, patients perceive sleep as an almost continuous state of wakefulness for still unknown reasons (American Academy of Sleep Medicine 2014). They feel as if their brain stays alert while the body is asleep and as if they had intense awareness of their surroundings throughout the night. Polysomnographic parameters are normal. The definition emphasizes the mismatch between objective and subjective findings.

Similar REM-Sleep Related Phenomena

REM-Sleep Behaviour Disorder

Whereas in SP, REM sleep muscle atonia persists into wakefulness, in REM sleep behaviour disorder (RBD) muscle tone persists into REM sleep phase and permits movements (Högl & Iranzo, 2017). Therefore, RBD is another parasomnia linked closely to REM sleep. ICSD-3 criteria require that sleep-related vocalizations or complex motor behaviors plus pathologically increased muscle tone during REM sleep on the polysomnogram (REM sleep without atonia) are present. Those complex motor behaviors and vocalizations are typically associated with dreaming and observers get the impression that dream content is enacted (Högl & Iranzo, 2017). Patients can be awakened easily and are usually quickly oriented (Högl & Iranzo, 2017). If dreamers are not awakened immediately after the episode, dream content is frequently forgotten. RBD is thought to arise from damage to the brainstem circuits that mediate REM sleep muscle paralysis (Fraigne et al. 2015).

Nightmare Disorder

Nightmare disorders (ND) include dream imagery that is negatively valenced, while content of SP hallucinations is usually but not always negative. During ND, conscious awareness of one's surroundings and atonia is missing.

Similar Non-REM Sleep Related Phenomena

Night Terrors

Night terrors occur during non-REM sleep and therefore more likely in the first hours of sleep. They typically last from 30 seconds to 5 minutes. The sleeper shows manifestations of intense fear, usually sitting up in bed and emitting a loud scream. Associated dream imagery may be reported and incoherent vocalizations can occur; the sleeper may bolt from the bed and run in an apparent attempt to avoid harm, thereby injuring themselves or others. It is more common in children. Night terrors are often only recalled vaguely, in contrary to nightmares and SP (Fleetham and Fleming 2014), and awareness of surroundings is lacking.

Similar Dissociative States during Wakefulness

Cataplexy/Narcolepsy

Despite being removed as a diagnostic criterion in the most recent edition of the International Classification of Sleep Disorders, SP is common in cataplexy. Cataplexy is a sudden and uncontrollable muscle paralysis that occurs during wakefulness and is precipitated by emotions. It has been theorized that cataplexy represents the to result from intrusion of REM sleep paralysis into wakefulness (Dauvilliers et al. 2014). And, like in SP, cataplexy affects all skeletal muscles aside from the diaphragm and extraocular muscles. Interestingly, hallucinations are more common in narcolepsy patients with cataplexy compared with narcolepsy without cataplexy (Leu-Semenescu et al. 2011). Furthermore, SP is more common in narcolepsy with cataplexy patients (53%) compared with narcolepsy without cataplexy (32%) (Bassetti, 1997).

Familial Periodic Paralysis Syndrome

Familial periodic paralysis syndrome is a rare condition, characterized by episodes of flaccid paralysis with loss of deep tendon reflexes and failure of muscle to respond to electrical stimulation. There are 4 forms: hypokalemic, hyperkalemic, thyrotoxic, and Andersen-Tawil syndrome. It may occur at rest and on awakening, however, the episodes usually last hours, are associated with carbohydrate intake, and are usually accompanied by hypokalemia (American Academy of Sleep Medicine 2014).

Dissociative States

The latest edition of the International Classification of Diseases (ICD-10) defines dissociation as " … a partial or complete loss of the normal integration between memories of the past, awareness of identity and immediate sensations, and control of body movements" (p. 151). In a study by Denis, French, and Gregory (2018), the frequency of dissociative experiences reported during the day was found to be related to both SP frequency and the frequency/intensity of hallucinations experienced during SP episodes. Both states imply a loss of control for the affected individual and might be related but distinct experiences.

Daytime Hallucinations

Waters et al. (2016) suggest that the above-mentioned sleep-related experiences show considerable overlap with hallucinations regarding subjective descriptions and underlying brain mechanisms. In both sleep-related phenomena and hallucinations, the thalamus is activated and the brain is focused inwards and selectively attentive to internal experiences. Limbic activity is thought to contribute to affective salience and emotive contents of both sleep-related perceptions and hallucinations in psychosis. During the short transition between sleep and wake, cortical inhibitory control is reduced but sufficiently sporadic to allow conscious awareness. Hypnagogia may, therefore, be an intermediate phenomenon between REM dreams and hallucinations. However, the experiences differ in the following: (1) Sleep-related perceptions are immersive and largely cut off from reality, whereas hallucinations are discrete and overlaid on veridical perceptions;

and (2) Sleep-related perceptions involve only a subset of neural networks implicated in hallucinations, reflecting perceptual signals processed in a functionally and cognitively closed-loop circuit. While multisensory experiences are common in sleep-related perceptions, hallucinations usually occur in a single modality (Waters et al. 2016).

Psychotic disorders and SP often share disturbing and unwanted hallucinations. However, hallucinations during SP are limited to sleep–wake transitions and gross reality testing is otherwise intact (B. Sharpless and Doghramji 2015).

Sleep Quality and SP

The occurrence and frequency of SP is closely related to the quality of sleep. Several studies (e.g., Hsieh et al., 2010; Ma et al., 2014; Munezawa et al., 2009) investigated this relationship with the help of subjective sleep measures such as the Pittsburg Sleep Quality Index (PSQI; Buysse, Reynolds, Monk, Berman, & Kupfer, 1989). One study in the UK showed that poor sleep quality (as measured by PSQI global score) was an independent predictor of SP. Denis, French, Schneider, & Gregory (2018) explored which of the seven components of the PSQI best explained SP and found that longer sleep latency and increased daytime dysfunction (trouble staying awake, problem to keep up enthusiasm) were significantly associated with SP. A period of heightened stress may cause a decline in general sleep quality, which in turn increases the risk of experiencing SP (Denis 2018).

The authors also found that heightened insomnia symptoms could predict SP and might be an even more important predictor than general sleep quality. Increased insomnia symptoms also predicted increased frequency of intruder and incubus hallucinations but was not associated with the intensity of hallucinations (Denis and Poerio 2017). Of the insomnia symptoms, difficulty maintaining sleep was found to be a significant independent predictor of SP occurrence in a sample of Chinese participants, whereas other insomnia symptoms such as difficulty initiating sleep and subjective insufficient sleep were not significant (Otsuka et al., 2018). Szklo-Coxe, Young, Finn, and Mignot (2007) found that self-reported insomnia (defined as waking up repeatedly during the night) occurring five or more times a month was associated with increased odds of experiencing sleep paralysis compared with experiencing insomnia less than five times a month. In addition, insomnia symptoms (but not a diagnosed insomnia disorder) have also been found to predict SP (Denis, 2018).

Increased cognitive and somatic presleep arousal have been linked to insomnia as well as SP (Denis, French, Schneider, & Gregory, 2018). The authors also linked faulty cognitions about sleep, such as being worried about losing control over one's own ability to sleep and being overwhelmed by thoughts at night and feeling that one has no control over these thoughts, to increased odds of sleep paralysis in one study. There is mixed evidence if bedtime and waking up time show any relationship with SP and if, what kind. Neither longer nor shorter durations of sleep have been found to be a predictor of SP since both have been observed in different studies (e.g., Ma et al., 2014; Munezawa et al., 2011, 2009). Studies that have used polysomnography (PSG), the gold standard for objective sleep recording, to identify potential differences in sleep architecture between those who report experiences of SP compared with those who do not, have found no significant differences in OSA patients with and without SP (Walther & Schulz, 2004). The same study found that REM sleep latency as well as sleep latency in general were significantly lower in individuals experiencing SP. Longer sleep latencies might be a sign for higher presleep arousal and heightened insomnia symptoms.

SP rates are elevated in case of obstructive sleep apnea (Ohayon et al. 1999). Excessive daytime sleepiness and sleep hygiene were shown to have an impact on SP as well (Hsieh et al. 2010; Ma, Wu, and Pi 2014; Munezawa et al. 2011; 2009).

Sleep paralysis has been shown to be heritable in twin studies and variation in the PER2 gene (single nucleotide polymorphism rs2304672) has been shown to increase the odds of having experienced SP (Denis et al. 2015). PER2 is a member of the Period (PER) family of genes and plays an important role in the regulation of circadian rhythms, such as the sleep–wake cycle. Therefore, dysregulation of the regular sleep–wake cycle may increase the risk of experiencing SP. However, the causal relationship between SP and sleep quality is not clear, and a reduction in sleep quality as a consequence of SP should also be considered, especially in cases where SP leads to significant levels of fear and postepisode distress (Sharpless et al., 2010). Individuals may avoid sleep due to fear of SP and frightening hallucinations, thus sleep may be further disrupted. Furthermore, Denis, French, Schneider and Gregory (2018) linked disruptive nocturnal behaviors such as nightmares of a traumatic experience, trouble sleeping due to general nervousness, and episodes of terror or screaming during sleep to the presence of SP. In general, SP often co-occurs with other 'anomalous' sleep experiences, like Exploding Head Syndrome, nightmares and night terrors.

Dreams, Nightmares, Lucid Dreams and SP

SP has been scientifically studied for about 125 years but continues to be confused with night terrors, anxiety dreams and sleeping nightmares. This might partly be due to an ambiguous use of vocabulary. Liddon (1967), for example, notes that the term *nightmare* originally referred to SP with hypnagogic or hypnopompic hallucinations. Therefore, it is necessary to differentiate between SP and similar phenomena.

Dreams

Dreams typically feature visual images of people, faces, objects, animals, and scenes that are rich in shape and full of movement, and like SP, can feature auditory and tactile percepts, and more rarely, smell, taste, pain, sexual sensations, and proprioceptive or kinesthetic sensations (Waters et al. 2016). Despite some similarities, SP and dreams do differ in a couple of important factors. During SP, subjects know they were asleep but have woken up, when dreaming, however, they usually do not know that they are dreaming (de Sá and Mota-Rolim 2016). Instead, the person is only aware of dream contents upon waking. Dreams may have bizarre contents that are usually readily accepted by the dreamer (Waters et al. 2016). Furthermore, SP experiences are usually more aggressive than a regular dream and there are four times more references to body parts (e.g., feelings of thorax pressure or paralyzed limbs) (Blackmore & Parker, 2002). While up to 90% of SP episodes are associated with fear (Cheyne, 2003), only about 30% of dreams are being rated as frightening (Schredl & Doll, 1998). Despite popular beliefs, dreams occur not only during REM sleep but also during all stages of non-REM sleep, whereas SP is linked to REM sleep.

Nightmares

Nightmares (repeated occurrences of extended, dysphoric, and well-remembered dreams that usually involve threats to survival, security, or physical integrity) may also occur in all sleep stages (Högl & Iranzo, 2017). Nightmares are defined as dream experience loaded with anxiety or fear (International Classification of Diseases ICD-10: F51.5; ICDS-2: 307.46) which mainly occur in the second half of the night during a long REM phase and are associated with slight vegetative arousals – an increase or fluctuations in the heart and respiratory rate – and cause the dreamer to wake up. Upon awakening, the dreamer is fully alert with little confusion or disorientation and is able to give a detailed account of the frightening dream usually involving threats to survival, security, or self-esteem (B. Holzinger, Klösch, and Saletu 2015). However, the awakening criterion in the definition of nightmares is still discussed (Blagrove and Haywood 2006) and is not included in the latest DSM-5. Nightmares in sleep terrors occur in N2 and N3 sleep and involve threats and the need to escape from something like a fire or a small room (Högl & Iranzo, 2017). Multiple studies have found a relationship between SP and nightmare frequency (Munezawa et al. 2011; 2009; Spanos et al. 1995).

Lucid Dreams

> "I used to dream that I was flying around the Earth in a rocket I had made from a garbage can. the bottom was glass and I had a lovely aerial view of the world as I flew wherever I wanted. When it was time to land (my rocket was not equipped for landing), during the descent I would tell myself, "Time to wake up," and I'd wake myself up. Though sometimes I would get perilously close to the ground, I was never afraid of the inevitable crash because I knew I was dreaming and could wake myself up at any time." (LaBerge and Rheingold 1991).

A lucid dream is defined as a dream during which one is aware of the dream state and able to volitionally control its content (e.g., (Brigitte Holzinger, LaBerge, and Levitan 2006; LaBerge 1981; LaBerge and Rheingold 1991; Van Eeden 1913). SP and lucid dreaming are thought to be related but distinct experiences and might be underlined by the similar neurophysiology (Dresler et al., 2012; Terzaghi et al., 2012; Voss et al., 2009). Both SP and LD can be conceptualized as dissociated REM states and as "hybrid states" between waking and REM-sleep (Mahowald and Schenck, 2005). Voss, Holzmann, Tuin, and Hobson (2009) found that to move from non-lucid REM sleep dreaming to lucid REM sleep dreaming, there must be a shift in brain activity in the direction of waking, just as in the transition from REM-sleep to SP. In LD, self-reflection arises and augments so that the dreamer recognizes that he is not awake but asleep, in contrary to non-lucid REM-sleep. The authors suggest that an increase in frontal lobe 40-Hz power and a global increase in EEG coherence might hold a functional role in the modulation of awareness across different conscious states. An increased activity in the right dorsolateral prefrontal cortex was found, an area thought to underlie a wide range of higher cognitive capacities, like self-focused metacognitive evaluation such as evaluation of one's own thoughts and feelings as well as task performance. Dresler et al., (2012) found the strongest increase in activation during lucid compared to non-lucid REM sleep in the precuneus, a brain region that has been implicated in self-referential processing, such as first-person perspective and experience of agency. They also found activation in the bilateral cuneus and occipitotemporal cortices, which might explain the exceptional brightness and visual clarity of the dream scenery often experienced during LD.

People are likely to enter SP directly from a lucid dream and vice versa (Emslie, 2014), the phenomenons often co-occur (Denis & Poerio, 2016). LD was found to be specifically related to those episodes of SP that contained vestibular-motor hallucinations and not intruder/incubus type episodes (Denis and Poerio 2017). This might be explained by the control factor: while individuals describe intruder episodes as having little or no control over the experience, LD is the clear opposite of this (Denis 2018). LD is typically positively valenced while SP is overwhelmingly negative. A further distinction is that SP involves full return to wakefulness during REM-induced muscle atonia, whereas LD involves the recovery of aspects of consciousness experienced during waking while the person remains asleep (Denis & Poerio, 2016). The authors also found that dissociative experience was the only common predictor of both SP and LD frequency, indicating that individuals who experience both unusual sleep experiences also experience greater dissociative experiences in daily life. SP frequency was associated with poorer sleep quality and greater stress and anxiety, whereas LD frequency was associated with positive constructive daydreaming and more vivid imagination (Denis & Poerio, 2016).

LD techniques might be useful for SP sufferers by enabling them to gain control of episodes, control the content, and turn them into positive experiences. In one study (B. Holzinger, Klösch, and Saletu 2015), subjects were randomly assigned to either Gestalt Therapy Group, or Gestalt and LD Group Therapy combined. Those attending to both interventions showed an increase in dream recall frequency and a more profound reduction of nightmare frequency than the controls. The intervention included conscious confrontation with frightening dream images, change of the dream plot and sharing of the dreams in group therapy, and in case of LDT additional education about LD, dream and nightmare education, changes of attitudes towards dreams, talking about and understanding dream content, sleep and dream journal, LD induction hypnosis, fantasizing about happier endings of the dream, and concentration exercises, relaxation techniques, meditation, and hypnosis, reading lucid dreaming literature. The significant reduction of nightmares, however, could only be measured in those who actually succeeded in lucid dreaming. LD is a learnable and effective strategy in coping with nightmares and can be applied while the frightening nightmare or hallucination is happening, not only in waking state (Holzinger, Klösch, and Saletu 2015). Therefore, LD may lead to the disappearance of frightening hallucinations and anxiety related to SP episodes, and increase ego strength as well as emotional safety.

CONCLUSION

In this chapter, we tried to give a comprehensive overview of the relevant criteria regarding sleep paralysis and which factors precipitate the occurrence. For people affected, it might be difficult to differentiate between related events such as nightmares, night terrors, or false awakenings, hallucinations and SP. Other diseases have to be ruled out to be able to diagnose SP. We therefore hope to make it clearer for everybody, what sets similar phenomena apart from SP and which neurological happenings are responsible for the occurrence. The investigation of neurophysiological correlates that can explain SP is still in its infancy and despite the phenomenon of SP always being a topic that provoked a lot of interest in many cultures around the world, still little is known about the experience of SP. Although the vast majority of episodes are associated with extreme fear and sometimes even lead to clinically significant levels of distress (Denis, 2018), SP is still an under-researched phenomenon that is relatively unknown in general population (Denis, French, & Gregory, 2017). Most people experiencing SP are uninformed about the phenomenon and education alone may provide relief, as individuals realize that those episodes are typically benign and experienced by a large part of the population (Wing, Lee, & Chen, 1994). Future research should focus on longitudinal designs (direction of effects). Since overall sleep quality could be linked with sleep paralysis, interventions need to take the patient´s general sleep quality into account. Lucid dreaming therapy has been shown to be successful in treating nightmares and could be helpful in treating SP by enabling those affected to gain control over the experience and decrease suffering. Studies testing this effect should be implemented in the future.

REFERENCES

[1] Abrams, Murray P., Ashlee D. Mulligan, R. Nicholas Carleton, and Gordon J.G. Asmundson. 2008. "Prevalence and Correlates of Sleep Paralysis in Adults Reporting Childhood Sexual Abuse." *Journal of Anxiety Disorders* 22 (8): 1535–41. https://doi.org/10.1016/j.janxdis.2008.03.007.

[2] Adler, Shelley R. 2011. *Sleep Paralysis. Nightmares, Nocebos, and the Mind-Body-Connection.* New Brunswick, New Jersey, London: Rutgers University Press.

[3] American Academy of Sleep Medicine. 2014. *International Classification of Sleep Disorders*. 3rd ed. Darien, Illinois.

[4] American Psychiatric Association. 2013. *Diagnostic and Statistical Manual of Mental Disorders (DSM-5®).* American Psychiatric Pub.

[5] Arzy, Shahar, Gregor Thut, Christine Mohr, Christoph M. Michel, and Olaf Blanke. 2006. "Neural Basis of Embodiment: Distinct Contributions of Temporoparietal Junction and Extrastriate Body Area." *Journal of Neuroscience* 26 (31): 8074–81. https://doi.org/10.1523/JNEUROSCI.0745-06.2006.

[6] Aserinsky, Eugene, and Nathaniel Kleitman. 1953. "*Regularly Occurring Periods of Eye Motility, and Concomitant Phenomena, During Sleep*" 118 (3062): 273-4.

[7] Awadalla, Abdelwahid, Ghenaim Al-Fayez, Michael Harville, Hiroko Arikawa, Marie E. Tomeo, Donald I. Templer, and Rocky Underwood. 2004. "*Comparative Prevalence of Isolated Sleep Paralysis in Kuwaiti, Sudanese, and American College Students*" 95 (1): 317–22.

[8] Bassetti, Claudio. 2003. "Narcolepsy, Cataplexy, and Sleep Paralysis." In *Sleep and Movement Disorders*, edited by Sudhansu

Chokroverty and Richard P. Allen, 373–94. Oxford: Butterworth Heinemann.

[9] Blagrove, Mark, and Suzanne Haywood. 2006. "Evaluating the Awakening Criterion in the Definition of Nightmares: How Certain Are People in Judging Whether a Nightmare Woke Them Up?" *Journal of Sleep Research* 15 (2): 117–24. https://doi.org/10.1111/j.1365-2869.2006.00507.x.

[10] Blanchard, D. Caroline, and Robert J. Blanchard. 1972. "Innate and Conditioned Reactions to Threat in Rats with Amygdaloid Lesions." *Journal of Comparative and Physiological Psychology* 81 (2): 281–90. https://doi.org/10.1037/h0033521.

[11] Brooks, Patricia L., and John H. Peever. 2012. "Identification of the Transmitter and Receptor Mechanisms Responsible for REM Sleep Paralysis." *Journal of Neuroscience* 32 (29): 9785–95. https://doi.org/10.1523/JNEUROSCI.0482-12.2012.

[12] Cheyne, James Allan. 2002. "Situational Factors Affecting Sleep Paralysis and Associated Hallucinations: Position and Timing Effects." *Journal of Sleep Research* 11 (2): 169–77. https://doi.org/10.1046/j.1365-2869.2002.00297.x.

[13] ——. 2003. "*Sleep Paralysis and the Structure of Waking-Nightmare Hallucinations*" 13 (3): 163–79.

[14] ——. 2016. "*Animal 'Hypnosis' and Waking Nightmares*" 16: 307–43.

[15] Cheyne, James Allan, and Todd A. Girard. 2007. "Paranoid Delusions and Threatening Hallucinations: A Prospective Study of Sleep Paralysis Experiences." *Consciousness and Cognition* 16 (4): 959–74. https://doi.org/10.1016/j.concog.2007.01.002.

[16] Cheyne, James Allan, and Gordon Pennycook. 2013. "Sleep Paralysis Postepisode Distress: Modeling Potential Effects of Episode Characteristics, General Psychological Distress, Beliefs, and Cognitive Style." *Clinical Psychological Science* 1 (2): 135–48. https://doi.org/10.1177/2167702612466656.

[17] Cheyne, James Allan, Steve D. Rueffer, and Ian R. Newby-Clark. 1999. "Hypnagogic and Hypnopompic Hallucinations during Sleep

Paralysis: Neurological and Cultural Construction of the Night-Mare." *Consciousness and Cognition* 8 (3): 319–37. https://doi.org/10.1006/ccog.1999.0404.

[18] Dahlitz, M., and J D Parkes. 1993. "Sleep Paralysis." *The Lancet* 341 (8842): 406–7. https://doi.org/10.1016/0140-6736(93)92993-4.

[19] Dauvilliers, Yves, Jerry M. Siegel, Regis Lopez, Zoltan A. Torontali, and John H. Peever. 2014. "Cataplexy—Clinical Aspects, Pathophysiology and Management Strategy." *Nature Reviews Neurology* 10 (7): 386–95. https://doi.org/10.1038/nrneurol.2014.97.

[20] Dement, William, and Nathaniel Kleitman. 1957. "Cyclic Variations in EEG during Sleep and Their Relation to Eye Movements, Body Motility, and Dreaming." *Electroencephalography and Clinical Neurophysiology* 9 (4): 673–90. https://doi.org/10.1016/0013-4694(57)90088-3.

[21] Denis, Dan. 2018. "Relationships between Sleep Paralysis and Sleep Quality: Current Insights." *Nature and Science of Sleep* Volume 10 (November): 355–67. https://doi.org/10.2147/NSS.S158600.

[22] Denis, Dan, Christopher C. French, and Alice M. Gregory. 2018. "A Systematic Review of Variables Associated with Sleep Paralysis." *Sleep Medicine Reviews* 38 (April): 141–57. https://doi.org/10.1016/j.smrv.2017.05.005.

[23] Denis, Dan, Christopher C. French, Richard Rowe, Helena M. S. Zavos, Patrick M. Nolan, Michael J. Parsons, and Alice M. Gregory. 2015. "A Twin and Molecular Genetics Study of Sleep Paralysis and Associated Factors." *Journal of Sleep Research* 24 (4): 438–46. https://doi.org/10.1111/jsr.12282.

[24] Denis, Dan, Christopher C. French, Melanie N. Schneider, and Alice M. Gregory. 2018. "Subjective Sleep-Related Variables in Those Who Have and Have Not Experienced Sleep Paralysis." *Journal of Sleep Research* 27 (5): e12650. https://doi.org/10.1111/jsr.12650.

[25] Denis, Dan, and Giulia L. Poerio. 2017. "Terror and Bliss? Commonalities and Distinctions between Sleep Paralysis, Lucid Dreaming, and Their Associations with Waking Life Experiences."

Journal of Sleep Research 26 (1): 38–47. https://doi.org/10.1111/jsr.12441.

[26] Drake, Christopher, Timothy Roehrs, John Shambroom, and Thomas Roth. 2013. "Caffeine Effects on Sleep Taken 0, 3, or 6 Hours before Going to Bed." *Journal of Clinical Sleep Medicine*, November. https://doi.org/10.5664/jcsm.3170.

[27] Dresler, Martin, Renate Wehrle, Victor I. Spoormaker, Stefan P. Koch, Florian Holsboer, Axel Steiger, Hellmuth Obrig, Philipp G. Sämann, and Michael Czisch. 2012. "Neural Correlates of Dream Lucidity Obtained from Contrasting Lucid versus Non-Lucid REM Sleep: A Combined EEG/FMRI Case Study." *Sleep* 35 (7): 1017–20. https://doi.org/10.5665/sleep.1974.

[28] Fleetham, John A., and Jonathan A. E. Fleming. 2014. "*Parasomnias*" 186 (8): E273–80.

[29] Fraigne, Jimmy J., Zoltan A. Torontali, Matthew B. Snow, and John H. Peever. 2015. "REM Sleep at Its Core - Circuits, Neurotransmitters, and Pathophysiology." *Frontiers in Neurology* 6 (May). https://doi.org/10.3389/fneur.2015.00123.

[30] Fuhrmann, Max, and Gerhard Mayer. 2016. "*Phänomenologie – Deutung – Coping* [*Phenomenology - interpretation - coping*]," no. 16: 275–306.

[31] Fukuda, Kazuhiko. 1989. "Preliminary Study on Kanashibari Phenomenon: A Polygraphic Approach." *Japanese Journal of Physiological Psychology and Psychophysiology* 7 (2): 83–89. https://doi.org/10.5674/jjppp1983.7.83.

[32] Fukuda, Kazuhiko, Robert D. Ogilvie, Lisa Chilcott, Ann-Marie Vendittelli, and Tomoka Takeuchi. 1998. "The Prevalence of Sleep Paralysis among Canadian and Japanese College Students." *Dreaming* 8 (2): 59–66. https://doi.org/10.1023/B:DREM.0000005896.68083.ae.

[33] González-Maeso, Javier, Rosalind L. Ang, Tony Yuen, Pokman Chan, Noelia V. Weisstaub, Juan F. López-Giménez, Mingming Zhou, et al. 2008. "Identification of a Serotonin/Glutamate Receptor Complex

Implicated in Psychosis." *Nature* 452 (7183): 93–97. https://doi.org/10.1038/nature06612.

[34] Grace, K. P., L. E. Vanstone, and R. L. Horner. 2014. "Endogenous Cholinergic Input to the Pontine REM Sleep Generator Is Not Required for REM Sleep to Occur." *Journal of Neuroscience* 34 (43): 14198–209. https://doi.org/10.1523/JNEUROSCI.0274-14.2014.

[35] Hetzel, Melanie D., and Thomas R. McCanne. 2005. "The Roles of Peritraumatic Dissociation, Child Physical Abuse, and Child Sexual Abuse in the Development of Posttraumatic Stress Disorder and Adult Victimization." *Child Abuse & Neglect* 29 (8): 915–30. https://doi.org/10.1016/j.chiabu.2004.11.008.

[36] Hinton, Devon E., Vuth Pich, Dara Chhean, and Mark H. Pollack. 2005. "'The Ghost Pushes You Down': Sleep Paralysis-Type Panic Attacks in a Khmer Refugee Population." *Transcultural Psychiatry* 42 (1): 46–77. https://doi.org/10.1177/1363461505050710.

[37] Hishikawa, Y, H Wakamatsu, E Furuya, Y Sugita, S Masaoka, H Kaneda, M Sato, H Nan'no, and Z Kaneko. 1976. "Sleep Satiation in Narcoleptic Patients." *Electroencephalography and Clinical Neurophysiology* 41 (1): 1–18. https://doi.org/10.1016/0013-4694(76)90210-8.

[38] Hobson, Allan J., and Robert W. McCarley. 1977. "*The Brain as a Dream State Generator: An Activation-Synthesis Hypothesis of the Dream Process*" 134 (12): 1335–1348. https://doi.org/10.1176/ajp.134.12.1335.

[39] Högl, Birgit, and Alex Iranzo. 2017. "Rapid Eye Movement Sleep Behavior Disorder and Other Rapid Eye Movement Sleep Parasomnias:" *CONTINUUM: Lifelong Learning in Neurology* 23 (4): 1017–34. https://doi.org/10.1212/CON.0000000000000489.

[40] Holzinger, B., G. Klösch, and B. Saletu. 2015. "Studies with Lucid Dreaming as Add-on Therapy to Gestalt Therapy." *Acta Neurologica Scandinavica* 131 (6): 355–63. https://doi.org/10.1111/ane.12362.

[41] Holzinger, Brigitte, Stephen LaBerge, and Lynne Levitan. 2006. "Psychophysiological Correlates of Lucid Dreaming." *Dreaming* 16 (2): 88–95. https://doi.org/10.1037/1053-0797.16.2.88.

[42] Hsieh, Sun-Wung, Chiou-Lian Lai, Ching-Kuan Liu, Sheng-Hsing Lan, and Chung-Yao Hsu. 2010. "Isolated Sleep Paralysis Linked to Impaired Nocturnal Sleep Quality and Health-Related Quality of Life in Chinese-Taiwanese Patients with Obstructive Sleep Apnea." *Quality of Life Research* 19 (9): 1265–72. https://doi.org/10.1007/s11136-010-9695-4.

[43] Hufford, David J. 2005. "Sleep Paralysis as Spiritual Experience." *Transcultural Psychiatry* 42 (1): 11–45. https://doi.org/10.1177/1363461505050709.

[44] Huot, Philippe, Tom H. Johnston, Tayyeba Darr, Lili-Naz Hazrati, Naomi P. Visanji, Donna Pires, Jonathan M. Brotchie, and Susan H. Fox. 2010. "Increased 5-HT $_{2A}$ Receptors in the Temporal Cortex of Parkinsonian Patients with Visual Hallucinations: 5-HT $_{2A}$ Receptors in Visual Hallucinations." *Movement Disorders* 25 (10): 1399–1408. https://doi.org/10.1002/mds.23083.

[45] Jalal, Baland, and Devon E. Hinton. 2013. "Rates and Characteristics of Sleep Paralysis in the General Population of Denmark and Egypt." *Culture, Medicine, and Psychiatry* 37 (3): 534–48. https://doi.org/10.1007/s11013-013-9327-x.

[46] Kometer, Michael, André Schmidt, Lutz Jancke, and Franz X. Vollenweider. 2013. "Activation of Serotonin 2A Receptors Underlies the Psilocybin-Induced Effects on Oscillations, N170 Visual-Evoked Potentials, and Visual Hallucinations." *Journal of Neuroscience* 33 (25): 10544–51. https://doi.org/10.1523/JNEUROSCI.3007-12.2013.

[47] Kotorii, Tsuyoshi, Tatayu Kotorii, Naohisa Uchimura, Yuji Hashizume, Shinichirou Shirakawa, Takeshi Satomura, Junji Tanaka, Youichi Nakazawa, and Hisao Maeda. 2001. "Questionnaire Relating to Sleep Paralysis." *Psychiatry and Clinical Neurosciences* 55 (3): 265–66. https://doi.org/10.1046/j.1440-1819.2001.00853.x.

[48] LaBerge, Stephen. 1981. "*Lucid Dreaming: An Exploratory Study of Consciousness during Sleep*."

[49] LaBerge, Stephen, and Howard Rheingold. 1991. *Exploring the World of Lucid Dreaming*. New York: Ballantine Books.

[50] Law, Samuel, and Laurence J. Kirmayer. 2005. "Inuit Interpretations of Sleep Paralysis." *Transcultural Psychiatry* 42 (1): 93–112. https://doi.org/10.1177/1363461505050712.

[51] Lebedev, A V, M Lövdén, G Rosenthal, A Feilding, David J. Nutt, and Robin L Carhart-Harris. 2015. "*Finding the Self by Losing the Self: Neural Correlates of Ego-Dissolution under Psilocybin.*" 36 (8): 3137–53. https://doi.org/10.1002/hbm.22833.

[52] Leu-Semenescu, Smaranda, Valerie Cochen De Cock, Valerie Dauriac Le Masson, Rachel Debs, Sophie Lavault, Emmanuel Roze, Marie Vidailhet, and Isabelle Arnulf. 2011. "Hallucinations in Narcolepsy with and without Cataplexy: Contrasts with Parkinson's Disease." *Sleep Medicine* 12 (5): 497–504. https://doi.org/10.1016/j.sleep.2011.03.006.

[53] Ma, Shengli, Tao Wu, and Guofu Pi. 2014. "Sleep Paralysis in Chinese Adolescents: A Representative Survey: Sleep Paralysis Epidemiological Survey." *Sleep and Biological Rhythms* 12 (1): 46–52. https://doi.org/10.1111/sbr.12045.

[54] Mahowald, Mark W., and Carlos H. Schenck. 2005. "Insights from Studying Human Sleep Disorders." *Nature* 437 (7063): 1279–85. https://doi.org/10.1038/nature04287.

[55] McNally, Richard J., and Susan A. Clancy. 2005. "Sleep Paralysis in Adults Reporting Repressed, Recovered, or Continuous Memories of Childhood Sexual Abuse." *Journal of Anxiety Disorders* 19 (5): 595–602. https://doi.org/10.1016/j.janxdis.2004.05.003.

[56] Mume, Celestine Okorome, and Innocent C Ikem. 2009. "Sleep Paralysis and Psychopathology." *South African Journal of Psychiatry* 15 (4): 4. https://doi.org/10.4102/sajpsychiatry.v15i4.185.

[57] Munezawa, Takeshi, Yoshitaka Kaneita, Yoneatsu Osaki, Hideyuki Kanda, Tadahiro Ohtsu, Hiroyuki Suzuki, Masumi Minowa, et al. 2011. "Nightmare and Sleep Paralysis among Japanese Adolescents: A Nationwide Representative Survey." *Sleep Medicine* 12 (1): 56–64. https://doi.org/10.1016/j.sleep.2010.04.015.

[58] Munezawa, Takeshi, Yoshitaka Kaneita, Eise Yokoyama, Hiroyuki Suzuki, and Takashi Ohida. 2009. "Epidemiological Study of

Nightmare and Sleep Paralysis among Japanese Adolescents." *Sleep and Biological Rhythms* 7 (3): 201–10. https://doi.org/10.1111/j.1479-8425.2009.00404.x.

[59] Ness, Robert C. 1978. "The Old Hag Phenomenon as Sleep Paralysis: A Biocultural Interpretation" 2 (1): 15–39. https://doi.org/10.1007/bf00052448.

[60] Ohaeri, Jude U., Abdel W. Awadalla, Victor A. Makanjuola, and Beatrice M. Ohaeri. 2004. "*Features of Isolated Sleep Paralysis Among Nigerians*" 81 (10): 509–19.

[61] Ohayon, Maurice M., and Colin M. Shapiro. 2000. "Sleep Disturbances and Psychiatric Disorders Associated with Posttraumatic Stress Disorder in the General Population." *Comprehensive Psychiatry* 41 (6): 469–78. https://doi.org/10.1053/comp.2000.16568.

[62] Ohayon, Maurice M., Jürgen Zulley, Christian Guilleminault, and Salvatore Smirne. 1999. "*Prevalence and Pathologic Associations of Sleep Paralysis in the General Population*" 52: 1194–1200.

[63] Otto, Michael W., Naomi M. Simon, Mark Powers, Devon Hinton, Alyson K. Zalta, and Mark H. Pollack. 2006. "Rates of Isolated Sleep Paralysis in Outpatients with Anxiety Disorders." *Journal of Anxiety Disorders* 20 (5): 687–93. https://doi.org/10.1016/j.janxdis.2005.07.002.

[64] Paradis, Cheryl, Steven Friedman, Devon E. Hinton, Richard J. McNally, Linda Z. Solomon, and Kelly A. Lyons. 2009. "The Assessment of the Phenomenology of Sleep Paralysis: The Unusual Sleep Experiences Questionnaire (USEQ)." *CNS Neuroscience & Therapeutics* 15 (3): 220–26. https://doi.org/10.1111/j.1755-5949.2009.00098.x.

[65] Parker, Jennifer D, and Susan J Blackmore. 2002. "*Comparing the Content of Sleep Paralysis and Dream Reports*" 12 (1).

[66] Preller, Katrin H., Marcus Herdener, Thomas Pokorny, Amanda Planzer, Rainer Kraehenmann, Philipp Stämpfli, Matthias E. Liechti, Erich Seifritz, and Franz X. Vollenweider. 2017. "The Fabric of Meaning and Subjective Effects in LSD-Induced States Depend on

Serotonin 2A Receptor Activation." *Current Biology* 27 (3): 451–57. https://doi.org/10.1016/j.cub.2016.12.030.

[67] Ramsawh, Holly J., Susan D. Raffa, Kamila S. White, and David H. Barlow. 2008. "Risk Factors for Isolated Sleep Paralysis in an African American Sample: A Preliminary Study." *Behavior Therapy* 39 (4): 386–97. https://doi.org/10.1016/j.beth.2007.11.002.

[68] Sá, José F. R. de, and Sérgio A. Mota-Rolim. 2016. "Sleep Paralysis in Brazilian Folklore and Other Cultures: A Brief Review." *Frontiers in Psychology* 7 (September): 1294. https://doi.org/10.3389/fpsyg.2016.01294.

[69] Sharpless, Brian. 2016. "A Clinician´s Guide to Recurrent Isolated Sleep Paralysis." *Neuropsychiatric Disease and Treatment* Volume 12 (July): 1761–67. https://doi.org/10.2147/NDT.S100307.

[70] Sharpless, Brian A., and Jacques P. Barber. 2011. "Lifetime Prevalence Rates of Sleep Paralysis: A Systematic Review." *Sleep Medicine Reviews* 15 (5): 311–15. https://doi.org/10.1016/j.smrv.2011.01.007.

[71] Sharpless, Brian A., and Monika Kliková. 2019. "Clinical Features of Isolated Sleep Paralysis." *Sleep Medicine* 58 (June): 102–6. https://doi.org/10.1016/j.sleep.2019.03.007.

[72] Sharpless, Brian A., Kevin S. McCarthy, Dianne L. Chambless, Barbara L. Milrod, Shabad-Ratan Khalsa, and Jacques P. Barber. 2010. "Isolated Sleep Paralysis and Fearful Isolated Sleep Paralysis in Outpatients with Panic Attacksb." *Journal of Clinical Psychology* 66 (12): 1292–1306. https://doi.org/10.1002/jclp.20724.

[73] Sharpless, Brian, and Karl Doghramji. 2015. *Sleep Paralysis. Historical, Psychological and Medical Perspectives*. Oxford: University Press.

[74] Spanos, Nicholas P., Stacey A. McNulty, Susan C. DeBreuil, Martha Pires, and Melissa Faith Burgess. 1995. "*The Frequency and Correlates of Sleep Paralysis in a University Sample*," 285–305.

[75] Spoormaker, Victor I., and Jan van den Bout. 2006. "Lucid Dreaming Treatment for Nightmares: A Pilot Study." *Psychotherapy and Psychosomatics* 75 (6): 389–94. https://doi.org/10.1159/000095446.

[76] Tagliazucchi, Enzo, Leor Roseman, Mendel Kaelen, Csaba Orban, Suresh D. Muthukumaraswamy, Kevin Murphy, Helmut Laufs, et al. 2016. "Increased Global Functional Connectivity Correlates with LSD-Induced Ego Dissolution." *Current Biology* 26 (8): 1043–50. https://doi.org/10.1016/j.cub.2016.02.010.

[77] Takeuchi, Tomoka, Kazuhiko Fukuda, Yuka Sasaki, Maki Inugami, and Timothy I. Murphy. 2002. "Factors Related to the Occurrence of Isolated Sleep Paralysis Elicited During a Multi-Phasic Sleep-Wake Schedule." *Sleep* 25 (1): 89–96. https://doi.org/10.1093/sleep/25.1.89.

[78] Torontali, Zoltan A., Kevin P. Grace, Richard L. Horner, and John H. Peever. 2014. "Cholinergic Involvement in Control of REM Sleep Paralysis: Perspectives." *The Journal of Physiology* 592 (7): 1425–26. https://doi.org/10.1113/jphysiol.2014.271304.

[79] Van Eeden, Frederik. 1913. "*A Study of Dreams. Proceedings SPR*." In, 26:431–61.

[80] Voss, Ursula, Romain Holzmann, Inka Tuin, and Allan J. Hobson. 2009. "Lucid Dreaming: A State of Consciousness with Features of Both Waking and Non-Lucid Dreaming." *Sleep* 32 (9): 1191–1200. https://doi.org/10.1093/sleep/32.9.1191.

[81] Waters, Flavie, Jan Dirk Blom, Thien Thanh Dang-Vu, Allan J. Cheyne, Ben Alderson-Day, Peter Woodruff, and Daniel Collerton. 2016. "What Is the Link Between Hallucinations, Dreams, and Hypnagogic–Hypnopompic Experiences?" *Schizophrenia Bulletin* 42 (5): 1098–1109. https://doi.org/10.1093/schbul/sbw076.

[82] William Moreau, Alexandre, Muriel Amar, Nicolas Le Roux, Nicolas Morel, and Philippe Fossier. 2010. "Serotoninergic Fine-Tuning of the Excitation–Inhibition Balance in Rat Visual Cortical Networks." *Cerebral Cortex* 20 (2): 456–67. https://doi.org/10.1093/cercor/bhp114.

[83] Wing, Yun-Kwok, Sharon Therese Lee, and Char-Nie Chen. 1994. "Sleep Paralysis in Chinese: Ghost Oppression Phenomenon in Hong Kong." *Sleep* 17 (7): 609–13. https://doi.org/10.1093/sleep/17.7.609.

[84] World Health Organization. 1992. *The ICD-10 Classification of Mental and Behavioural Disorders : Clinical Descriptions and Diagnostic Guidelines.* Geneva: World Health Organization.

[85] Wróbel-Knybel, Paulina, Michał Flis, Rafał Dubiel, and Hanna Karakuła-Juchnowicz. 2018. "What Do We Know about Sleep Paralysis?" *Current Problems of Psychiatry* 19 (3): 174–84. https://doi.org/10.2478/cpp-2018-0014.

In: Parasomnia Dreaming
Editor: James Frederic Pagel

ISBN: 978-1-53617-815-9

Chapter 5

TRAUMA AND PTSD NIGHTMARE CONTENT

Robert Hoss*

Dream Science Foundation,
International Association for the Study of Dreams

ABSTRACT

Most victims of trauma and/or those diagnosed with PTSD report nightmares related to the trauma that last for weeks, months or years afterwards. While actual and repetitive replays of the traumatic memory are a common occurrence in these nightmares, they occupy about half the nightmares and thus are not the rule. The studies on PTSD subjects discussed herein, as well as the 45 years of post-combat dreams in the Vet Content Study, illustrate the variable nature of post-traumatic and PTSD nightmares. The variability is a function of many factors: the nature and duration of the trauma; the person's susceptibility; the person's history of like past traumas; changing situations in the person's life; the nature and phases of treatment and the ability for the normative function of dreaming to process the trauma over time. In some cases, the traumatic event is replayed with no resolution, in others it is only the emotional impact that

* Corresponding Author's Email: bob@dreamscience.org.

is played out or pictured *metaphorically*, in others there is a *mixture* which might combine past and recent traumatic events, and in many the dreamer attempts to *master* the threat – all possible indications of the state of recovery.

Keywords: dreams, nightmares, PTSD, post-traumatic dreams, content analysis

INTRODUCTION

Nightmares are vivid and emotionally disturbing dreams typically involving threats to the dreamer's emotional or physical sense of safety that often awaken the dreamer due to the uncontained emotional surge. They most frequently, but not always, occur during rapid eye movement (REM) sleep. They are characterized by emotional intensity, bizarreness, aggressions, failures/helplessness, unfortunate endings and emotions such as fear, anger, sadness, helplessness, confusion, disgust, and guilt (Roberts and Zadra, 2014). They are often a result of psychological trauma, which is an assault on the human mind that disrupts the ordinary functioning often leading to a constellation of disorders that do not seem to mend, such as persistent anxiety and depression (Krippner, 2016; Krippner and Taitz, 2017). At the extreme is Post-traumatic stress disorder (PTSD). Classified under Trauma and Stress-Related Disorders in the DSM-5 (APA, 2013), PTSD is considered a mental health problem resulting from direct or indirect exposure to a traumatic or stressful event. The estimated prevalence of PTSD in the general population is 1-9%; among victims with significant trauma the figure is 20–45% and among veterans it is 15–20% (Khazaie, et al., 2016).

Sleep disturbance is also a core feature of PTSD and the related nightmares are perhaps some of the most emotionally intense and can include flashbacks bearing actual or varying degrees of resemblance to the traumatic even, some being repetitive without resolution or progress. Up to 72% of individuals with PTSD report post-traumatic nightmares, and up to

91% report problems maintaining sleep (Casement and Swanson, 2012). Casement and Swanson found that post-traumatic sleep disturbance also predicts PTSD onset and severity and is independently associated with daytime distress and functional impairment, contributing to feelings of anxiety, helplessness, and heightened distress. All such nightmares are not replays of the event. Some feature the disturbing complex of emotions pictured in a representative symbolic way (metaphorically), which vividly play out the emotional impact of the trauma and/or the broader unconscious impact to the self. Over time, as the patient receives therapy and/or in other ways gains a degree of *mastery* over the effects of the trauma, the nightmares and dreams also change, becoming more of a mixture of reduced replication and increased metaphor as recovery occurs.

Veteran Content Study

An objective of this chapter is to present some of the typically reported content, characteristics and hypotheses regarding trauma and PTSD related dreams (from various studies) and compare those findings with long-term nightmare and dream content reports from a Vietnam war veteran who recorded over a 47-year period following combat. He provided these dreams for researchers, in a searchable dreambank.net databased at UC Santa Cruz. Although this is only a single person study, it is unique given the period of time over which the dream records were recorded and the changing content profiles which roughly support various hypothesis regarding intervention, the role of the dream, and stages of recovery. As far as can be determined, there are no other nightmare journals from PTSD victims, over such a post-event lifelong period of time, that have been made available to dream researchers.

While only a very few waking life influences (noted herein) are offered by the Vet on his life over that time, it is not critical to this comparison since the objective is not to speculate on the dream-to-life connection, but rather to compare these long-term dream content reports with reports and hypothesis from other PTSD dream content studies, most of which consisted of reports collected at one point in time or over shorter periods. Relating the content of his dreams and nightmares to like observations reported in other PTSD nightmare studies, can tell us much about general recovery and transition stages that he may have been going through. According to G. William Domhoff (2003) content analysis can provide a good indication of the mental profile of the dreamer. He states, "75 to 100 dreams from a person give us a very good psychological portrait of that individual."

Mark Levy (the "Vet") left combat in 1970 and recorded his nightmares and significant dreams at various times (in roughly four distinct periods) through 2017. The source of the information, for what we will call the "Vet Content Study" herein, are 139 post-combat dreams and nightmares recorded by Mark over a 45 year period (1970 to 2015) that G. William Domhoff worked with him to enter into the UC Santa Cruz dreambank.net database for use in content research. The database dreams are also available, with some added information, in his two books Dreams Vietnam and Other Dreams. The nightmares and dreams were recorded in four groups which reveal their changing nature from combat, through stages of treatment and a few influencing events, to the more recent time of his life.

A summary of Mark's history, as related in the dreambank.net database and his books, is that he was a medic in the Cavalry Division in the Vietnam war from 1969 to November 1970. Three weeks a month were spent on jungle patrol. Once a month his company pulled guard on remote fire bases. His area of operation included: Bien Hoa, phuc Binh, Phuc Vinh, An Loc, Loc Ninh, Quan Lot, Song Be, Tay Ninh, and Bu Dop. The company saw most of its action and took most of its casualties during the Cambodia invasion, where his "Delta" company helped defend a site called LZ Ranch when it was overrun. The Veterans Administration diagnosed him in 1972 as having a fairly mild case of PTSD, but over the next several years it became worse and then eventually improved. He was hospitalized in the late 1970s and then returned to everyday life. He traveled in Asia in 1995, which he reported to aggravate the condition. He wrote, "prior to entering a seven-week stay in a VA PTSD unit in 1996, I was rated 100%." Symptoms at home included: avoiding sitting or sleeping with his back to an open room or door, a heightened startle reflex and strong emotions which often follow. When asked about the frequency of nightmares after Vietnam, he wrote as follows (from dreambank.net): "During the first twenty-five or so years after coming home, I had fairly frequent war or war related dreams. They've tapered off, especially in the last ten years, to one or two a month, and they are less vivid and frightening."

His dreams were recorded in four time-periods with some long breaks between them. He stated, "I do not recall how or why the journal began." He also didn't offer a reason for the long breaks between each period – although the first two were timeframes in which/when he was hospitalized for short periods.

- Period 1 - mid-1970 to 1973; 10 nightmares one within Cambodia just before leaving combat. He reported, "from 1972 to 1974 he slept with a loaded twenty-five caliber automatic pistol under my pillow and kept a canteen of water nearby."
  - Interim 3 yrs. 1974 to 1977: He spent 2 months in a Jungian psychiatric clinic in Zurich in the late 1970s then returned to everyday life. He stated that "for various reasons" not all war nightmares were recorded between July 30, 1971, and October 6, 1981.
- Period 2 - 1977 to 1981: 23 dreams and nightmares plus one reported from within the clinic in 1970.
  - Interim 7 yrs, 1981 to 1998: He graduated from Hunter School of Social Work and practiced from 1984 to 1994 then in 1994 worked at a hospital in New Zeeland. He stated that his nightmares continued from 1981 to January 1999 (but published only one entry in 1998). Then in February 1999 he resumed entries as war-related dreams occurred. From 1983 to 1989 he reports sleeping with a meat cleaver under his pillow and from 1997 to 2000 with a machete close at hand. In 1995 he read a book by a North Vietnam vet Bao Ninh about their experience and backpacked in Asia in 1995, which he wrote appeared to "aggravate the condition." Prior to entering a seven-week stay

in a New York VA PTSD unit in 1997, he states his PTSD was rated at100%. In late 1998 briefly corresponded with, and then 2 years later (at a writers workshop), met and befriended Bao Ninh.

- Period 3 - 1998 to 2002; 65 dreams, some in 2 parts and one from 2008 his book making 70 in this study.
  - Interim 13 years 2003 to 2015: no dreams recorded and not explained. This was a period in which he wrote his first book *Dreams, Vietnam* (Levy, 2016), the content of which was almost entirely the recorded dreams from periods 1 through 3.
- Period 4 – 2015: 34 dreams and nightmares (included only in the dreambank.net database) representing a more recent stage at age 66.
  - Post Period 4 – 2016 and 2017: In the 2 years following the 4th period he recorded 463 dreams in 2016 and 2017, including them in the dreambank database and his second book *Other Dreams* (Levy, 2018). These records were not included in the study herein as they were recorded from a different basis ("It occurred to me that writing down all my dreams, not just war nightmares, seemed the logical next step"). Upon examination, many of the dominant content factors in 2016 and 2017 were much like those in 2015 (period 4) which provided an adequate sampling for this more recent timeframe.

What Is a Nightmare?

According to the Diagnostic and Statistical Manual of Mental Disorders, or DSM-5 (American Psychiatric Association, 2013), nightmares are defined as extremely dysphoric and well-remembered dreams that typically involve threats to an individual's survival or someone's emotional or physical sense of safety.

The nature of nightmarish experiences tend to fall along a *disturbed dreaming* continuum (Levin and Nielsen, 2007) where those on the milder end might be classified a "bad dream" and on the more sever end as a "nightmare disorder" a "post-traumatic nightmare" or a "PTSD nightmare." They suggest that the two are variations of the same trait but differ in how effective or ineffective they are in regulating affect loads and surges (emotion). With a nightmare the emotional surge can overwhelm the ability of the dream process to dampen it, such that the surge itself wakes the dreamer. Bad dreams are extremely negatively toned dreams that are recalled upon awakening and do not of themselves cause the awakening. DSM-5 defines a Nightmare Disorder as repeated awakenings with recollection of terrifying dreams, usually involving threats to survival, safety or physical integrity. To be considered a "disorder," nightmares also need to cause significant daytime suffering and distress (Zadra, 2019).

Post-Traumatic Nightmares, following a traumatic event, are considered an unconscious reflection of our attempts to overcome the trauma, sometimes but not necessarily a sign of pathology or PTSD (Siegel, 2001). Domhoff (in Levy, 2018) states that "unlike other dreams, trauma driven nightmares are dramatizations of enactments of our personal concerns based on our general understandings of the situations in which we find ourselves." At the extreme, PTSD Nightmares are unique in that they replicate, in whole or in part, the terrifying and often-excruciating experiential details of the trauma (Mellman and Pigeon, 2005). According to Hartmann (1984) such nightmares occur when a trauma victim reaches an impasse in their recovery and the failure to cope is reflected in nightmares that *replicate* the traumatic event, a condition he refers to as *encapsulation.* In a study by Wittmann, Schredl, and Kramer (2007), 50% of PTSD patients report such *replicative* nightmares. In various other studies (Esposito et al., 1999), nightmares of PTSD subjects were found to feature settings, characters, and objects similar to the traumatic experience, along with high threat, low contemporaneity, and low distortion (more a memory than a dream creation). As this chapter will demonstrate, PTSD related nightmares can vary, however, depending on the nature of the trauma, vulnerability, passing time, and recovery and begin to contain sequences that retain elements of the traumatic event but change or omit some details, becoming more *metaphoric* or dream-like over time, yet preserving or amplifying associated emotions.

All can be distinguished from Night Terrors and REM Disorders which are disturbed sleep parasomnias which involve vocalizations, sleep walking, and thrashing movements during sleep, sometimes startled or panicked awakenings, and exhibit a terror or fight/flight response. Night Terrors typically arise from non-REM sleep, and do not always result in awakenings and may or may not be accompanied by a recalled dream or experience. REM based parasomnias, or REM sleep disorders, involve non-atonia conditions whereby the dreamer in essence acts out a vivid often threatening dream (e.g., walking, talking).

The Nature of Post-Traumatic Nightmares

Increased awareness and research into trauma and PTSD over the last few decades have revealed that the devastating and life-altering impact of trauma is not limited to war veterans (Zadra, 2019). Extreme trauma may stem from child abuse and neglect, domestic violence, terrorism, political persecution, loss, an all forms of accidental or natural disasters such as car or plane crashes, fires, explosions, earthquakes, floods or hurricanes (e.g., Duval and Zadra, 2010). However, Mellman, et al., (1995) and Neylan et al., (1998) found disturbing dreams about combat experience and war-zone exposure to be more specifically associated with PTSD than disturbing dreams about other traumatic events. Van der Kolk et al. (1984) reported that Vietnam veterans who met criteria for PTSD were more likely to state that their nightmares exactly or almost exactly replicated an actual event, than lifelong nightmare sufferers without combat experience. Moreover, far from being infrequent or isolated occurrences, recurrent about a traumatic event can persist for years and even decades after the trauma.

Content Variability

Victims of trauma and/or those diagnosed with PTSD report nightmares and dreams related to the trauma that last for weeks, months or years afterwards. While repetitive replays or representations of the traumatic memory are a common occurrence in such nightmares, they are not the rule. Post-traumatic dreams vary in nature depending on many factors both personal and involving the nature of the trauma. From most studies, it appears the traumatized person may dream first about the actual trauma or some close semblance of it (though not always), but then selectively process the dominant emotions (Stickgold and Walker, 2013), often picturing it in a *metaphoric* manner (Malinowski and Horton, 2015) or as Hartmann (1998) termed it, *contextualizing* the emotional concern. Kramer et al. in a study of eight Vietnam combat veterans with PTSD (1987), noted that the veterans are vulnerable to developing an acute PTSD in response to a current stressors in their environment which facilitate the return of dreams about Vietnam which are *mixed*, being expressive of their current problems as well. Barrett (2018) also noted two variations from actual replay in post-traumatic nightmares: a) horrors that were only a threat or imagined in waking life can materialize in the nightmare and b) the actual traumatic event can become more surreal or dreamlike (*metaphoric*). An example given of the first is, A World Trade Center survivor dreams of barely escaping, as in real life, but then seeing his wife among the dead (she actually worked in the building but was reunited safely with him within the hour). An example of the second is that a human attacker becomes a literal monster in the dream re-creation.

As will be discussed in more detail below, the normal dreaming process is thought to serve a self-regulating emotional problem-solving function of emotional down regulation, *contextualization*, adaptive learning and extinction. As differentiated from typical dreams and nightmares, some post-traumatic nightmares and most PTSD nightmares are resistant to this function. The trauma (emotional memory) is too overwhelming to be dampened by this self-regulating process (Levin and Nielsen, 2007) or to be interrelated with the remainder of life's experiences and memories that might blend and soften the overwhelming emotion (Krippner, 2016). Thus new contexts do not replace the existing threatening context and are therefore ineffective at extinguishing the fear memories. Their repetitive nature may reflect failed attempts at extinction. Such dreams are for the most part incapable of regulating emotion (affect load), and in fact may increase affect load further preventing the extinction of fear memories (Levin and Nielsen 2007) and further may activate previously encoded fear memories, leading to greater waking emotional distress and subsequent increase in nightmares.

Hartmann, however, stated that repetitive PTSD dreams and seeming replays, often turn out to be new creations, stating that, "in all the cases I have examined in detail, the dream is not 'just the way it was' - there's at least one important change" (Hartmann, 2011). An example of this concept (which also supports the notion of a sequential emotional processing function within dream cycles) was captured in the REM state awakening reports of one of the Vets in a sleep lab study of eight combat Vets with PTSD by Kramer et al. (1987). One of the vignettes (described further below) involved reports from five subsequent REM awakenings. Two initial periods contained discrete vivid replays of two different traumatic combat events (with minor distortions), followed by a REM report that appeared to integrate memory fragments from those two, then followed by a REM report of combat/non-combat memory integrations (including elements of a childhood trauma) *mixed* in a fictive metaphoric manner.

Content Studies on Post-Traumatic and PTSD Nightmares

There have been many studies related to the nature of and content of post-traumatic and PTSD nightmares. Table 1 summarizes a few. It should be noted that the results aren't totally comparative due to the differences in design, nature of the trauma, population size and recovery status of the subjects. The table, however, can provide some common features which may be useful in understanding the general content of post-traumatic nightmares in comparison with the Vet Case Study.

Table 1. Summary findings of various post-traumatic nightmare studies

Condition	Content	% or Comment	Study
PTSD Patients	• Replication of traumatic event • Mixture of elements of much earlier traumatic and adverse events, also repetitious non-traumatic elements	• About 50%	Wittmann, Schredl, and Kramer, 2007
PTSD nightmares	Replications with modified details, but that preserve or amplify associated emotions	General observation	Hartmann, Zborowski, Rosen, and Grace, 2001; Mellman and Pigeon, 2005
18 Vietnam combat Vets with PTSD using a PTSD dream rating instrument	• Dreams contain features characteristic of combat • Moderately to highly threatening • Contained "distorted" elements (vary with regard to replication of trauma and elements normally associated with dreaming).	• About 50% • 83% • 79%	Hall van de Castle, 1966(b)
Combat Vets	• Reported dreams containing threats • Contain explicit combat details	• 83% • About 50%	Esposito, Benitez, Barza, Mellman, 1999
304 War Vets w/combat nightmares n = 359	• realistic combat dreams, • plausible war sequences not actually experienced • alluded to the war but also included fantastical and everyday elements • killing, being killed, death, dying, wounded, committing atrocities • being chased, animals, looming danger; cycle of service: approaching combat, engagement, returning home and longing to return to Vietnam	• 50% • 21% • 26% • Most common • Generally common	Wilmer (1996)
Hospitalized Accident or Assault	• replicated the traumatic event • dissimilar to the traumatic event but containing high levels of distress	• 46% • Remaining (54%)	Mellman et al., 2001
Survivors of Hurricane Iniki	• Dreamed about the hurricane itself • Especially stressful life experience • Only general stressors	• 13% • 67% vs. 37% in the control • 74% vs. 48% in the control	Pagel, Vann, and Altomare, 1995

Table 1. (Continued)

Condition	Content	% or Comment	Study
Combat vets	• One's own comrade's death, wounded, witnessing killing and destruction, • Realistic combat refences, factual, emotionless, mundane reporting	Dominant content contained in the nightmares	Siegel (2001)
Nam Vets w/PTSD and depression	Vets with depression and PTSD more likely to be set in the past vs. reports from depressed groups without PTSD.	Reports from Lab Awakenings	Dow, Kelsoe, and Gillin (1996)
Vietnam veterans	Sleep Lab awakenings: dreams containing military references	47% (vs in-frequent in control)	Kramer, Schoen, and Kinney (1984)
8 Vets w/PTSD	Sleep Lab awakenings illustrated fragmented replays followed by metaphoric mix in subsequent stages	Memory fragmentation & integration	Kramer, Schoen, and Kinney (1987)
19 Vietnam combat veterans with PTSD (diverse races and ages from 41 to 61)	Using DRS (dream rating scale for combat-related PTSD): • Settings, characters, objects, weapons similar to the combat experience • Reference to Enemy • Atrocities • Moderately to highly threatening • Set at least partially in the preset • Contained distorted elements • Having exactly replicated an event similar to the traumatic experience • Vary as to replication and elements normally associated with dreaming	 • 42% to 47% • 16.3% • 2.6% • 83% • 53% • 79% • 21% • 79% w/distorted elements.	Esposito, Amparo Benitez, Lydia Bana, and Thomas Mellma (1999)
PTSD Nightmares (Overview)	• Exact replays are rare • Rarely employ metaphoric content that represents beneficial working through of negative affect as observed in normative dreams	Observations specific to PTSD nightmares	Davis (2009)
Children in war zones or w/life threats (using DTS rating scale)	• Very high proportion of threat simulations in their dreams that were severe and realistic; vs children from safe environments with simulations that contained more mild threats.	• 80% (vs. 30% to 50% in the safe environ. control groups)	Valli et al. (2005)

What these studies above and in the table demonstrate is that post-traumatic and PTSD nightmares have, in general, a number of unique characteristics:

- Post-traumatic or PTSD nightmares can be highly variable: from near exact *replications*; to highly *metaphoric* dreams carrying through the emotional theme; some with present and past traumatic memory fragments *mixed* with fictive activity.
- Exact replays in PTSD nightmares are not the rule; only about half appear to replicate the terrifying and often excruciating experiential details of the trauma.
- When exact and/or repetitive replays occur in PTSD nightmares, they exhibit a resistance to the *metaphoric* content that might represent a beneficial working through of negative affect.
- Upon close examination even apparent replays may not be "just the way it was" but can include distortions, notable missing elements, and a mixing of memory fragments from other past or present traumatic experiences and waking life stressors. This evolution from replay to *metaphoric* distortion can even occur in subsequent REM stages within a single night.
- Post-traumatic nightmares and dreams, of a trauma survivor over time, show increasing dissimilarity to the traumatic event (in the 50 to 80% range) but tend to maintain high levels of distress and preserve/amplify the associated emotions.
- Over half of combat related PTSD nightmares can also depict vivid, realistic and actual combat scenes and references, characters, and objects similar to the traumatic experience.
- PTSD nightmares generally contain high threat levels (in the 80% range).
- Post-traumatic nightmares can picture intensified representations of the actual or unexpressed emotions: horrors that were only a threat or imagined in waking life can materialize in the nightmare, and the actual traumatic event can become more surreal.

Sleep Parasomnias and PTSD Nightmares

Sleep parasomnia (sleep walking, talking, night terrors, REM disorders) do not necessarily imply a psychiatric disorder or PTSD, however sleep parasomnias have been found to be a sleep disturbance feature of PTSD.

Arnulf (2019) indicates that most studies in patients with sleep parasomnias show no psychiatric disorders but that parasomnias can involve nightmares. In a study of 73 adults with sleepwalking and night terrors, 53 percent reported nightmarish mental content, where the sensation of threat to survival and the need to escape co-occurred with motor episodes. The source of the dream experiences is sometimes obvious to the patients. In this 73-subject series with sleepwalking and night terrors, four patients (5.4%) replayed a recent event that they had encountered during the day, either in real life or on TV; this as opposed to <2% per normative dream studies (Fosse, 2000; Malinowski and Horton, 2014). With night terrors, the contents of mental imagery was often recurrent (e.g., being buried alive, insects and snakes, or facing a life-threatening event) across nights and across individuals, and they are often followed by a flight response.

Arnulf found that NREM parasomnias (sleepwalking and night terrors) are most often associated with dreams of misfortune and result in a flight response. 80% were considered negative. The notable emotions are: friendliness 6%; failure 37%; aggression against the dreamer 26%, misfortune 54%, apprehension 84% and threat 70%. For REM Behavior Disorders, she found: 10% friendliness; 20% failure; and 60% threat.

A study by Mellman et.al (1995) concluded that PTSD results in sleep intrusions (particularly REM sleep) with more highly aroused behaviors and states. The study included: 58 combat veterans with and without PTSD; sleep diary records from 52 hospitalized combat veterans with PTSD; and overnight polysomnography recordings from 21 medication-free combat veterans with PTSD and 8 healthy comparison subjects not exposed to combat. They found that recurrent awakenings, threatening dreams, thrashing movements during sleep, and startled or panicked awakenings or anxiety related to threatening dreams, represented the most prevalently reported sleep-related symptoms.

Mysliwiec et al. (2014) also found disruptive nocturnal behaviors (DNB) consisting of excessive movements, autonomic hyperarousal, abnormal vocalizations and complex motor behaviors, and nightmares, which contained replays of traumatic experiences, are frequently reported sleep disturbances in combat veterans and trauma survivors with and without PTSD. Mysliwiec proposes the term "Trauma associated Sleep Disorder" (TSD), to describe a unique sleep condition that represent trauma related parasomnia associated with PTSD, encompassing such clinical features as: a) disruptive nocturnal behaviors (DNB) in both REM and NREM (thrashing movement, startle/panic-like awakenings, acting out dreams); b) PSG characteristics such as altered REM latency, REM without atonia, and any EMG activity that is normal to increased and c) nightmares that are frequently reported that follow and replay traumatic experiences.

THE VARIABILITY OF POST-TRAUMATIC NIGHTMARES

As noted above and in Table 1, post-traumatic and PTSD nightmares are not simply repetitive replays but vary dramatically in their nature. Hartmann (1998, 2001) determined, in studying dreams of traumatized persons over time, early on the person may (but not always) dream about the actual trauma or a close replica (most often with some small variation), the dreams then soon appear to deal with and picture the dominant emotion in a *metaphoric* or *contextualized* manner. In the Wilmer (1996) study of over 300 Vietnam War veterans, they found that while over half of the veterans reported realistic combat dreams, 21 percent reported plausible but not actual war scenes, and 26 percent reported dreams with war references, but *mixed* with fictive *metaphoric* and everyday elements. Mellman et al. (2001) also noted this variability in his study of accident of assault patients where only 46 percent of reported dreams were found to replicate the traumatic event; the rest were either dissimilar to the traumatic event but contained high levels of distress with little to no elements of the event. In the Kramer study (1987) of Vietnam Vets, one of the PTSD subject's dreams varied from two replays with missing elements, to a *metaphoric* mixture of multiple combat and early life traumatic memory fragments, in subsequent REM stages in a single night.

Variability over Time in the Vet Case Study

Aside from these studies, most of those listed in Table 1 can tell us much about the general nature of the post-traumatic and PTSD nightmare content, but do not necessarily provide data related to variation in content over time as recovery progresses. Studying the dreams of combat veterans with the variations in mind can provide evidence of how the human psyche appears to cope in different ways with overwhelming trauma. This involves not only a variation in *replicative* versus *metaphoric* content, but also the role the trauma survivor began to play in the nightmares over time (passive victim versus attempts at controlling, containing or *mastering* the threat within the dream itself).

In order to quantify this variability, a useful set of parameters can be borrowed from some standard rating scales used to quantify dream content for both common and combat related dreams. Hall Van de Castle's scale (1966a) is structured with the following major indicators: Characters, Aggression, Friendliness, Sexuality, Activities, Success/Failure, Misfortune/Good Fortune, Emotions, Settings, Objects, Modifiers, Negative, Temporal, Food. Exposito et al. (1999) developed a Dream Rating Scale (DRS) for combat related PTSD, incorporating many of the Hall and Van de Castle parameters but bundled them with a combat focus: a) Setting - characteristic of combat or military reference vs. *mixture* of combat/noncombat vs. civilian, no military references; b) Threat: catastrophic or death possible; discomfort; friendly or comfortable; c) Contemporaneity (temporal measure): past or future vs. *mixture* of past/present vs. present; and d) Distortion: replication of actual event vs. did not occur but plausible; vs. bizarre, magical. Tables 3 through 6 is a result of re-grouping these parameters to best reflect the combat related content that dominates the Vet Content Study. Table 3 displays Setting content and the nature of Threat; Table 4 expressed Emotions; Table 5 "distortion" or replay versus *metaphoric* representation and Table 6 *Mastery* (a separate category oddly not included in the above studies) aimed at observing how the dream-self initiates attempts to take charge and contain or gain control over the threat situation within the dream.

Setting and Threat

The contents from the Vet Content Study is summarized in Table 3 for the four periods where the nightmares and dreams were recorded. These are roughly compared with the results of other studies representing a spectrum from common dreams or "norms" to common bad dreams and nightmares then combat nightmares for subjects with PTSD. Note that as with Table 1, the comparison with other studies is only a relative perspective due to the differing methodological approaches and definitions between each study. For the Vet Content Study however the definitions used for the 23 content parameters have been held consistent and thus comparison between periods is possible. This table contains all of the recorded Vet dreams, mostly nightmares but also those he considered important enough to record. This time dependent series highlights the subtle and not so subtle changes in the dreams as the Vet struggles through what might represent four relative stages in his recovery over a 45-year period (from post-combat to near present day). Content data related to the other categories (emotion, variability and *mastery*) follows in later tables and discussions.

Table 3. Comparative content of the Vet Content Study dreams over 4 periods in 45 years, with other classes of dreams or nightmares (% of dreams in the study containing that content)

Content	(1) Norms Common Dreams	(2) Common Bad Dreams and Nightmares	(3) Vets with Combat Nightmares	(4) Vet Content Study PTSD (% of all dreams recorded)			
Source	a) HVDC Males b) HVDC Males c) DTS M and F	a) Zadra '04 b) Robert and Zadra '14 c) Schredl '10 d) Other	a) Esposito '99 (PTSD) b) Wilmer '96 (Nightmares)	1st 3 yr '70 - '73 n=10	Period #2 '77-'81 post-therapy n =25	Period #3 '98-'02 post-therapy n = 70	Period #4 2015 n = 32
SETTING CONTENT							
Combat Scene		5.4% [b]	42-47%[a]	90%	36%	39%	0%
Actual Loc.	1.2%[b]		6.5% [b]	80%	20%	23%	0%
War Reference	3.7%[b]		~ 83%[b]	100%	72%	83%	28%
Shoot, Stab	7.3%[b]		9.5%[b]	50%	72%	44%	6%
Atrocities			2.5-6%[a] 8.5%[b]	0%	8%	11%	0%
Service Cycle			2.2%[a] 2.2%[b]	10%	16%	1%	0%
Weapons	15%[a]			100%	68%	46%	9%
Enemy (stated)	2.4%[b]		16.3%[a]	60%	16%	23%	3%
Family (his)	13.8% [b]			0%	20%	9%	6%
Female (all)	24.4%[a] 43%[b]			10%	12%	31%	31%
Sexuality	11.2%[a]			0%	4.7%	7.9%	15.6%
Threat Content	77%[c]	63% [d]	83%[a] ~65%[b]	100%	96%	83%	75%
Victim of	67.4%[c]	12.2%[b] 25%[d]	29.6%[a] 40%[b]	100%	72%	63%	56%
Life and Death	17.7%[c]		~ 68%[b]	100%	68%	51%	19%
Others Dying			~ 27%[b]	70%	48%	36%	9%
Self Dying			3.3%[b]	30%	8%	10%	6%
Wounded (any)	8.6%[a] (self)		4%[b]	0%	36%	17%	3%

Content	(1) Norms Common Dreams	(2) Common Bad Dreams and Nightmares	(3) Vets with Combat Nightmares	(4) Vet Content Study PTSD (% of all dreams recorded)			
Source	a) HVDC Males b) HVDC Males c) DTS M and F	a) Zadra '04 b) Robert and Zadra '14 c) Schredl '10 d) Other	a) Esposito '99 (PTSD) b) Wilmer '96 (Nightmares)	1st 3 yr '70 - '73 n=10	Period #2 '77-'81 post-therapy n =25	Period #3 '98-'02 post-therapy n = 70	Period #4 2015 n = 32
Aggression	47%[(a)] 33.5%[(c)]	26%[(a)] 31.5%[(b)]		100%	92%	70%	53%
Aggressor	11.7%[(a)]			20%	48%	21%	19%
Victim	18%[(a)]			100%	68%	64%	56%
Chased or Escaping	12%[(a)] 6.9%[(b)]	<5%[(a)] 7.6%[(b)] 26%[(c)] 27%[(d)]	4%[(b)]	0%	12%	17%	3%
Accident	17%[(c)]	<5%[(a)] 6.3%[(b)] 17%[(d)]		0%	12%	6%	3%
Inter-personal		9%[(a)] 29.5[(b)]		0%	20%	17%	31%

Legend:

1) Normative Dreams: (a) Hall van de Castle norms dreamresearch.net (Schneider and Domhoff, n.d.) (b) Hall van de Castle norms in dreambank.net word search w/applicable synonyms of 491 male dreams; (c) Valli et al. (2007) Dream Threat Scale (DTS), 248 reports, 50 Male and Female (at least 1 threatening event).

2) Nightmares and Bad Dreams of those with no reported disorder: (a) study by Zadra, Duval, Begin, and Pilon (2004), 125 nightmares reported by 125 women. (b) study by Robert and Zadra (2014) of 331 adults, 253 nightmares and 431 bad dreams. (c) Schredl (2010) 1,022 adults reporting multiple nightmares per year, using a retrospective questionnaire representing 23 nightmare themes. (d) Other: Cason and Desjardins and Zadra in Robert and Zadra (2014)

3) Combat-related PTSD: (a) 18 combat Vets with PTSD evaluated using DRS rating scale with Hall and Van de Castle content analysis parameters (Esposito et al., 1999). (b) Wilmer (1996) 304 Vets w/combat nightmares (~ notates the mean computed from other measured parameters)

4) Vet Content Study PTSD All dreams and nightmare records, 4 periods, dreambank.net database

Emotion

Chart 1a illustrates expressed emotions in all dream records over time. Emotional expression remains fairly high (100% at first and about 60 to 70% across all remaining periods). When all recorded dreams are taken into account, negative emotions begin very high but then reduce (varying dramatically over the remaining periods) whereas positive emotions increase almost linearly from periods 2 to 4 as he begins to record more non-nightmarish dreams. When nightmares (disturbing dreams with threats to self) alone are taken into account (Chart 1b and Table 4): negative emotions are high (70%) in period 1 where combat memories are *replicated*. Negative emotion drops to 48% in the 2nd period which seems to correspond to a greater degree of *mixed* or *metaphoric* nightmares (Table 6) and as discussed later (Table 5) a period when he attempts to contain (or *master*) the threat. Negative emotion then increases in period 3 (to 66%), which followed what he reported as an "aggravation of his condition" and subsequent treatment. However as will be discussed later, the content in period 3 suggests it may be inner turmoil from abandoning the "combat self" and re-establishing a new self-image. In the more recent period 4, the dreams are nearly all *metaphoric* with little to no combat references, as negative emotion decreases again into the 44% range (about the same as the norms for common male dreamers per Hall van de Castle).

In addition to variation in total or positive and negative emotion, the specific emotions in the Vet Content Study varied over time. The specific emotions closely aligned with those found in a disturbed dreaming study by Robert and Zadra (2014) of 331 adult volunteers who reported 253 nightmares and 431 bad dreams (using the Hall Van de Castle method). He found eight content categories to account for approximately 90% of emotions reported in disturbed dreaming: fear, anger, sadness, confusion, disgust, guilt, frustration plus all positive emotions. In a dream by dream review in the Vet Content Study, two other frequent emotions were noted separately: Anxiety/Apprehension (similar to fear but more impending and less immediate) and Helplessness (inability to avoid the threat); Although more emotions seemed apparent from the dream plots, only "expressed" emotions were captured for inclusion in the Table (where the Vet stated the feeling in words or a synonym for it). The emotions are sorted top to bottom by percent of dream records containing those emotions.

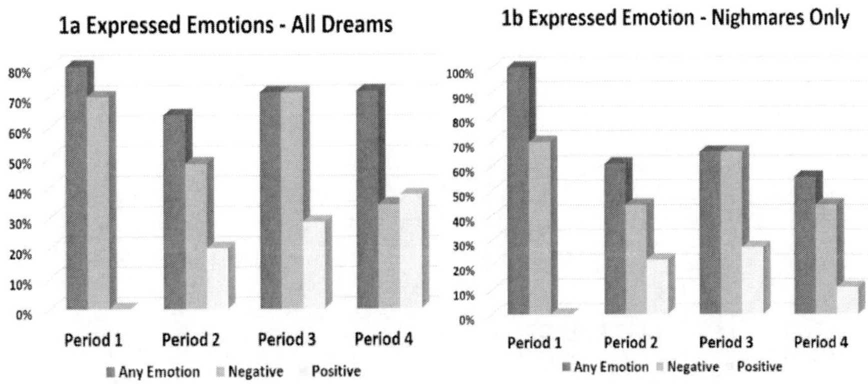

Chart 1a and b. Variation in Expressed Emotion in All Recorded Dreams over Time. (% of dreams expressing that emotion).

Note that in Table 4 the most commonly expressed emotions over the full 45 years of nightmares in the Vet Content Study were fear, anxiety, confusion, anger and sadness. These were also the highest percentages in the Robert and Zadra non-PTSD nightmare study (2014). It should also be noted how the emotions vary over time as the Vet transitions through various stages of apparent recovery. Note how Helplessness and Confusion steadily decreased over time as attempts at *mastery* (control over the threat) increased (Table 6). Anger and Sadness also decreased as combat memories decreased (Table 3). Aside from the immediate post-combat period of memory replays, total expressed emotion (negative plus positive) remained fairly constant, and roughly at the level of normative dreams (column 1) which along with the increasing *metaphoric* content (Table 5) might suggest a growing influence of the hypothesized normative emotional problem-solving function for dreaming.

Table 4. Emotional dream content over time

Emotion	(1) Norms Hall V D Castle (n = 491 Male of 891)	(2) Bad Dreams and Nightmares Robert and Zadra	(3) Veterans Content Study - Combat Related Nightmares (% w/that expressed emotion)				
			Period #1	Period #2	Period #3	Period #4	All Yrs
Expressed	56%[(c)]		100%	61%	66%	56%	67%
Negative	45.4%[(c)]		70%	44%	66%	44%	58%
Positive	6.1%[(b)] – 11%[(c)]	4.2%	0%	22%	30%	11%	21%
Fear	16.1%[(b)]	52.2%	40%	11%	32%	28%	28%
Anxiety	11.3%[(c)]	11.4%	10%	6%	16%	6%	11%
Confusion	7.5%[(b)] – 12.3%[(c)]	5.4%	10%	17%	9%	6%	10%
Anger	6.7%[(b)] – 9%[(c)]	6.1%	0%	11%	9%	6%	8%
Helpless	7.5%[(a)]		30%	11%	2%	0%	7%
Sadness	2.2%[(b)] – 5.2%[(c)]	10.1%	0%	17%	7%	0%	7%
Frustration	1.8%[(a)]	3.5%	0%	0%	5%	0%	2%
Guilt	1.4%[(a)]	3.1%	10%	6%	0%	0%	2%
Disgust	2.9%[(a)]	4.0%	0%	0%	0%	0%	1%

Legend:

1) Hall van de Castle Norms: a) dreambank.net database search of Hall van de Castle files for % of male dreams with those expressed emotional terms and synonyms; b) Bulkeley 2009 dreambank.net search; c) Schneider and Domhoff dreamresearch.net

2) Common Nightmares and Bad Dreams: Robert and Zadra, 2014; 331 Adults (n=253 nightmares, 431 bad dreams)

Classification of Nightmare Variability

As mentioned briefly above, Harry Wilmer (1996) worked extensively with Vietnam combat veterans, at one point conducting a study of 359 dreams from 103 veterans. Some of the more consistent content he observed in such studies is listed in Table 1. Out of this research Wilmer, observed three stages of impasse and resolution in the dreams which he suggested coincided with each stage of recovery that the veterans had reached.

Category 1 – 'Actual' (Replications)

Wilmer's category one dreams he termed "actual" (perhaps more commonly called "flashbacks," "replays" or herein *replications*) were characteristic of the men who suffered the most severe symptoms of PTSD. These dreams were more realistic and less "dream-like." They tended to replay horrific events (perhaps a single horrific event) witnessed by the dreamer (such as a buddy getting 'blown away' in his presence). The content of such dreams almost exactly *replicate* the actual event (or realistic aspects of it) and have nothing to do with day residue, or the dreamer's past or present life. Category 1 dreams accounted for 53% of the total in Wilmer's study. In the Vet Content Study, such *replications* in period 1 (first 3 years post-combat) accounted for about 70%; although all records contained war references and 80% were in an actual named combat location.

Hartman believes that soldiers who lose a person with whom they were extremely close are the most vulnerable, with the related nightmares often occurring months or years later when, a subsequent loss, rejection or trauma triggers the painful memories that had not been totally dealt with.

Some repetitive *replications* can even come from an early forgotten trauma as in this report provided to me from a person at a lecture I was giving.

> The woman reported that she had recurrent nightmare all her life from the time she was a child where she always sees the same airplane diving down and shooting at her. It comes so close she could see the face of the pilot sneering at her.

She was reluctant to tell her mother, who she felt did not put much importance in dreams, but finally did so later in her adult life. When she told her the dream, her mother said, "Oh you can remember that? It actually happened when you were 18 months old." As a child in Europe during WWII, while trying to run for shelter, she and her mother had been attacked by an enemy airplane. Once her mother revealed the story, her mind could now place the memory in a new nonthreatening context, and nightmarish replay never returned. Note that, as in the above example, where a dominant feature was, "*...the face of the pilot sneering at me*" post-traumatic nightmares often feature an odd peripheral aspect or fragment that was a part of the overall sensory experience.

Category 2 – 'Variable' (Mixed)

Wilmer found that as veteran's were recovering, either from treatment or spontaneously, the imagery began to transform whereby the nightmares still involve an intrusion of the war experience, but the content begins to be mixed with events from the present, past and other areas of the person's life. These dreams, which are often referred to as "distortion" or *mixed* (as referred to herein), accounted for about 21% of the dreams in Wilmer's study.

In the first 3 years of the Vet Content Study, although 70% appeared as actual or plausible memory *replications*, with all containing war references, a couple were "variable" or *mixed*. The final dream, as did all, depicted a realistic combat scene, but the bunker he was cowering in was from his home:

> "The bunker is actually the bar in the basement of 39 Ivy Street!"

He stated that before the dream he had been going through an album of family pictures. After spending 2 months in treatment, the 25 nightmares recorded in the next 5-year period there were almost no *replications* and about 1/3 were *mixed* (realistic combat scenes mixed with past or present memory references or symbolic metaphors).

Category 3 – 'Hallucinatory' (Metaphoric)

Wilmer's category 3 dreams (which will be herein termed *metaphoric*) are more like common bad dreams and nightmares. They change details of the trauma in a story-like *mixture* of past, present and fictitious imagery, but their emotional and thematic relationship with the original trauma is ever present - preserved and perhaps amplified in the actions and symbolism of the dream (Hartmann, et al. 2001 Mellman and Pigeon, 2005). They tend to place the events and emotional impact of the original trauma into the broader picture of the victim's self-image and world view. In Wilmer's study, 26% were of this type. In the Vet Content Study in period 2 (following his first reatment), almost all dreams (96%) contained highly *metaphoric* content, although 36% took place in an actual named combat location or similar setting.

This variability between Category 1 *replications,* and *metaphoric* representations as in Categories 2 and 3, is seen in other studies (generally noted as *mixtures*, dissimilar or distortions in Table 1). In a study by Mellman of patients hospitalized following an accident or assault, 46% of reported dreams were found to replicate the traumatic event and the rest were either dissimilar to the traumatic event while containing high levels of distress or showed little similarity to the trauma or distress (Mellman et al., 2001). Kramer et al. (1984) who reported that 47% of dreams elicited in the sleep laboratory from combat veterans with disturbed dreaming contained military references: 21% were rated as having exactly replicated an event and 79% demonstrated some degree of distortion. A study by Pagel et al. (1995) of 22 survivors of Hurricane Iniki reported dreaming only rarely about the hurricane itself (13.6%), but more about an especially stressful life experience (67% vs. 37% for the control) or about general stressors (74% vs. 48% for the control). The control consisted of 265 questionnaires from male, female and cross-cultural populations.

VARIABILITY IN RELATION TO RECOVERY

Wilmer considered the emergence of a stage three dream after prolonged unchanging nightmares as an important sign and part of the healing process. Observing the lessening or ending of their brutal repetitive nightmares helps the veterans to feel understood and to observe their inner response to the therapy and healing process. Seigel (2001) stated that post-traumatic nightmares are like graphic memories of the trauma that gradually fade as the person recovers, giving way to dreams less focused on the trauma and more mixed with other concerns. Ernest Hartmann (1984) who studied trauma dreams over periods of recovery found the first dreams after the trauma can often directly portray memories of the actual event but over time it is the dominant emotions (especially fear or vulnerability) that become *contextually* pictured, often in a powerful dream (Hartmann liked to use the example of a tidal wave). This is then followed by a whole series of dreams which appear to be connecting memories of the actual trauma, *metaphoric* pictures of the emotion, and pictures of similar past traumas or other related events that have some emotional relationship to the new one. He states that as recovery progresses the dreams usually gradually return to the pattern they had before the traumatic event.

But not all severe traumas or even those that might be classified as PTSD create nightmares that occur in this specific sequence. In some cases, the initial dream pictures only the emotional impact to the individual *metaphorically*. Hartmann considered the imagery in dreams, particularly the overpowering ones, to "picture" the emotional state of the dreamer, often using a "tidal wave" as an example of how a dream might picture the overpowering emotion of fear. Hartmann (2011) developed what he called the Contextualizing Image or Central Image (CI) as a means of recognizing and measuring the emotions contained within a dream narrative and to reflect how active and unresolved the current emotional concern is. He found that those suffering from trauma or an extreme emotional event, may not dream of actual elements from the event, but may have dreams with a mean CI score that is higher than the mean CI score of a control group. For example, Hartmann (2003) performed a study on the effects of the attacks of September 11, 2001, comparing 880 dreams recorded before and after the event for 44 persons around the United States who had recorded their dreams for many years. The results were that not a single dream actually pictured planes hitting tall towers or anything close to that, but they did show a highly significant increase in CI scores.

Sometimes the dreams begin as *mixtures* relating combat fragments to other lifetime memories. In the Vet Case Study, although most of the nightmares recorded in the first few years appeared as *replications* of his vivid and realistic combat memories, the first dream recorded right out of Cambodia was a *mixture* of a seemingly repetitive and replicated combat scene combined with a *contextualizing* metaphor of experiencing combat from the perspective of a vulnerable child:

> "The enemy is above, in the sky. We are not soldiers but school children. A voice from everywhere begins to warn us, 'Seeds are dropping.' Then I hear the faint echo of the tubes popping. There is an underground bunker; most of us run for the school bus. The fear of realizing death is pacing through my mind."`

Alan Siegel (2001) stated that post-traumatic nightmares tend to blend the pain of past losses with present traumatic anxieties – which creates odd mixtures of past and present imagery or references. Pagel et al. (1995), in a study of those that survived Hurricane Iniki, found that only 13.6% of the subjects responded that they sometimes dreamed of the hurricane, while 67% reported that their dreams were related to an especially stressful life experience. Siegel found that unresolved past trauma can play into post-traumatic nightmares. If the person hadn't been able to work out the troubling emotions from the earlier trauma, they will have a more difficult time working out the parallel emotional stumbling blocks caused by the recent trauma. Among the thousands of survivors of the Oakland/Berkeley firestorm he studied 215 of 42 subjects, some who lost their homes and other who's homes in the burned area were spared. He found that those with more profound backgrounds of trauma were more severely affected and slower to recover – some dreaming of long dead parents, the death of a child, divorces and other past tragedies.

Siegel (2003) also observed that "for some, emotional regressions occurred only after substantial progress had been made and signaled a *breakthrough* in the ability to face the most abhorrent memories and emotions of the firestorm." It is uncertain but this *breakthrough* factor may have contributed to the phenomenon observed in Chart 2 of the Vet Case Study when comparing the third period content with that of the second. In the second period, after treatment, combat memory *replications* all but disappeared giving way to *mixed* nightmares containing fictive combat situations and *metaphoric* imagery. It appeared from the dream content alone that progress had been made. In the third period, however, we see a significant increase in combat memory *replays*, war references, fear and negative emotion. The third period came after a visit back to the war zone which aggravated his condition, but also after a second hospitalization and what he called a "turning point" (turmoil of a *breakthrough*?) as he befriended the former enemy, a North Viet Nam veteran Bao Ninh.

Table 5 and Chart 2 illustrates the resultant variation in the nature and content of the Vet's nightmares between the initial four periods in which he recorded his nightmare, over the 45 years since leaving combat.

- Period 1 - the initial 3 years following combat 10 nightmares were recorded. Seven appeared to be Category 1 (*replications* of combat memories) and three appeared to be Category 2 (realistic combat scene *mixed* with *metaphoric* content).
- Period 2, after some time and his first treatment, *replications* all but disappeared, being replaced with fictitious but realistic thematic intrusions of his combat experience *mixed* 60% with other life experiences, settings and *metaphoric* representations.
- Period 3 came after correspondence with a former enemy Vet (a NVA Vet), traveling in Asia, (which may have "aggravated the condition") and in 1996 entering a seven-week stay in a VA PTSD unit (when "rated 100%" on the PTSD scale). Here *replications* and references to combat (including combat locations) increased as *metaphoric* representations decreased. But as will be discussed later, the reflections on the war and his combat role appeared to take on a

new perspective, more from a whole-life context and the turmoil of moving from a combat self to a new self.

- Period 4 – the more recent period in 2015 *replications* of combat memories and actual locations dropped to zero, with war references remaining but dropping substantially as most of the dreams appeared as normative *metaphoric* dream-like creations dealing with internal conflict as opposed to combat memories.

Table 5. Variability: nightmare replications versus metaphoric representation

Content	Vet Content Study Progression			
	Period 1	Period 2	Period 3	Period 4
Replication or Similar (Cat. 1)	70%	6%	20%	0%
Mixed (Cat. 2)	30%	44%	36%	6%
Metaphor or Fictive (Cat. 3)	0%	56%	44%	94%
Actual or Like Combat Scene	90%	39%	45%	0%
Fictive/Mixed Combat Scene	20%	33%	36%	17%
Home/Non-Combat Loc.	0%	56%	20%	83%
War Reference	100%	79%	84%	33%

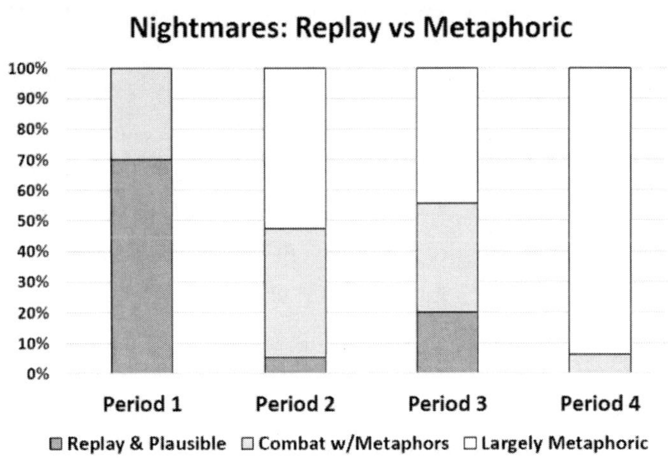

Chart 2. Variability over the 4 Periods in the Vet Content Study.

VARIABILITY IN RELATION TO DREAM FUNCTION

From a literature review, Cartwright (2012) concluded that during sleep, dreamers review emotive experiences from the day, defuse that emotional impact with similar experiences from memory, update an organized sense of self, and rehearse new coping behaviors. Even repetitive nightmares may play a role in recovery (Hartmann, 1998) and as researchers Brewin and Holmes (2003) suggest the unfit, provocative, overwhelming and traumatic experience must be encountered, remembered and processes as long as it is not neutralized. Only when it is neutralized and integrated will it cease. Some of the theories and observations surrounding the role of the dream itself in recovery are discussed below.

Emotional Regulation, Extinction and Adaption

Emotional Regulation and Extinction

Levin and Nielsen (2007) propose that the engagement of a fear extinction processes might be one of the default functions of REM sleep. They propose what they call the AMPHAC/AND model of disturbed dreaming, which involves emotion processing, and a memory-fragment activation and recombination function for sleep and dreaming. In normative dreams the process begins with memory element activation – most dreams do not represent coherent episodic memories (2009) but rather activate selective emotional memories of an event along with various associated memories from multiple sources in time and context – extracting the "meaning" or "gist" (Wamsley, 2016, Payne, 2009). Then memory elements are recombined and continuously assemble into a coherent flow of imagery, which provides a new context for highly arousing (positive and negative) memory elements (2009), that are incompatible with existing fear memories. Stickgold and Walker (2013) propose that memory activation begins with a "memory triage" to determine which information to select and integrate into existing memory networks, looking for common patterns; perhaps as Stewart and Koulack put it (1993) activating habitual defense mechanisms then matching them with the present stress situation and integrating the two. Ernest Hartmann (2011) contends that it is the emotional concern that is "illuminated" and *contextualized* (depicted it in less threatening form that can be more easily processed) and connected with other memories, so that it is no longer so unique or extreme—so that the next time something similar or vaguely similar occurs, the connections will already be present, and the event will not be quite so traumatic.

In the case of PTSD nightmares, the dream attempts to perform these functions but fails to do so – the emotional memory being too overwhelming to enable access to associated memory elements that create new contexts to replace the fearful context. If incapable of regulating emotion (affect load), they may in fact increase affect load further preventing the extinction of fear memories (Levin and Nielsen 2007). If fearful memories are not able to be broken down and recombined, a cycle of disturbed processing begins, in which emotionally laden waking memories trigger more disturb dreams and dreaming reactivates those memories further – some people being more vulnerable than others.

The fear extinction function is supported by a network of limbic, paralimbic, and prefrontal regions (Levin and Nielsen, 2007). This includes the amygdala (the "control center" for affect load) and the anterior cingulate and medial prefrontal cortex (which mediate extinction by downregulating impulsive emotional expression). These are understood to be a site of plasticity/learning for inhibition of fear responses and constitute the control center for the expression and regulation of emotion, both in sleep and in wakefulness. The hippocampus plays a crucial role in processing emotion by encoding and consolidating memories as well as re-mapping and representing stimuli in novel contexts to enhance its effect on limbic structures, thus shaping emotional imagery during dreaming (Krippner and Taitz, 2017).

Emotional Problem-Solving

Kramer (2019) considers this process to be one of emotional problem-solving, and observed two patterns: (1) a progressive-sequential pattern, in which an emotional problem is stated figuratively, worked on, and resolved, and (2) a repetitive-traumatic pattern, in which the emotional problem is simply restated in different images or metaphors and no progress toward resolution occurs. He suggests a mechanism of "emotional problem-solving" or failure to "problem solve"— related to the success or failure to contain the emotional surge (1991). These patterns account for variability in how a person feels (mood) in the morning. If the dream is unsuccessful, the dreamer may recall a disturbing dream or nightmare. Kramer (1984, 1993) suggests that with repetitive nightmares, the emotional surge overwhelms the ability of the dream process to dampen it - they lack the "emotional problem-solving" function to bring about recovery. As the repetitive dream attenuates the surge, however, through gradual integration of the traumatic experience into existing memory schemes, it eventually becomes possible to create dream narratives that accommodate themes of *mastery*, consoling and strength.

Adaptive Learning

Emotional problem-solving may rely on a process referred to as adaptive learning which introduces and tests creative resolution scenarios (Levin, Stewart and Koulack, Stickgold, Coutts, Hartmann, McNamara in Hoss, et al., 2019). Antti Revonsuo (2000, 2016) proposes that dreams have an adaptive function (with an evolutionary basis) to help us better adapt to life. He proposes that dreams are "simulations" of threatening events or social situations that provide an illusory feeling of reality motivating the dreamer to act on it as if it were a real, this affording us the ability to rehearse threat perception and avoidance and social perception and interaction in a harmless environment. Coutts (2008) also proposed that dreams improve our ability to meet waking need by adapting or rejecting scenarios depending on their outcome; retaining those that appear adaptive and discarding those appearing maladaptive.

This process often includes the introduction of *counterfactuals* or "what-if" mental simulations of what might happen if an alternative approach is tried (to that previously considered the behavioral norm or stress response). In two pilot studies (McNamara et al., 2000) one with 52 pain patients and another with 34 dream reports from subjects in an ongoing neuropsychological health study, McNamara found that 97% of the reports in both cases contained *counterfactuals* and 47% to 44% respectively contained attempts to correct the situation (the "norm violation"). The dream plot then tests these alternatives and observes/processes the outcome as potential resolutions. An example of such *counterfactuals* can be found as the Vet begins to *master* the threats in his dreams – to try something new, a "what if I did this instead." For example, in the first 3 years after combat, in the majority of the nightmares the Vet appeared only to take a defensive posture, to hide or protect himself from inevitable doom. But in the last few recorded nightmares he switched at the last moment and took action to eliminate the threat, *I open up with three short bursts and literally rip the dink gunner in half.*

Given the time period and multiple, sequential REM/NREM cycles over which dream sleep affords, a fear processing or memory reconsolidation process might be involved. Studies by Nader, Schafe, and LeDoux in 2000 concluded that even strongly consolidated memories (traumatic emotional reactions or conditioning) can be made unstable and readily open to change when we recall them. The synapses involved become temporarily labile (unlocked or deconsolidated) for about four to five hours before relocking (reconsolidation). If a critical new learning experience takes place during that reconsolidation "window," the old learning can be permanently altered by the new learning. In a review of animal and human waking-state studies from 2004 to 2009, Ecker (2012) found four common process elements involved, which he termed the "transformation sequence": 1) reactivate (deconsolidate) a memory/learned reaction; 2) create a mismatch schema, an experience contradictory to that original learning; and 3) juxtapose the two in a new learning experience during the reconsolidation window (dream simulation in this case) such that, 4) the memory reconsolidates with the new learning. There appears to be a reward-based processing involved in dreams as well (Perogamvros and Schwartz, 2012) which may be integral to the reconsolidation of the new more adaptive memory (Hoss, et al., 2019).

The creative introduction and planning of alternative schema may be in part initiated by a number of REM active reward based planning centers (Hoss, et al., 2019): the caudal and ventral medial orbitofrontal cortex; the medial prefrontal cortex (involved in plan generation) and the anterior cingulate which plays a role in imagining or observing an activity, selecting a resolution scenario, mediating action to choose between conflicting perceptions, and adapting the response to changing conditions. The basal ganglia is also a locus for reward-based adaptive action planning and learning, motivating us to seek eventual rather than immediate reward and selecting which response to make or inhibit. Emotional reinforcement may be the natural dream based expression of these reward-based centers.

Variability between Sleep Cycles in One-Night of Dreaming

The emotional processing function of the dream and resultant variability (from replay to memory fragmentation, integration and *contextualization*) can even be observed within subsequent REM cycles in one night of dreaming. In one of the cited PTSD cases, in the sleep lab study by Kramer et al. (1987), information was at first collected on four traumatic combat experiences as well a traumatic non-combat events from the veteran's earlier life that involved him being unfairly blamed for a boy being burned with gasoline that he tried to prevent. The dreams from 5 REM state awakenings in a single night were then recorded. The 1st and 3rd REM state dreams respectively, were close but fragmented replays of two of the traumatic combat events. The 4th REM dream however appeared as fragments of the memories from the first two, combined in a fictive story. Then in the 5th REM cycle, the dream contained *mixed* fragments from all traumas including the emotionally related one from childhood.

Three of the combat traumas were described as follows: 1) He and his buddy were riding in a jeep, went over a land mine, and were blown up. He was thrown 20-30 feet from the explosion. He. ran to his buddy but found he was dead; 2) He was walking down the street in Saigon and saw a number of marines gathered around a little boy with a basket. Suddenly there was an explosion, and when he looked over at the marines, they were lying dead. He went over to help them and saw their insides had fallen out. He then ... saw another little boy with a basket...and ran over and choked him to death…later discovering that the boy had nothing dangerous in the basket, just some bread. He felt very guilty; 3) On the way to Saigon he saw saw women hanging by the side of the road impaled on tall poles.

The dream reports from the first and third REM periods were found to be near replays of the first two events with some interesting omissions:

> In the first REM Period: "I was driving a truck in Vietnam and was getting ready to get blown up. … I have had this dream many times… There were no other people in the dream. I got blown up. I found myself in the rice paddies. …. I felt confused." In the third REM period: "I was walking down the streets of Saigon. There were six marines on the other side of the street. A boy came up behind them with a basket selling things out of it. There was an explosion. I looked and the marines were lying on the ground. I ran over. I wound up killing this boy by choking him to death. I was mad, after I killed the boy (another boy). I was confused."

Then in the 4th REM period:

> It happened in Saigon. Me and another guy riding in a jeep. I heard something tinkle in the back seat …It was a grenade, and I picked it up and threw it … killing four people. I just kept driving. I felt mad …

The first two REM reports were near replays of the two traumatic events, but with some interesting omissions: in the first his buddy who was killed is absent; and in the second the gore and emotion of guilt was absent (now anger and confusion). Further fragmentation and recontextualization is suggested in the fourth REM period which creatively integrated elements of the first two: the explosion was a grenade (which he could act on and *master*) rather than a land mine (helpless against); his "buddy" is now just "another guy" and finally they were unharmed and continued to drive off.

In the 5th REM report he sees girls hanging upside down slit open with "intestines hangin out" plus a guy in the village that was burned alive …someone had poured gasoline on him. Here we see the hanging women from the third atrocity, mixed with the guts, guilt and blame from the second trauma, along with a non-combat event from his childhood trauma perhaps related to the second by emotional memories of guilt and blame.

Mastery

Siegel (2001) observed *mastery* as an indicator of resolution. *Mastery* is often pictured as the creative attempt at an alternative course of action by the dreamer to try to overcome or eliminate the threat. As noted in the above discussions, the dreaming mind may at times creatively introduce what-if-scenarios or *counterfactuals* that change the plot in order to resolve the threatening situation. Therefore, the dream process itself may contain a self-initiated *mastery* capability which works in continuity with therapy and supportive waking life experiences. Sometimes *mastery* is the direct decisions and actions of the dreamer, sometimes magical or *metaphoric* elements introduced by the dream, and in others changes in one element from dream to dream (as in the Kramer 1987 study where the dream changed the origin of the explosion from a land mine, over which the dreamer had no control, into a grenade which he could toss away).

In the Vet Content Study, in period 1 prior to treatment, attempts at *mastering*, or taking charge of the situation, were made in four of the ten nightmares, where half failed and half succeeded. *Mastery* appeared as switching from a reactive defense response (*I recall huddling in a corner, shifting positions, as I anticipate a direct hit amidst us*...), to deciding to switching from a fearful defensive posture to taking charge, going on the attack in an attempt contain or counter the threat (*We gather about and form a defensive perimeter. A machine gun team of dinks stalks right up to us. It seems everyone is too scared to fire. I open up with three short bursts and literally rip the dink gunner in half.*)

In the second period after having been treated for two months, attempts at *mastery* increased to 78% of the threatening nightmares, most (64% of those) succeeding. In a number of cases it was characterized by switching from defense to attack, bringing in elements from his past, or the appearance of an unusual or surprising out of context element. In one a surprising new element appears:

> I am in hand to hand combat with a man….. With a sinister leer he taunts me. Twisting his knife through the air he is easily the more skillful. I am lightly scarred several times but somehow, as my knife becomes a fork, I spear him in the throat, causing him a mortal wound.

In another it is the intervention of something magical:

> Attacking soldiers find me. … The second aims and shoots just as the atmosphere begins to freeze. It forms a protective shield in front of me. The bullet punctures the ice, then falls to the floor. The solider outside suffocates from lack of air. And in another the Vet invents something intended to alter the outcome, I help to invent a catapult that hurls a half dozen grenades at a time.

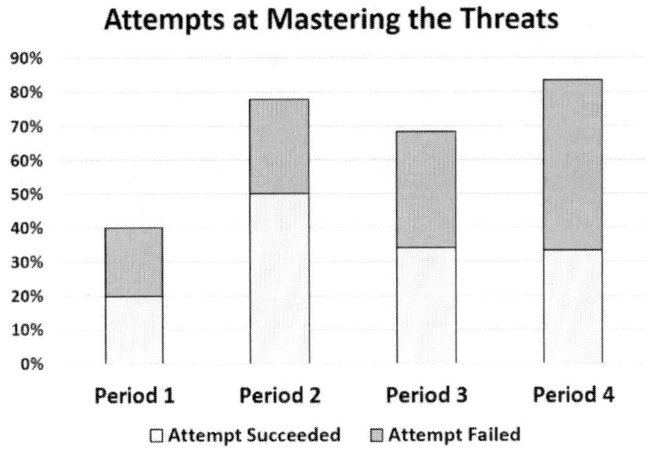

Chart 3. Variation in Attempts and Success at Mastering the Threat in Nightmares Over Time.

Table 6. Mastery (in the nightmares - dreams imposing a threat to the dreamer)

	Comments	Vet Content Study			
Period of the Dream Records =		#1	#2	#3	#4
% of recorded dreams considered Nightmares (threat to dreamer)		100%	72%	63%	56%
Nature of *mastery* in those nightmares					
No Attempt	No attempt made to take charge other than flight or avoidance; the threat remained.	60%	22%	32%	17%
Try but Fail	Attempt made to take charge and change the anticipated outcome - failed.	20%	28%	34%	50%
Try and Succeed	Attempt to take charge and change the anticipated outcome - succeeded.	20%	50%	34%	33%

Integrating the Fragmented Self

Carl Jung (founder of Analytical Psychology), promoted a theory that the role of the dream goes beyond emotional problem-solving but involves attempts at re-integrating a fragmented self – facing and attempting to accommodate previously suppressed experiences or rejected parts into an internal model of self (Jung, 1971). The objective is for the existing ego personality and these unconscious aspects of self to interact in order to find a way to *integrate* or work unison, as opposed to the suppressed material emerging in explosive or dysfunctional ways (as is often the case with PTSD). Jung was not alone, Fritz Perls (1974), who founded Gestalt Therapy, also considered dreams the "royal road to integration," striving to bring suppressed, alienated fragments into the "whole" self or "Gestalt."

Jung called this the "transcendent function" which results from an "interactive tension" between opposites that eventually leads to a resolution. A *metaphoric* example of this can be found in an interesting negotiation sequence in one of the Vet Content Study nightmares:

> I talk with the opposing force. A young man, friendly, gifted in mathematics, offers to end the fighting by settling monetary debts. He

shows me how much we owe in relation to what is owed us. Quick calculations reveal the debts are equal, I accept his peace offer.

Jung (1973) wrote that dreams bring this about through another action he called *compensation* - providing content that opposes the ego identity or viewpoint (much like the "opposing schema" or *counterfactuals* discussed previously). Punamäki (1997) found evidence in traumatized children for a compensatory dynamic, whereby children who responded with passive, withdrawing and emotion-focused coping modes in daytime treats showed active, heroic and problem-focused coping responses in their dreams. An example of *compensation* leading toward the adoption of a new identity can also be seen in the Vet Content Study (further discussed below): upon seeing his "*old hat*" being incompatible with his expectation he rejects it, exclaiming, "*This hat cannot be mine!*" He then lines up in the stairwell of a cellar where they will be issued "*new hats.*"

Carl Jung observed that this function is pictured in dreams by symbolic patterns representing basic mental processes, which he called *archetypes*. (Jung, 1973). The rejected fragments of the personality can appear as unknown, sometimes shadowy, undesirable, opposing characters that the dreamer is conflicted over. The integration or re-integration therefore often appears as a symbolic union of opposites in various forms. One common form is the symbolic union or pairing of the dream ego with a typically unknown (although sometimes know) figure of the opposite sex (Jung, 1973, 1993) – symbolizing the integration of masculine and feminine attributes within the dreamer. It can appear as male and female attraction, partnering, sexual union, marriage etc.

In the Vet Content Study, the world of combat might demanded that the Vet take on and strongly maintain qualities that are traditionally considered "masculine," in order to operate and survive in that environment (for example: aggression, emotional and physical strength, directive logic, an external attack or survival focus) and perhaps suppress qualities traditionally considered "feminine" which are less compatible with combat (for example: empathy, compassion, sympathy, nurturing, introspection). At first most all characters in the Vet's dreams, and those he most identified with, were male (usually platoon or combat buddies and old friends as well as fictive male companions). Over time this changed as his dream themes (see Table 3) evidenced rejection of a combat role, reduction in combat references, introspection and sadness about atrocities of war, followed by an increase in females (mostly unknown at first) and a near linear increase from rejection to sexual attraction. The female roles also evolved from dying, unattractive and undesirable at first, to guides, nurses and doctors, then alluring and sexually attractive at the end of the 3rd period; where (as discussed later) in the last dream he made love to an alluring unknown female who, "*has promised me a place to live.*"

The Healing Nightmare

The "turning point" within a dream or series of dreams might be when *mastery* begins to occur in the nightmares in a manner which correlates with general improvements in their waking condition. Wilmer (1996) called this "the *healing nightmare*." Although it may still present itself as a nightmare the healing or *mastery* relationships become reversed in the dream – instead of say a medic helping the victim, the victim becomes the healer – instead of being rescued, the dreamer becomes the rescuer – instead of being attacked and defeated the dreamer goes on the attack and gains control over the threat. In an example from the Vet Case Study he flips from confident cunning aggressor to medic:

> I am full of confidence and cunning. Outmaneuvering the other fellow, I stab him in the back with a knife and say, "Give up or I'll kill you." He yields and I help him outside and dress his wounds. In another: I am in Vietnam and we kill many soldiers. It is a recollection from the war. The dead lay clumped in heaps... Then the situation is reversed, as in the dream I imagine what it would be for an NVA to come across American dead. I wake up anxious, angry and fearful.

There were a number of other, perhaps more profound, turning point dreams or *healing nightmares* recounted in the Vet Case Study. Many were clustered around an event that the Vet stated was a "turning point" in his healing - when he wrote to and eventually met (at a writers workshop) and became friends with a former enemy, a North Vietnam (NV) veteran named Bao Ninh. The dreams that followed show this NV veteran not as the enemy, but as a guide, a healer, a Zen master. In one dream he states:

> Bao Ninh tends to me in hospital. … Like a Zen master he is deliberate and clear in all he does. I complain and grow anxious, become uncooperative. Bao Ninh ignores or engages me depending on my awareness of what is correct. I wake up still much moved by the force of his personality. In another dream of Bao Ninh: I am forced to accept the conflicts of my own nature and to understand them. The shock of recognition is painful.

THE VET CONTENT STUDY

Key elements of the content of the Vet's dream records over the four recording periods was provided above in relation to various theories and other studies related to post-traumatic and PTSD nightmares. In this section the data will be put in chronological order highlighting some of the apparent and evolving themes over the four periods. Although the Vet did not reveal much about his life during these periods in relation to his dream records, themes apparent in the content of his dreams are suggestive of the state and elements of recovery – particularly when compared with findings from other post-traumatic studies. The dream records that illustrate these themes are for the most part abbreviated to conserve space, but can be reviewed in full in the www.dreambank.net database and Levy's books (2016, 2018).

Period #1: June 1970 to March 1973 Immediately Post Combat, 10 Nightmares Recorded

Ten nightmares were recorded, one in 1970 while in Cambodia the remainder after he was released. As Table 3 shows all nightmares in this period were combat related. All contained aggression against himself and a life and death threat conditions, nine were combat scenes, eight occurred in an actual combat location, six with the actual enemy, and all contained the actual weapons. Dominant emotions were fear/anxiety and helplessness. Seven appeared as actual or plausible memory *replications* and the remaining three pictured realistic combat action or *mixed* with *metaphoric* representations of childhood or home environments (see Table 5).

In his first dream, while still in Cambodia he records:

> In the dream we are being overrun by sappers who have got past the Night Defensive Perimeter trips and claymores and now crawl forward.

The guarded sense from the dream continues into waking where, upon "seeing a boot tread close to his face" he draws his gun to shoot then stops when he recognizes that it is his buddy.

His first recorded nightmare after leaving combat was the *mixed* one mentioned previously of experiencing the combat from a child's perspective:

> The enemy is above, in the sky. We are not soldiers, but school children. … There is an underground bunker; most of us run for the school bus. … The mortars are dropping close by… The fear of realizing death is pacing through my mind. …All at once there is a sharp explosion. I am hit, though I scream for only a moment. It is such a helpless feeling. The concussion has shattered the teeth in my mouth, and I am going into shock.

> The next was more of a replication: The mortars begin to drop; I hear the explosions very clear. This part of the dream seemed the most real. I recall huddling in a corner, shifting positions, as I anticipate a direct hit amidst us. The mortars continue to fall.

These two examples are much like six out of the ten where the dreamer remained as fearful or helpless victim attempting to avoid the threat. About a year later, however, in the final five recorded nightmares, he attempts to *master* the threat four times. Twice the attempts fail as in this one:

> ... I am behind a wall of some kind, the danger of being hit is very great. In a fury I grab a silver object, which seems to resemble an RPG, and hurl it towards the foes. It explodes and immediately the entire area is shrouded in flames, even I am enveloped. I awaken with the all-too-familiar feeling that I notice each time I have a dream that ends with my own death.

But among his final recorded nightmares he begins to succeed. In this one, as discussed previously, his actions and the outcome is reinforced:

> We gather about and form a defensive perimeter. A machine gun team of dinks stalks right up to us. It seems everyone is too scared to fire. I open up with three short bursts and literally rip the dink gunner in half. A while later, a single dink walks up to us, unmolested. I sense that no one is going to fire and spatter him wickedly with my M-16. It is a good feeling.

Period #2: 1978 to 1981, 25 Dreams and Nightmares Recorded

This period of dream records continued after 1977 when the Vet spent 2 months at Klinic am Zurichhberg in Switzerland a Jungian psychiatric center. It appeared to be a period where he was gaining a greater degree of *mastery* over the combat memories, and an apparent shift away from his combat identity. Life threatening themes and combat references dropped dramatically (Table 3). Of the 25 recorded nightmares and dreams 72% contained threats to himself and only 68% were life and death situations, (both down from 100% in period 1). Only one dream appeared to be an actual or plausible memory *replication*; war references dropped from 100% to 72% with combat scenes reducing to 40% (only half being actual Nam locations). The remaining dreams contained combat actions or conflicts but were in fictional combat settings or non-combat/home settings. Interpersonal conflicts (as opposed to war) appeared in 20% of the nightmares, perhaps representing the emergence of broader internal struggles versus combat memories alone. The dreams begin associating combat experiences more broadly to his life, as references to his family and childhood begin to appear about midway through this period (about 20% of dreams). Negative emotions decreased from 70% to 48% while positive emotions began to appear and were expressed in 20% of the dreams. Dominant emotions were confusion and sadness then fear/anxiety, helplessness and guilt as the dream content evidenced a degree of reflection.

From the onset he began gaining a greater degree of *mastery* over the combat situations in his nightmares, with more successes (52%) than failures (20%). Being the victim dropped to 68% while becoming the aggressor (mostly fighting back) more than doubled to 48%. Most of the successful *mastery* actions were matter of fact, typical combat action, with no emotion expressed, but a couple introduced surprising or magical elements. In one the dream repeats:

> This time I do the killing with a laser beam. In another: I am in hand to hand combat with a man … Twisting his knife through the air he is easily the more skillful… but somehow, as my knife becomes a fork, I spear him in the throat, causing him a mortal wound.

Role confusion or reversals begin to emerge. *I go to Vietnam. I am with my old company, but I am not accepted. I become a loner, an outcast. Roaming about, I shoot down American helicopters.* Another is the dream described previously where the dream begins where he is a confident cunning aggressor who stabs a man shouting, *give up or I'll kill you,* then he flips to his medic role once the man yields, *I help him outside and dress his wounds*. He reflects on how things are not always the way they seem: *I see the familiar jungle that hides the enemy. I know only too well this paradox of beautiful jungle and the enemy waiting within.*

In another, the alternative of viewing his combat memories in the context of being no longer threatening, is reinforced:

> A group of ten or fifteen men with fixed bayonets sallies towards us. They appear to be Oriental, though at the last moment I see that it is all a sham. These are our own men. I have been the object of a great joke. Rolling on the ground, David and I howl with laughter. We begin to sing a popular song. Our harmony and rhythm are surprisingly good.

His dreams also begin to reinforce leaving his combat soldier role and moving on.

> I am with my old unit in Vietnam. The order comes over the radio that we are now finished with jungle patrols … As the others have walked off, I throw the gun down and follow after them. There is a feeling of complete freedom. In very high spirits…I am aware that I do not wear my steel helmet.

In another, pride in his Veteran identity is reinforced:

> I am in a large classroom. The teacher is taking roll. He calls, "Veterans." No one answers. A second time the name is called. I trumpet out, "Don't you know that's me, you big baboon!" Everyone in the class begins to laugh. Even the teacher … breaks into a grin. The laughter … grows and grows, until it is followed by loud and sustained applause. … These people are all familiar to me, they are from my past … depicted as children. … I am born aloft by the many hands and arms of the great crowd and deposited at the head of the class, alongside the teacher.

His final entry during this period appears to be a *metaphoric* attempt to salvage something that has been damaged (himself?) and sense of not yet being in full control but a healthy sense of encouragement.

> I seem to be with someone else; we are flying low over the ground, surveying the land. We decide to salvage a damaged plane. … It is unclear if I am flying the damaged plane or the one used for surveying. In any event I am at the controls and telling the other person with me that I have very little or no feeling of control. This other person, patient and willing, remains calm and gives me encouragement.

Period #3: 1998 to 2002; 70 Dreams Recorded

Prior to this period of records, in 1995, after backpacking in Southeast Asia, he purchased a paperback copy of *The Sorrow of War*, by the North

Vietnamese army veteran and novelist Bao Ninh, which impacted him stating, “I fell into a trance, felt as if I were floating above my bed. What was happening, I did not know.” “At the sight of Ninh's dust jacket photo, I saw again the faces of enemy soldiers my platoon had hunted, and who had hunted us in return. That night, I dreamt my platoon was trapped, out of ammo, and trying desperately to escape.” All this appeared to "aggravate the condition" and in 1996, when his PTSD was rated at100%, he was again hospitalized; a seven-week stay in a VA PTSD treatment unit.

From the dream content it appears that a significant degree of reflection and turmutuous change was taking place. War reference increase to 83% and memory *replications* (actual or plausible) which were practically nil in the last period, begin to increase again and intrude on 20% of his nightmares. Although overall threats decreased slightly (10% or so), his ability to *master* those threats began to decline. No-attempts doubled to 32% with successes declining from 50% to 34%, (now roughly equal to failures), which was reflected in his role as aggressor which dropped more than half to 21%. Expressed emotion increased (61% to 74%), mostly negative (from 44% to 66%) but some positive emotional increase as well (22% to 27%). On the surface it would appear to be a stage of regression – but upon closer examination, the events which interleaved, and the content and themes of the dreams appear to tell a different story.

In 1998, the Vet sent Bao Ninh a letter and two months later received a warm Christmas and New Year greeting. Then 2 yrs later (2000) he decided to attend the Annual William Joiner Institute Writer's Workshop, located at U Mass Boston. Quite by surprise Bao Ninh was there and they met. “We rushed to each other’s open arms. Ninh pulled me close, clapped my back once, twice, three times, as if I were a long-lost friend. Overcome by deep feelings, I sobbed uncontrollably Ninh took my hand, led me away. 'No... I'm all right." I said, still sobbing.” They talked for three hours. That night, an old recurring nightmare returned.

> An invisible spirit grabs my feet. Yanks them straight up. I can't move. I scream and struggle and wake with dread. But this time, this night, I fight

> back. Defeat the demon. This time, upon waking, I'm calm, free, euphoric. This was a genuine turning point. Something inside me had changed.

The Vet referred to these events as a "turning point." Alan Siegel (2019) discusses dreams and nightmares that occur at a "turning point" in people's lives as including, "vivid and compelling dreams that often cluster before, during, and after expectable life transitions." Thus a "turning point" is often a period of increased turmoil. While the dreams in this period appeared cycle between progress and regression in relation to conflicted situations, there appeared to be in general a positive evolution taking place.

After his interaction with Bao Ninh, in mid-1999 to mid-2000 there were a cluster of dreams of himself (mostly in a soldier or combat role) either dying or being killed just as he awoke. These were then followed by a period of increasing reflection and introspection and acceptance of a new role (or as the dreams put it, adopting a "new hat" figuring out "which room is his" and being given "a new place to live"). This was accompanied by dreams of accepting, friendly and intimate interactions with other dream characters, particularly with unknown females.

As Alan Siegel (2019) put it, "changes in dream content can parallel improvements in the post-trauma reaction and thus may serve as indicators of clinical progress in trauma victims' response to trauma treatments." Some of the leading themes (discussed below) throughout this period include: a) reflecting on the war and the atrocities (17%); b) reflection on his present condition; c) dealing with issues of respect (10%); d) and establishing a new sense of self or world view (25%). The ratio of 80 percent *metaphoric* content to 20 percent memory *replications* remained high, suggesting the possibility that some normative restorative function might be operating in his dreams and nightmares.

Reflection on the War and Atrocities

In about 17% of the nightmares he begins to reflect on the war and atrocities, whether replicated or simulated.

> In one: I am witnessing the execution of individuals at close range. … There is very little sound or movement, no struggle or attempt to flee. The shot is fired, the body falls.
>
> Others picture a reflection on the cruelty: I sit with a young boy who is looking through a thick book about war … the section about Vietnam. I look at the pictures with him. Several NV A soldiers have dragged two dead GI's into a street and are castrating them. Then the images become real, and we are caught up in an NV A attack. And another: I watch as prisoners are beaten senseless by members of the Central Intelligence Agency. The victims are dragged off and thrown into piles, condemned to be shot. The bloodied, half-conscious men plead for their lives. The cruel agents ignore them.
>
> The perspective is switched in another: I am in Vietnam and we kill many soldiers. It is a recollection from the war. The dead lay clumped in heaps, like pictures from the civil war. Then the situation is reversed, as in the dream I imagine what it would be for an NVA to come across American dead. I wake up anxious, angry and fearful.

Some reflective nightmares released deep emotion to the point of weeping:

> With third platoon I rushed up to watch as the men picked off the enemy soldiers. Their legs were blown off. They screamed to each other, firing their AK's wildly. I knelt and wept at the slaughter. In another: I am with Jamie H. on a trail in Vietnam. We are in a hostile environment.... I begin to tell him about war, about Chicom's and combat and death. … I am fearful of going out on ambush. The next moment I begin to weep and sob deeply. A third appears to express a very deep release: I am with a large black woman … She listens while I speak about war. Suddenly a low resonant howl escapes me. The black woman says, "I know what that is." I begin weeping.

Reflection on His Condition

Some appear as possible reflections on feelings about his current condition; not in control, regressing, lost at sea, a wounded soldier running for help – no longer the doctor. In one, as a pilot of a crashed helicopter, the 'grunts' on board twice ignore his orders.

> This becomes the basic tension of the dream: Am I or am I not in control? It would appear that I am not. In another, I am with a group of emotionally disturbed people … lying on the cold floor. … I am aware of giving way, to regressing, to curling up like an animal. Some begin to picture his attempts at gaining help: "A man is lost at sea." I help to rescue him.

In another he pictures a contrast with who he was in combat:

> On a cancer ward all the patients are men, …On the wall hangs a photograph of a male doctor at war. His pants are torn and dirty. His knees are wounded; he is running for help. I say, "What does he know? I was the medic. They all came to me." I begin weeping.

The theme of respect begins to appear in about 10% of the dreams. In one dream when a man is sent out to challenge his command, *I shoot him in the leg*. In another many in his patrol have injured feet and are marching in the cold hard snow to an aid station where they overwhelm the staff, who complain. *I am sad because no one cares, no one understands how we suffer.* In another he and his brother appear young.

> His brother throws a heavy silver ring at him. He tells his brother, "If you ever treat me with disrespect again, I'll kill you," eventually doing just that as his brother continues to question him.

A New Sense of Self

The nightmares and dreams begin to picture themes of reflection and releasing the old combat image of self and movement toward or acceptance of a new non-combat sense of self; a "new hat" a new "place to live," as his dreams *metaphorically* present it. The dreams might be seen to follow the transformative integration process which Carl Jung hypothesized (discussed previously). Jung noted that unresolved conflict is often followed by giving in - pictured by what he called "death of the ego" imagery. This is followed by an integration of suppressed or conflicted aspects, so as to establish a new more balanced or "whole" self, evidenced as a new attitude or world view (Jung, 1973). It generally comes at a point where internal conflicts cannot readily be overcome. Some relevant themes in the dream content in this period of note is as follows:

- **Conflicted Sefl:** At the beginning of this period, after communicating with the North Vietnamese Vet Bao Ninh, he dreams:

  I am in a classroom setting. Bao Ninh is critiquing my work. His voice is warm yet authoritative. He commends what is good, denotes what is imperfect…. I am forced to accept the conflicts of my own nature and to understand them. The shock of recognition is painful.

  The theme of confusion begins to appear. In one dream:

  I am confused. Which room is mine? In another he is torn between following orders and doing what he considers right: Opposite me … in another trench, are several Americans. Lt. Sharp tells me to fire on them. I don't want to kill them. Waking up from the dream I am half shouting, "Get down."

  In another he switches roles from aggressor to medic:

  A Viet Cong attempts to steal American weapons … we fight. Each time I stab his belly he says, "Kill me." I feel terrible. It is as if I knew this man. When he weakens, I take him in my arms and call for help. … I am crying. I am saying, "Oh God.... Oh God." …. We march to a hospital.

- **"Death" of the Combat Self:** a cluster of 7 dreams mid-1999 to mid-2000 picture him (mostly in a combat role) dying or being killed as he wakes. After this cluster, combat related conflicts becomes increasingly replaced by interpersonal conflicts. More positive emotions arise about this time and become more frequent afterwards.

- **Emergence of a new Self:** Toward the end of the "death" cluster, his interaction with Bao Ninh (which he called the "turning point") is pictured as a rescue and a healing. The dream begins:

> A man is lost at sea. I help to rescue him. It continues: There is war. I am hit with shrapnel in the back. Bao Ninh tends to me in hospital. He is clearly in charge of himself and the matter at hand. Like a Zen master he is deliberate and clear in all he does. I complain and grow anxious, become uncooperative. Bao Ninh ignores or engages me depending on my awareness of what is correct. I wake up still much moved by the force of his personality.

In a later dream he recognizes his combat hat is no longer his and is to receive a new hat:

> An old officer says my boonie hat is worn out. The flimsy material has lost its shape. My CMB pinned in front is missing, as are the encircling love beads, the Cav patch sewn on top, the grenade rings clipped round the brim. There is no ground-in muck, no accumulated sweat. This hat cannot be mine! We … line up in a stairwell, which leads to a cellar where we will be issued new hats.

- **Integration of Suppressed Material:** About mid-way through the "death" cluster the concept of a balance is introduced: I talk with the opposing force.

> A young man, friendly, gifted in mathematics, offers to end the fighting by settling monetary debts. He shows me how much we owe in relation to what is owed us…. Quick calculations reveal the debts are equal, I accept his peace offer….

During and following the "death" cluster a notable trend began in the number and roles of unknown females. The number begin to increase (to 31%, up from 12% in the previous period) along with a change in role from injured or dying, ugly or undesirable to healing roles of nurses, doctors and guides. The male to female relationship also changes as women who were rejected at the beginning, become more alluring to sexually attractive at the end of the period. Although associated waking life influences are unknown, if these themes are considered purely on the basis of Jung's archetypal theories (discussed previously) which picture the integration of suppressed material as an integration of opposites, an interesting picture emerges. There may be a balancing of "masculine" attributes demanded by combat (which the Vet may have up to now identified with), with a re-emergence of inner "feminine" attributes (empathy, introspection, etc.) which may have been suppressed out of necessity during this period. At the onset, females are considered undesirable in a number of dreams, some examples being:

> I walk next to a woman who is old and smelly, who has a heart condition. In another: A pretty American girl with long hair makes her way past. I have no desire for her. It is a good feeling to be with my men.

But eventually the females are pictured as nurses and guides helping injured males:

> One nurse … attends to the man, who is mortally wounded; or in a dream of being in a cancer ward, My doctor is female. She is dedicated and loving.

Eventually they become alluring and sexually attractive as in this example where the final dream in the 3rd period ends with a symbolic and literal union; metaphorically "a new place to live":

> I am at a dance with an attractive woman who has promised me a place to live…. Afterwards we walk down a snowy moonlit street. We are happy and young. She puts her arm around my shoulder. "I like you," she says. "I really like you." Then we sit in her car and neck. Our kisses are sensual,

giving, tender…. Outside, a small wood sign is tacked to a streetlamp. Its handwritten letters state, "Kit Karson Scouts." I raise up and embrace her. "That's gold," I say. "Absolute gold." Then we make love.

Note: The Kit Carson Scouts belonged to a special program created by the U.S. Marine Corps during the Vietnam War involving the use of former Viet Cong combatants as intelligence scouts for American infantry units - who defected and rallied behind the Saigon government. They became known as Hồi Chánh Viên loosely translated as "members who have returned to the righteous side." An apt metaphor picturing re-integration.

- **Emergence of an Altered world view:**

Principles of life are being explained. I am told to think of life as a daily game, to imagine luck or good fortune. I agree, but question why this works, since it does. In another: It is the present and I am with Bao Ninh in Vietnam. He leads me about, showing me different sites and soldiers. There is a certain intimacy, a certain understanding between us all. We communicate without language. … There is no sense of war.

Period #4: 2015, 32 Dreams Recorded

The 13-year delay in reporting dreams was unexplained however this next period of reports began the year before publishing his 2016 book *Dreams, Vietnam* (Levy, 2016) which contained the stories from the first 3 periods. Only one dream during this interim period (from 2008) was released in that book which again reflects the theme of respect which surfaced in period 3. His brother in waking life sends him a copy of an anti-war film *Living With War*. In the dream which takes place in a Vietnam jungle where he is "breaking in" his brother who is new to Vietnam, he asks his brother for help at one point:

> My brother laughs at me. I take out my 45 pistol. I say, “If you do that again, I’ll kill you. My brother laughs. I shoot him point blank in the chest. I wake up saying, "You didn't listen, so I killed you."

He states that his nightmares have reduced to two or less per month. Threats in general in all dreams dropped (Levy, 2018). In just the nightmares, life threats dropped to a third of those in period 3 (from 62% to 19%) and war references reduced by half (from 84% to 39%). Negative emotions in all dreams dropped by half to 34% (being fear/anxiety, confusion and anger) and positive emotion increased slightly. Most all of the content factors (with the exception of specific war references) seemed to fall within what might be considered normative dreams (see tables 4 and 5).

The nature of conflict in his dreams shifts from combat to interpersonal (which increased from 17% to 31%) mostly with an unknown non-war related figure (for example a kidnapper, tiger, teacher, policeman, bar owner, friend, his brother, childhood punk, and even the wind or an industrial oven). None of the dreams contained an original combat location or *replication* of a combat memory – 84% were non-combat or home locations and 84% were *metaphoric*. Only 25% contained any references to his combat experience and those just brief or general (NV, an MP, in the Army, defending a position, a password). His most intense nightmares resulted from such content as: a man trying to kill him; a gun; being stabbed, a pending attack, an ominous presence; being paralyzed; no escape from tear gas; or a repetitive one of feeling suffocating after telling someone about an element of the war. The only theme that immediately stood out, other than being under attack at the onset, is a notable shift in the nature and roles of males and females.

The number of females remained at about 31% but their nature and roles differed. As opposed to mostly unknown, the females in this period are mostly all known past or recent girlfriends. Also, although there was an increase in sexual attraction, most of the relationships evolved toward rejection or loss – as opposed to period 3 where the relationship evolved from rejection to extreme attraction. For example:

> I become affectionate toward her, but she has trouble accepting my feelings.

In another the female he is in love with is leaving:

> It's our final time together. It's just understood.

In others the female begins to be replaced by males:

> She [the teacher] is gone .... an old man in a suit with wild hair, takes charge; and in another, I look for her [a former girlfriend] and only find her brothers.

There is also notably changing theme with male figures. In period 4 the males are mostly unknown, as opposed to prior periods where they were generally army buddies or friends (past or recent). At the onset of the period, in most of the dreams (7 in the first month or 22% for the period) he appears to be being attacked, captured, harmed or impeded by unknown males - whereby he fights, resists or ignores them (in Jungian terms perhaps in conflict with "masculine" attributes). The following example illustrates such a conflict perhaps suggesting the rejection of his combat/vet identity, giving nothing to it and moving past it:

> An army vet [wanting something from me] blocks my way out....I get up and deliberately walk toward him and tell him, "No," I have nothing to give him... I'm not intimidated by him. I'm going to walk right past him.

He also resists and "knocks down" the male intrusion on his regained relationship with a female he once was in a union with:

> A woman that I had dated twenty years ago... says she still loves me. I feel the same toward her. A [male] guard arrives... He puts me under arrest, but I knock him down.

Then the male/female relationships become conflicted. In one dream the male/female roles flip four times:

> I meet a woman and want to sleep with her, but she declines.

There is an ongoing conflict with another group

> ...one of their men [has] been executed. I meet another woman.... I feel doubtful toward her and lose interest. Then I meet an exotic looking young man ... I begin to cuddle him.

Midway in this period the males begin to take on a more positive helpful role: [while combating a large industrial oven]

> ... the other man grabs me. I say to them, "Get me away, boys." And they lift me out of danger.

In another:

> There's an old man just behind me. We're both good at finding our way through this thicket; and he himself takes pride in leadership, I'm the leader, and I want people to see me, to know that.

Without knowing what is taking place in his waking life it is not possible to correlate these shifts with waking events. We might speculate that this was a time of reflection since he was writing his first book at this point. We might also speculate on the *archetypal* significance (Jung, 1973) as a conflict and shift in the balance in masculine and feminine personality attributes – from the content alone we can't be sure.

Post Period #4: 2016 and 2017, 463 Dream Records

This period of records came after his conscious decision, to write down all of his dreams, not just war nightmares, which he stated, "seemed the logical next step" to understand "how often did the war raise its chthonic head?" These he reported in his second book *Other Dreams*. Upon examination the themes and primary content parameters did not change appreciably (period 3 appeared to be a good representation of recent times) and also due to the changed basis for recording, these dreams were not bundled with period 4 to include in this quantitative Vet Content Study herein. Some summary observations, however. Fragments of war references (a name, location, a weapon, a word, a similar scene, etc.) remained about the same as period 4 (at 24%), as did the *metaphoric* nature of the dreams. Examining the continuity of emotional themes (primary emotions being fear/anxiety, confusion and anger), fear/anxiety dropped about half, with confusion remaining about the same, and other expressed emotions remained less than 2% - all in the range which might be considered normative. The male to female relationship shifts seen in 2015 did not appear to continue into 2016 and 2017, but rather returned to more of a balance. The males in this period were observed about equally in neutral and conflict roles. The females were pretty much a mix of unknown and known and most of the encounters were positive, or helpful and the women were for the most part considered attractive, desirable or involved love making.

CONCLUSION

Although only a singular case, the Vet Content Study illustrates that when researching the nature and content of PTSD nightmares, the design should not ignore time period and stage of recovery factors. The Vet Content Study and others cited herein, demonstrate that dream content is definitely not static over time. The post-traumatic nightmares and dreams will progressively take on elements that indirectly relate to the trauma and that increasingly picture how the dream has woven and accommodated the emotional remnants of the trauma into their overall personality and totality of self. This variability is a function of many factors: the nature and duration of the trauma; the person's susceptibility; the person's history of like past traumas; changing situations in the person's life; and the nature and phases of treatment. Evidence also suggests a natural emotional problem-solving and extinction process taking place in dreams. It is one that can be overwhelmed by the emotions that surge from the traumatic memory, but at some point the dream or nightmare may attempt to fragment and select emotional memories to process in order to interleave them with other associated life experiences so as to dampen the overwhelming emotion and place it in a less threatening context that is incompatible with the existing fear memory. Alternative resolution scenarios (counterfactuals) might be introduced as attempts by the dreamer to *master* the threat. The process may never entirely eliminate all of the fragments of the trauma from surfacing but working with the dreams as an integral part of therapy can be, as Alan Siegel states, "a key to recovery from traumatic events."

References

APA, 2013. *Diagnostic and statistical manual of mental disorders: DSM-5*. American Psychiatric Association. Arlington, VA.

Arnulf, I. 2019. *Dreams: Understanding the Biology, Psychology and Culture,* edited by Hoss, R., Valli, K, Gongloff, R ABC-CLIO.

Barrett D. 2018. "*The Special Nature of Nightmares After Trauma*." Article on IASD website for Brain Awareness Week, March 12-18, 2018, https://www.asdreams.org/articles/.

Brewin, C. and Holmes, E. 2003. "Psychological theories of posttraumatic stress disorder." *Clinical Psychology Review*, 23, 339-376.

Bulkeley, K. 2009. "Seeking patterns in dream content: A systematic approach to word searches." *Consciousness and Cognition;* www.elsevier.com/locate/concog.

Cartwright, R. 2012. *The twenty-four hour mind: The role of sleep and dreaming in our emotional lives*. New York, NY: Oxford University Press.

Casement, M., Swanson, L. 2012. "A Meta-analysis of Imagery Rehearsal for Post-trauma Nightmares: Effects on Nightmare Frequency, Sleep

Quality, and Posttraumatic Stress." *Clin Psychol Rev.* 2012 Aug; 32(6): 566–574).

Coutts, R. 2008. "Dreams as modifiers and tests of mental schemas: An emotional selection hypothesis." *Psychological Reports, 102*(2), 561–574.

Davis, J. 2009. "Overview of trauma and post-trauma nightmares." In *Treating post-traumatic nightmares,* edited by J. L. Davis, 1-23. New York, NY: Springer.

Desjardins S, Zadra A. 2004. "Threat simulation in nightmares." *Sleep.* 2004;27:A61.

Domhoff, G. W. (n.d.). *dreambank.net, a database of over 30,000 dream reports with access provided to the author for the purposes of content analysis*, courtesy of G. William Domhoff, PhD.

Domhoff, G. W. 2003. *The scientific study of dreams: Neural networks, cognitive development, and content analysis* (1st ed.). Washington, D.C. American Psychological Association.

Duval, M., Zadra, A. 2010. "Frequency and content of dreams associated with trauma." *Sleep Medicine Clinics, 5*, 249–260.

Ecker, B., Ticic, R., and Hulley, L. 2012. *Unlocking the Emotional Brain,* 21–26. New York, NY: Routledge.

Esposito, K., Benitez, A., Bana, L., Mellma, T. 1999. "Evaluation of Dream Content in Combat-Related PTSD." *Journal of Traumatic Stress*. Vol. 12. No. 4. 1999

Fosse, R. 2000. "REM mentation in narcoleptics and normals: An empirical test of two neurocognitive theories." *Consciousness and Cognition* 9:488–509.

Hall, C. and Van de Castle, R. 1966a. *The Content Analysis of Dreams*, Appleton-Century-Croft, New York

Hall, C., and Van de Castle, R. I. 1966b. "Evaluation of dream content in combat-related PTSD." *Journal of Traumatic Stress,* 12, 681–687.

Hartmann, E. 1984. *The Nightmare: The Psychology and Biology of Terrifying Dreams*, Basic Books, NY.

Hartmann, E. 1998. “Nightmare after trauma as paradigm for all dreams: a new approach to the nature and functions of dreaming.” *Psychiatry*. 1998 Fall; 61(3): 223-38.

Hartmann, E. 2001. *Dreams and Nightmares*, Perseus, New York, 2001.

Hartmann, E., Zborowski, M., Rosen, R., Grace, N. 2001. “Contextualizing Images in Dreams: More Intense After Abuse and Trauma.” *Dreaming* 11, 115 (2001).

Hartmann, E. and Basile, R. 2003. “Dream Imagery Becomes More Intense After 9/11/01.” *Dreaming,* Vol. 13 No. 2, June 2003.

Hartmann, E. 2011. *The Nature and Functions of Dreaming*. New York, NY: Oxford University Press.

Hoss, R., Valli, K, Gongloff, R 2019. *Dreams: Understanding the Biology, Psychology and Culture,* 340-349, 365-413. ABC-CLIO.

Jung, C. G. 1971. “The Transcendent Function.” *The Portable Jung - Part 9,* Edited by Joseph Campbell, The Viking Press, N.Y.

Jung, C. G. 1973. *Man and His Symbols*, Dell Publishing Co. NY, NY.

Jung, C. G. 1993. “On the nature of the psyche.” In *The basic writings of C. G. Jung*, edited by V. S. de Laszlo, 53–61. New York, NY: Random House.

Khazaie, H., Ghadami, M., Masoudi, M. 2016. “Sleep disturbances in veterans with chronic war-induced PTSD.” *J Inj Violence Res*. 2016 Jul; 8(2): 99–107.

Kramer, M., Schoen, L. S., Kinney, L. 1984. “Psychological and behavioral features of disturbed dreamers.” *Psychiatric Journal of the University of Ottawa, 9,* 102-106.

Kramer, M., Schoen, L., and Kinney, L. 1987. "Nightmares in Vietnam veterans." *Journal of the American Academy of Psychoanalysis, 15*, 67-81.

Kramer, M. 1991. “The nightmare: A failure in dream function.” *Dreaming,* I, 277-285.

Kramer, M. 2019. “Mood regulation and emotional problem-solving.” In *Dreams: Understanding the Biology, Psychology and Culture*, edited by Hoss, R., Valli, K, Gongloff, R, 349-354. ABC-CLIO.

Krippner, S. 2016. In *Working with Dreams and PTSD Nightmares 14 Approaches for Psychotherapists and Counselors*, edited by J.E. Lewis and S. Krippner, 251-268. Santa Barbara, CA: Praeger.

Krippner, S., and Taitz, I. 2017. "Psychotherapeutic approaches for post-traumatic stress disorder nightmares." *International Journal of Dream Research, 10*(2), 101–109.

Levin, R., Tore A. Nielsen 2007. "Disturbed Dreaming, Posttraumatic Stress Disorder, and Affect Distress: A Review and Neurocognitive Model." *Psychological Bulletin 2007.*

Levin, R., Nielsen, T. 2009. "Nightmares, bad dreams, and emotion dysregulation a review and new neurocognitive model of dreaming." *Current Directions in Psychological Science,* 18(2), 84-88.

Levy, M. 2016. *Dreams, Vietnam.* Winter Street Press, Salem MA.

Levy, M. 2018. *Other Dreams.* Telegraphos Press, San Bernadino, CA.

Maher M., Rego S., Asnis G. 2006. "Sleep disturbances in patients with post-traumatic stress disorder: epidemiology, impact and approaches to management." *CNS Drugs*. 2006; 20(7): 567–90.

Malinowski, J., and Horton, C. L. 2014. "Evidence for the preferential incorporation of emotional waking-life experiences into dreams." *Dreaming*, 24, 18-31.

Malinowski, J., Horton, C. 2015. "Metaphor and hyperassociativity: The imagination mechanisms behind emotion assimilation in sleep and dreaming." *Frontiers in Psychology, 6*, 1132. 2015, August 18.

McNamara, P., Dowdall, J., Auerbach, S. 2002. "REM sleep, early experience, and the development of reproductive strategies." *Human Nature, 13*(4), 405–435

Mellman T., Kulick-Bell R,. Ashlock L., Nolan B. 1995. "Sleep events among veterans with combat-related posttraumatic stress disorder." *Am J Psychiatry*. 1995;152:110–5.

Mellman, T., David, D., Bustamante, V., et al. 2001. "Dreams in the acute aftermath of trauma and their relationship to PTSD." *Journal of Traumatic Stress, 14*(1), 241–247.

Mellman, T., Pigeon, W. 2005. "Dreams and nightmares in posttraumatic stress disorder." In *Principles and practice of sleep medicine* (4th ed.), edited by M. H. Kryger, T. Roth, and W. C. Dement. 573-578. Philadelphia, PA: Elsevier.

Mysliwiec V, O'Reilly B, Polchinski J., et al. 2014. "Trauma associated sleep disorder: a proposed parasomnia encompassing disruptive nocturnal behaviors, nightmares, and REM without atonia in trauma survivors." *J Clin Sleep Med* 2014;10(10):1143-1148.

Nader, K., Schafe, G., LeDoux, J. 2000. "The labile nature of consolidation theory." *Nature Reviews Neuroscience* Volume 1, December 2000, 216-217.

Neylan T., Marmar C., Metzler T. et al. 1998. "Sleep disturbances in the Vietnam generation: Findings from a nationally representative sample of male Vietnam veterans." *American Journal of Psychiatry*. 1998;155:929–933.

Nielsen, T., Levin, R. 2007. "Nightmares: A new neurocognitive model." *Sleep Medicine Reviews, 11*(4), 295–310.

Pagel, J., Vann, B., Altomare, C. 1995. "Reported association of stress and dreaming: Community background levels and changes with disaster (Hurricane Iniki)." *Dreaming: Journal of the Association for the Study of Dreams*. Vol 5(1) 43-50, Mar 1995.

Payne, J., Schacter, D., Propper, R., et al. 2009. "The role of sleep in false memory formation." *Neurobiology of Learning and Memory, 92*(3), 27–334.

Perls, F. 1974. *Gestalt Therapy Verbatim*, 27-76. Bantam Books, Real People Press.

Perogamvros, L., Schwartz, S. 2012. "The roles of the reward system in sleep and dreaming." *Neuroscience and Biobehavioral Reviews, 36*(8), 1934–1951.

Punamäki, R. 1997. "Determinants and Mental Health Effects of Dream Recall Among Children Living in Traumatic Conditions," *Dreaming*, Vol. 7 No. 4, 1997.

Punamäki, R. 2007. "Trauma and Dreaming." in *The New Science of Dreaming*, edited by D. Barrett and P. McNamara, 211–251. Westport, CT: Praeger Greenwood.

Robert, G., Zadra, A. 2014. "Thematic and Content Analysis of Idiopathic Nightmares and Bad Dreams." *Sleep.* 2014 Feb 1; 37(2): 409–417.

Revonsuo, A. 2000. "The reinterpretation of dreams: An evolutionary hypothesis of the function of dreaming." *Behavioral and Brain Sciences, 23*(6), 877–901.

Revonsuo, A., Tuominen, J., Valli, K. 2016. "The simulation theories of dreaming: How to make theoretical progress in dream science: A reply to Martin Dresler." In *Open MIND,* edited by T. Metzinger and. M. Windt, 1341–1348. Cambridge, MA: MIT Press.

Schneider and Domhoff. "The quantitative study of dreams"; *dreamresearch.net,* University of California, Santa Cruz.

Siegel, A. 2001. "Dreams of Firestorm Survivors." in *Trauma and dreams*, edited by Barrett, D., 159. Cambridge, MA: Harvard University Press.

Siegel, A. 2003. *Dream Wisdom: Uncovering Life's Answers in Your Dreams*. Berkeley: Celestial Arts: 2003.

Siegel, A. 2019. "A Lifetime of Dreaming." In *Dreams: Understanding the Biology, Psychology and Culture*, edited by Hoss, R., Valli, K, Gongloff, R, 497. ABC-CLIO.

Stewart, D., and Koulack, D. 1993. "The function of dreams in adaption to stress over time." *Dreaming: Journal of the Association for the Study of Dreams, 3*(4), 259–268.

Stickgold, R. and Walker, M. 2013. "Sleep-dependent memory triage: Evolving generalization through selective processing." *Nature Neuroscience, 16*, 139–145.

Valli, K., Revonsuo, A., Pälkäs, O., et al. 2005. "The Threat Simulation Theory of the Evolutionary Function of Dreaming: Evidence from Dreams of Traumatized Children." *Consciousness and Cognition* 14 (1):188-218.

Valli, K, Lenasdotter, S., MacGregor, O., Revonsuo, A. 2007. "A test of the Threat Simulation Theory - Replication of Results in an Independent Sample." *Sleep and Hypnosis*, 9:1, 2007.

Van der Kolk B, Blitz R, Burr W, Sherry S, Hartmann E. 1984. "Nightmares and trauma: a comparison of nightmares after combat with lifelong nightmares in veterans." *Am J Psychiatry*. 1984 Feb;141(2):187-90.

Wamsley, E. 2016. “Memory consolidation in dreams and waking thought.” Paper presented at the *IASD Online Research Conference*, June 2016. Retrieved from www.iasdreamresearch.org.

Wilmer, H. 1996. “The healing nightmare: War dreams of Vietnam veterans.” In *Trauma and Dreams,* edited by D. Barrett, 85–99. Cambridge, MA: Harvard University Press.

Wittmann, L., Schredl, M., Kramer, M. 2007a. “Dreaming in posttraumatic stress disorder: a critical review of phenomenology psychophysiology and treatment.” *Psychother Psychosom*. 2007;76(1):25-39.

Wittmann, L., Schredl, M., Kramer, M. 2007b. “Understanding posttraumatic nightmares: An empirical and conceptual review.” *Sleep Medicine Reviews*, 10, 19–31.

Zadra, A., Duval, M., Begin, E., Pilon, M. 2004. “Content analysis of nightmares.” *Sleep,* Supplement, 27, A64.

Zadra, A. 2019. “Nightmare therapies.” In *Dreams: Understanding the Biology, Psychology and Culture*, edited by Hoss, R., Valli, K, Gongloff, R, 479-490. ABC-CLIO.

In: Parasomnia Dreaming
Editor: James Frederic Pagel

ISBN: 978-1-53617-815-9

Chapter 6

DREAMING IN REM SLEEP BEHAVIOR DISORDER

Isabelle Arnulf**, MD, PhD
Sleep disorder Unit of Pitié-Salpêtrière Hospital,
Sorbonne University, Paris, France

ABSTRACT

Rapid eye movement sleep behavior disorder (RBD) correspond to violent dream-enacting behaviors. REM sleep is normal except for imperfect blockade of muscle tone and movement. The dreaming features in RBD are less studied than its motor aspects or than isolated RBD as heralding Parkinson's disease. Dream content during RBD is often recalled afterwards. The question whether RBD is a dreaming disorder or just normal REM sleep with motor release, remains open. Indeed, RBD-associated dreams differ from normal dreaming by a more active or violent content, and by frequent agressions from humans or animals. However, these reports may be biased toward remembering scenarios that awakened

* Corresponding Author's Email: isabelle.arnulf@aphp.fr.

or injured the sleeping couple. Plus, quiet, mundane dreaming (e.g., eating, smoking, lecturing) is also observed with RBD. When dream content is systematically collected after provoked awakenings during RBD movements and during quiet REM sleep in RBD patients, its content is not different from that of subjects without RBD. RBD behaviors are quite concordant with dream recall (or at least part of it, as behaviors are present during only 9% of REM sleep), as reported in numerous case-reports by history and by direct observation in the sleep laboratory. This concordance has been tested in a single controlled study. Eventually, the rare, scenic RBD behaviors can be viewed as "solid" parts of dream. The gestures, speeches and facial expressions, and their isomorphism with the dream recall offer a window to overtly approach the physiology of dreaming during REM sleep.

Keywords: REM sleep behavior disorder, status dissociatus

INTRODUCTION

Rapid eye movement behavior disorder (RBD) is often defined as dream enacting behaviors. However, more attention has been paid to the motor aspects (because of risk of injuries) and to RBD as prodromal neurodegeneration than to its dreaming aspects. The first RBD animal model was developed in cats in the 1960's by Michel Jouvet in France (M Jouvet, 1962), around twenty years before the discovery, in 1986 of the human RBD by Carlos Schenck and coll. (C. H. Schenck, Bundlie, Ettinger, & Mahowald, 1986). Notably, the feline RBD was named 'oneiric behavior' ('oneiric' meaning 'dream' in ancient Greek language) because these animals displayed apparent dream-related behaviors (leaping, chasing and fighting) during REM sleep. The model was even used to determine "what does a cat dream about?," illustrating how much these pioneers believed in the dream-action hypothesis (M. Jouvet, 1979; Sastre & Jouvet, 1979).

Here, we will describe the nature of RBD, its similarity (to the exception of movements and muscle tone) with normal REM sleep, and its dreaming aspects. Plus, RBD constitutes a unique window to study the dreaming process from a point of view external to the dreamer. Indeed, behaviors, facial expressions and verbal utterances are in accordance with the dream reports obtained upon awakening. This condition (named isomorphism) brings strong, unbiased evidence that dreaming occurs during sleep and is not built upon awakening or reconstructed afterwards by the sleeper to please the investigator/clinician. One fascinating aspect of RBD is that the observer has, for the first time, the feeling of seeing "solid" mental images.

The Various Types of REM Sleep-Associated Parasomnias

The international classification of sleep disorders includes in this category the abnormal phenomena occuring during full blown REM sleep (REM sleep behavior disorder [RBD], nightmare disorder), at the transition between wakefulness and REM sleep (sleep-associated hallucinations and exploding head syndrome, recurrent isolated sleep paralysis) and, by analogy to RBD, status dissociatus (which is a rare disorder characterized by continuous dreamlike or mundane activities during apparent sleep, in absence of sleep EEG figures). Sleep talking is considered as a normal variant both in NREM and REM sleep. Cataplexy is a cardinal symptom of narcolepsy and not an isolated phenomenon. Lucid dreaming in REM sleep is not mentionned in the classification as it is not considered as a disorder. Here we will focus on RBD.

What is REM Sleep Behavior Disorder?

Diagnosis

REM sleep behavior disorder has been identified as a parasomnia in 1986 by Carlos Schenck and coll., long after arousal disorders were identified (C. H. Schenck et al., 1986). It is characterized by violent behaviors during REM sleep, usually corresponding to enacted nightmares and active dreams, that may injure the patients and their bed-partners or disturb sleep (American Academy of Sleep Medicine, 2014). The usual patient is a sixty- or seventy-years old man, refered to the sleep specialist because he fell from bed, or had an hematoma or a fracture when bumping into the wall or the bedside table during the night. He afterwards often recalled having had a nightmare, during which he had to fight an agressor. RBD is associated with abnormally enhanced muscle tone during REM sleep, a sleep stage during which muscle tone should normally be abolished (atonia). The diagnosis requires a medical history (ideally including the bed-partner, as some behaviors are unknown to the patient), a neurological examination (to look for mild parkinsonism, mild cognitive impairment, cerebellar syndrome or orthostatic hypotension which are conditions often associated with RBD) and video polysomnography (including chin and the limbs electromyography) to document abnormal behaviors (gesturing with eye closed, often kicking and boxing, vocalizing, yelling or speaking) and/or an excess of muscle activity during REM sleep (Figure 1). Abnormal loss of atonia is usually defined when 15-18% of REM sleep epochs contain a muscle tone twice greater than the lowest muscle tone of the noght during more than half of the epoch. Scoring RBD sleep requires a large experience plus a concomitant video analysis, otherwise some epochs with muscle artefacts on EEG could be scored as wakefulness. The video-polysomnography also helps excluding other mimicking disorders in the differential diagnoses of RBD, including NREM parasomnias (sleepwalking, night terrors, confusional arousal, sexsomnia, sleep-related eating disorders), REM sleep-associated severe apnea, periodic leg movement disorder, nocturnal confusions and hallucinations (frequent in dementia with Lewy bodies), seizures, non-pathological dream enacting behaviors and dissociative disorders.

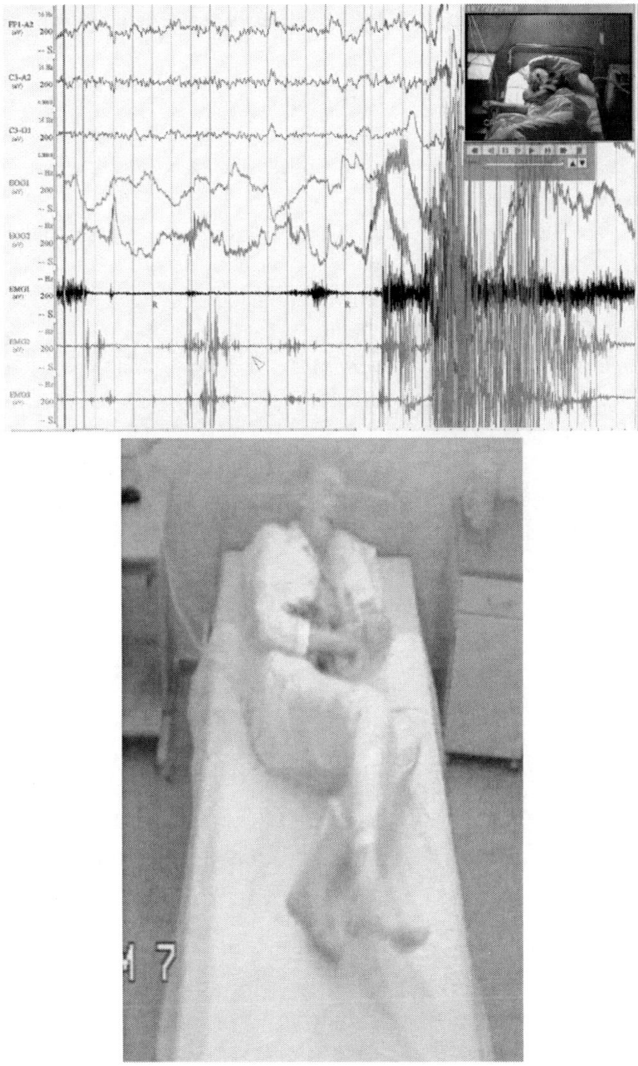

Figure 1. Videopolysomnography during an RBD event. n this 30 s epoch, the EEG (FP1-A2, C3-A2 and C3-O1, 3 first black lines) with theta rhythm and rapid eye movements (EOG1 and EOG2) are typical of a REM sleep episode. However, the muscle activities on the mentalis (EMG 1) and right (EMG2) and left (EMG3) anterior tibialis are not totally abolished, as they should be in REM sleep, with occasional phasic activities in the first part of the epoch, and major tonic and phasic activities in the last part of the epoch, which are concomitant with shouting and kicking on the video-clip.

Table 1. The RBD screening questionnaire

1. I sometimes have very vivid dreams. yes/no
2. My dreams frequently have an aggressive or action-packed content. yes/no
3. The dream contents mostly match my nocturnal behavior. yes/no
4. I know that my arms or legs move when I sleep. yes/no
5. It thereby happened that I (almost) hurt my bed partner or myself. yes/no
6. I have or had the following phenomena during my dreams:
6.1. speaking, shouting, swearing, laughing loudly yes/no
6.2. sudden limb movements, "fights" yes/no
6.3. gestures, complex movements, that are useless during sleep, e.g., to wave, to salute, to frighten mosquitoes, falls off the bed yes/no
6.4. things that fell down around the bed, e.g., bedside lamp, book, glasses yes/no
7. It happens that my movements awake me. yes/no
8. After awakening I mostly remember the content of my dreams well. yes/no
9. My sleep is frequently disturbed. yes/no
10. I have/had a disease of the nervous system (e.g., stroke, head trauma, parkinsonism, RLS, narcolepsy, depression, epilepsy, inflammatory disease of the brain), which?

One point per positive answer. A total score greater than 5/13 is abnormal (Stiasny-Kolster et al., 2007).

Some questionnaires have been developed to screen RBD in the general population (Li et al., 2010; Postuma et al., 2012; Stiasny-Kolster et al., 2007). An example of the most frequently used questionnaire is shown in Table 1. All questionnaires have been built following two dimensions: i) abnormal dreaming, and ii) dream-enactment. They have excellent sensitivity in the general population, except in patients with NREM parasomnia who have scores similar to those of RBD patients, illustrating that patients with NREM parasomnia also report dream-enacting behaviors, although during NREM sleep (Haridi et al., 2017). It is therefore recommended to be cautious using these questionnaires in young adults and children (who are at highest risk of NREM than REM parasomnia), and when ambulation (which is exceptional in RBD and frequent in NREM parasomnia) is present. When there is a history of dream-enacted behaviors in a context suggestive of RBD (e.g., in patients with Parkinson's disease) and absent polysomnography, the term of 'probable RBD' is used, in opposition to 'definite' RBD when the polysomnography has been performed and shows excessive muscle tone in REM sleep or behaviors.

Epidemiology

The prevalence of RBD is estimated to be 0.5-2% of the population aged 60 years or older, with 1.15% having idiopathic RBD (Kang et al., 2013), and up to 6.8% of those older than 70 years may have probable RBD (Boot et al., 2012). If not associated with other obvious neurological or psychiatric diseases, it used to ve named "idiopathic RBD," a term that had been recently replaced to "isolated RBD" (iRBD). There are more men (around 80%) than women with iRBD, whereas the sex ratio is 1 in neurodegenerative and autoimmune RBD. RBD can be also associated with neurodegenerative (including Parkinson disease, dementia with Lewy bodies and multiple system atrophy), auto-immune (narcolepsy, Guillain-Barre syndrome, limbic encephalitis, Morvan syndrome) or vascular, autoimmune and tumoral lesions of the pons or medulla (I. Arnulf, 2012) (Table 2). Among neurodegenerative diseases, those characterized by alphasynuclein deposits in the neurons are strongly associated with RBD, wheras those characterized by tau protein and amyloid deposits (eg, Alzheimer disease, supranuclear palsy, or corticobasal degeneration) are exceptionnally associated with RBD. Consequently, the presence of RBD has a great diagnostic value (in favor of prodromal dementia with Lewy bodies and not Alzheimer disease) in presence of mild cognitive impairment or dementia.

Higher than expected comorbid RBD is found in some psychiatric disorders (mostly major depressive disorder and more rarely post-traumatic stress disorder), and affected subjects are younger and more frequently female than those with iRBD or Parkinson disease-associated RBD (Husain, Miller, & Carwile, 2001; Wing et al., 2018).

Table 2. Frequency of RBD in various neurological and psychiatric disorders

Disorder	Frequency	Remarks
Neurodegenerative disorders		
Isolated RBD	100%	The most severe form of RBD, with frequent and violent behaviors
Parkinson's disease	40-60%	RBD starts before PD onset in 1/3 patients. Dementia and hallucinations develop earlier in patients with than without RBD
Dementia with Lewy bodies	50-86%	RBD starts before dementia onset in 64%. Status dissociatus is possible.
Multiple systemic atrophy	90%	RBD is more severe at the beginning than at the end of the disease
Spinocerebellar ataxia	56%	Type 3: A rare genetic disorder
Autoimmune brain disorders		
Narcolepsy	45-60%	More in narcolepsy type 1 than type 2. Often mildler form of RBD than neurodegenerative- and iRBD
Autoimmune encephalitis	20-100%	Linked to LGi-1, Caspr2, DPPX and IgLON5 antibodies. Status dissociatus is possible.
Metabolic disorder		
Wilson disease	15%	Genetic disorder with copper accumulation in the brain and liver
Psychiatric disorders		
Post-traumatic stress disorder	Unknown	Enacted reviviscent nightmares in NREM and REM sleep
Depression	5-8%	More frequently women

The suspected mechanism of RBD is a lesion of the REM sleep atonia system, which is located in the pontomedullary brainstem (Ehrminger et al., 2016; Luppi et al., 2011), in parallel with several models of dreamlike behaviors obtained by damaging these area in cat and rat (Sastre & Jouvet, 1979; Valencia Garcia et al., 2017). The behavioral disorder is alleviated with treatments such as melatonin, clonazepam and rivastigmine and worsened by antidepressants. Some spouses or RDb patients resort to sleeping in another bed so as not to be harmed.

Isolated RBD Often Herald Parkinsonism or Dementia

Most patients with iRBD will eventually develop a neurodegenerative disorder (mostly a synucleinopathy), within a median time of 6-7 years (Postuma et al., 2019). In addition, these patients may present with various mild symptoms, including motor (mild bradykinesia and hypertonia), autonomic (constipation, dysuria, dyserection, orthostatic hypotension), cognitive (mild non-amnestic cognitive impairment, decreased verbal fluency, altered visuospatial abilities), sensorial (hyposmia, impaired color vision) and psychological (apathy, anxiety, emotional dysregulation) symptoms (Arnaldi, Antelmi, St Louis, Postuma, & Arnulf, 2017). They may have indices of neurodegenerative lesions in the striatum (altered DAT scan or Fluoro-DOPA scintigraphy), substantia nigra (decreased melanin signal), and in the locus coeruleus/subcoeruleus, as well as in the neurons of the gut, skin and heart (Antelmi, Donadio, Incensi, Plazzi, & Liguori, 2017; Ehrminger et al., 2016; Knudsen et al., 2018). This latency period provides a time window to observe how the symptoms develop and potentially intervene to modify the course of neurodegeneration. However, since no drug or intervention can yet prevent the development of parkinsonism or dementia, how to annouce the diagnosis and present the pronostic raises some ethical issues, and may generate anxiety (Arnaldi et al., 2017).

Are RBD Dreams Occurring During Genuine REM Sleep?

Is REM Sleep in RBD Different from Normal REM Sleep?

Because visually elaborate dreams are closely associated with normal REM sleep, the report of complex dreams, congruent with the observed sleep behavior, in patients awakened from an RBD episode constitutes convincing evidence that RBD is a manifestation of normal REM sleep (apart from the motor dyscontrol). This question will be key to determine whether observations on dreaming in the RBD model could apply to normal REM sleep dreaming or not, exactly as the model of lucid dreaming in REM sleep may slightly differ from normal REM sleep in terms of eye movements, EEG, area activated in the brain, muscle tone, EKG and respiratory activity (Dodet, Chavez, Leu-Semenescu, Golmard, & Arnulf, 2015).

First, REM sleep time is unchanged in RBD patients, compared to non-RBD, showing that the enacted behaviors rarely interrupt REM sleep (De Cock et al., 2007; Ferri et al., 2017; Leclair-Visonneau, Oudiette, Gaymard, Leu-Semenescu, & Arnulf, 2010). Regarding rapid eye movements (REMs) during REM sleep, their density (ratio between 3 sec-miniepochs containing REMs and the total of mini-epochs of REM sleep), index (number of REMs per min of REM sleep) and complexity (isolated, clustered, or burst REMs) was similar in patients with RBD vs. controls matched for age (Leclair-Visonneau et al., 2010). This suggests that there is no supplementary central blockade on eye movements during normal REM sleep, otherwise it would be unmasked during RBD (higher number of REMs). Regarding EEG during REM sleep, the analyses are restricted to "clean" periods (without movements or with distal, minor movements), as movements frequently artifact the EEG. The EEG power spectrum in REM sleep is characterized by a lower degree of difference from that of stage 2 (vs. normal controls) and by a higher degree of instability in patients with iRBD (Ferri et al., 2017), including a less pronounced REM-related decrease of power in all bands with frequency <15 Hz than controls (indicating a "slower" EEG activity). Because these subtle changes are not found in animal RBD models (Peever, Luppi, & Montplaisir, 2014), they are interpreted as early amrkers of cortical neurodegeneration, rather than changes secondary to the disinhibited muscle tone. Although the autonomic system is early affected in iRBD and in parkinsonism, heart rate variability during REM sleep without movements is similar in patients with iRBD vs. controls (Sorensen, Mehlsen, & Jennum, 2013). However, it has been early noticed that during RBD movements, patients had not the expected reactional tachycardia that one should have when displaying gross body movements during wakefulness (Ferini-Strambi, Oldani, Zucconi, & Smirne, 1996). This may indicate in fact some "normal" REM sleep processes, as REM sleep is a state of reduced autonomic control, and reduced inputs and outputs.

Is Consciousness during REM Sleep in RBD Different from That in Normal REM Sleep?

The patients with RBD have no reflexive consciousness when they exhibit movements during the RBD episodes, meaning that they have no mere perception that they really move in the bed. This absence of reflexive consciousness is illustrated by one of our patients with narcolepsy, monitored while his wife was present in the room (Oudiette, Leclair-Visonneau, & Arnulf, 2010). During an RBD episode in the early morning, he whistled and seemed to draw and take measures (he dreamt that he was building a stair). His wife, thinking that he was awake and eager to be unhooked from the electrodes, told him 'Don't move! Just wait for the nurse, she will come and unhook you'. The patient needed several seconds before reacting; then he woke up and answered 'What?' [she repeated her remark] 'I did not say anything, I was sleeping'. His wife concluded: 'Oh, just I thought… So you were dreaming!' Thus, the patient was able to qualify his previous state as sleep and his present state as awake, strongly suggesting that RBD is a within-sleep state phenomenon. We also observed a patient snoring during RBD behaviors, and another having a penile erection associated with a fighting behavior (whereas penile erection is an autonomic marker of REM sleep) (Oudiette et al., 2010).

All in all, these eye movements, EEG, respiratory, cognitive and autonomic clues support the concept that RBD occurs within genuine REM sleep, indicating that RBD is not an intermediary state between wakefulness and REM sleep. If this assumption is right, RBD (and its overt motor, autonomic, emotional, verbal and cognitive features) could be used as an original model to study some mechanisms of normal REM sleep. For example, the penile erection associated here with overt fighting in RBD provides additional evidence that REM sleep-associated penile erection are autonomic automatisms unrelated to sexual dreams.

Motor Behaviors during RBD Events

The variety of movements, vocalizations, facial expressions and behaviors observed during RBD is infinite. The most frequent motor events are minor, and include some fine fumbling fingers and hand movements, which purpose is often impossible to guess from the observator's point of view, despite they are suggestive of shaking, writing or 'catching flies'. Sudden body jerking is also common. Compared to the wakefulness movements, the RBD movements have however a specific motor signature, observed in all disorders associated with RBD (Oudiette et al., 2012). They are brief, faster, more often repeated, jerky and pseudohallucinator, not self-centered, and they rarely involve the environment in an appropriate manner. Despite they seem purposeful, they often look like "super twiches," as they might include a phasic muscle activity in a limb, performed against a tonic atonia. Notably, parkinsonism disappears during RBD movements, even in patients with severe parkinsonism during wakefulness (Cochen De Cock et al., 2011; De Cock et al., 2007). No tremor, bradykinesia, dysarthria or amimia are observed, in sharp contrast with movements, speech and facial expression during arousals and awakenings, raising the possibility that the altered basal ganglia (which modulate and slow movements awake in Parkinson's disease) are bypassed during RBD movements. Scenic behaviors are however exceptional in the sleep laboratory. Standing up and walking are exceptional during RBD, and more suggestive of dementia with Lewy bodies or of parasomnia overlap syndrome (a combination of NREM and REM sleep parasomnia). All in all, the motor events represent only 9% of REM sleep time, which means that only a reduced part of the dreamt behaviors are enacted (Leclair-Visonneau et al., 2010).

All facial emotions are observed, including fear, disgust, laugh and smile, surprise, and pain (Cle et al., 2019). However, the main difference with awake expressions is that the eyes are almost always closed. Sleep talking during RBD includes nonverbal utterance (silent speeches, mumbles, shouts, whispers, and laughs) in 61% of the utterance, and distinct words and sentences (with correct syntax and subordinate clauses) in 39% (I Arnulf et al., 2017). As many as 16% of clauses contain profanities. Verbal abuse lasts long and is mostly directed toward insulting or condemning someone. Men sleep-talk more than women and used a higher proportion of profanities. Apparent turn-taking in the conversation respects the usual language gaps. Coverbal gestures are observed, in coherence with the enonced sentences.

Dream Content during RBD Events

The vocalizations, face expressions or behaviors during RBD often correlate with simultaneously occurring dream mentation, leading to the frequent report of "dream-enactement." The behaviors usually manifest as attempted enactments of unpleasant, action-filled and violent dreams or nightmares in which the individual is being confronted, attacked, or chased by unfamiliar people or animals (Figure 2).

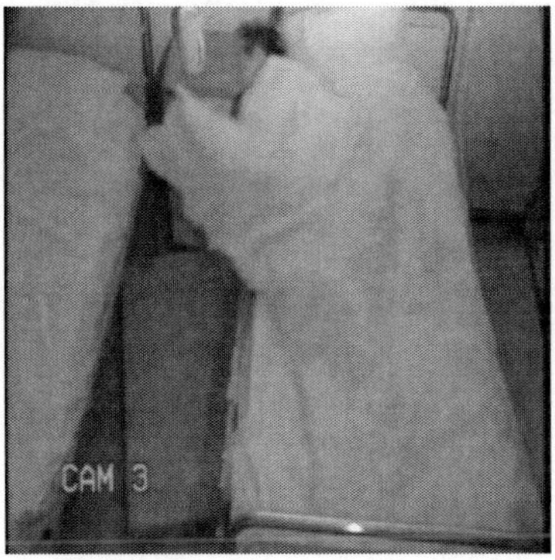

Figure 2. Examples of RBD violent behaviors. Patients are asleep in REM sleep and display violent movements of kicking (left) or boxing (right), reproduced with patients' permissions.

Typically, at the end of an episode, the individual awakens quickly, becomes rapidly alert, and reports a dream with a coherent story. The dream action corresponds closely to the observed sleep behaviors (American Academy of Sleep Medicine, 2014). The evidence supporting the presence of dreams upon awakening from an RBD episode, the evidence supporting the concordance between the action in dreams and in reality, and the evidence for/against predominance of violence in RBD dreams will be examined below.

Is There Always a Dream Recall upon Awakening from REM Sleep Behavior Disorder ?

The recall of a dream upon awakening from REM sleep in normal subjects is frequent, but 20-23% of REM sleep awakenings do not elicit any dream recall (Siclari et al., 2017). Similarly, if most patients report a dream upon awakening from an RBD episode, this is not a universal finding. Among patients with RBD and Parkinson's disease 40.3% of patients reported 'often or always' (vs. 'sometimes' in 22.6% and 'never' in 35.5%) experiencing spontaneous awakenings from the RBD (Scaglione et al., 2005) Among them, 66.1% were always/often readily awoken, whereas it was only sometimes the case in 19.4% and never the case in 3.2%. The orientation upon awakening was frequently good in most (88.7%) patients, but occasional or absent in 6.4%. Dream content recall upon awakening was often present in 59.7% patients, sometimes in 33.9% and never in 1.6%. When serially awakened in REM sleep (52 times in total in 9 patients with Parkinson's disease and RBD), the mean dream recall rate was 57% in RBD patients, vs. 65% in 6 patients with Parkinson's disease and no RBD, a non different rate (K Valli et al., 2016). Dream recall was equally frequent after spontaneous (50%) and forced awakenings (55%) in NREM and REM sleep in both groups. In another series, 25 of the total 35 awakenings (71.4%) in REM sleep presented with dream recall (K. Valli et al., 2012). All in all, it suggests that a dream recall is obtained after an RBD event as often as (and not more often than) after any usual REM sleep awakening.

Violent Dreams Are Prominent in RBD

Most clinicians report the forceful and violent aspect of the RBD-associated motor behaviors, which are usually associated with vivid, unpleasant and active dreams (C. H. Schenck & Mahowald, 2002). The dreams associated with RBD are usually different from those experienced by patients before the RBD onset, although this assertion is difficult to prove, as none completed a dream diary prior to RBD onset. When they are asked about the dreams they remember in the last month, 60 patients with RBD report enacted dreams containing more elements of aggression and animals than control subjects, in sharp contrast with an absence of aggressiveness duirng their daytime personality (Fantini, Corona, Clerici, & Ferini-Strambi, 2005). In 58 patients with Parkinson's disease plus RBD, the most commonly associated dream is fighting in response to danger (91%), whereas pleasant activity is reported in 20% of patients and daily activity in 22% of patients (Scaglione et al., 2005). The thematics of unpleasant dreams reported in 188 patients with iRBD are indicated in Table 3 (Fernández-Arcos, Iranzo, Serradell, Gaig, & Santamaria, 2016). Dreams containing were reported by patients. In a group of 66 patients with Parkinson's disease and RBD, all reported they had at least once some violent behaviors (Oudiette et al., 2009). Our colleagues from Japan reported patients with RBD reproducing the Samourai Seppuku (suicide by honor), thankfully without any weapon, during sleep.

In systematic study conducted in 24 subjects with RBD, 60% of RBD dreams (i.e., those associated with a behavior) involved a threat, mostly an aggression by a human or an animal (Uguccioni et al., 2013). Most (75%) patients with RBD counterattacked when assaulted. They defended themselves, and less frequently their family from attackers (mostly human strangers). They rarely were the first attacker in the dreamt fight (6%). The threats and their responses were compared with those in adult patients with night terrors/sleepwalking: the sleepwalkers mostly fled from a disaster. These major differences in the type of threat and in the dreamer's response were reminiscent of the fight- or-flight response to threats. The RBD-enacted dreams involved more aggression when retrospectively collected over a lifetime span than when prospectively collected on the morning following the sleep monitoring, suggesting a recall bias. Indeed, dream recall likely is enhanced when the dream-enacted behaviors lead to an awakening or injuries, which more frequently occurs when the dream content is violent.

Table 3. Content of unpleasant dreams recalled by 188 patients with iRBD (Fernández-Arcos et al., 2016)

Unpleasant dreams	Frequency of patients
Attacked by someone	76.8%
Arguing with someone	63.5%
Chased by someone	55.7%
Falling from a cliff	47.8%
Attacked by an animal: dogs, snakes, lions, bulls, horses, insects, cats, rats, tigers, pigs, wolves, crocodiles, cows, moles, piranhas, wild boars (in descending order of frequency)	39.9%
Children in danger	12.8%

Non Violent Dreams May also Occur during RBD

Non-violent elaborate behaviors occur in 18% of patients with Parkinson's disease and RBD (coexisting in this case with violent behaviors within the same or other nights), as well as in patients with iRBD and RBD associated with narcolepsy or with dementia with Lewy bodies (Oudiette et al., 2009) (Figure 3).

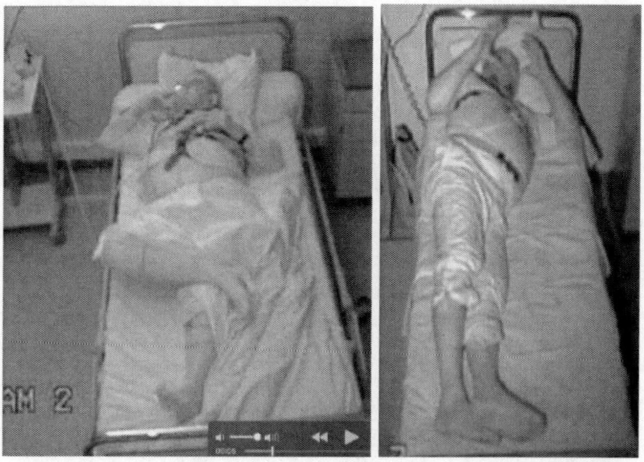

(a) (b)

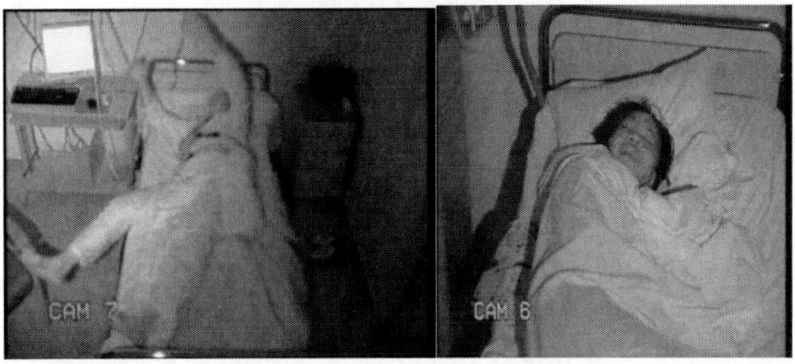

(c) (d)

Figure 3. Examples of non violent behaviors (supporting non violent dreams) during RBD. These 4 images from video-clip during RBD illustrates a patient with narcolepsy smoking a fictive cigarette (upper left), another patient with narcolepsy taking a picture (upper right), a patient with iRBD scoring a goal and rasing hands (lower left), and a patient with iRBD smiling asleep (lower right).

In some cases, the recalled dream is available and translates this same feeling of non violent content (Table 4). Most behaviors are learned behaviors in accordance with the cultural and social context of the patient. Singing and whistling are possible with correct musicality, and the local dialect is maintained (I Arnulf et al., 2017; Oudiette et al., 2009). In 203 patients with idiopathic RBD (Fernández-Arcos et al., 2016), action-filled sports are present in 15.8% of RBD dream content, including soccer (81.3%), then boxing (6.3%), skiing, basketball, motorcycling and cycling (3.1% each). Love (kissing in three cases), giving a political speech (three cases), teaching a lesson (one case), shuffling, picking thinks and riding are also reported (Fernández-Arcos et al., 2016). In one patient, a behavior resembling sexual intercourse with an imaginary partner and accompanied by a disgusting comment occurred on a single night, as reported by his wife. Patients who experienced these nonviolent behaviors also displayed aggressive behaviors during the same or different nights. The enactment of nonviolent dreams is also observed in patients with Parkinson's disease, with or without RBD (Borek, Kohn, & Friedman, 2007).

Table 4. Examples of non violent behaviors observed during RBD, and, if obtained, the corresponding dream recall, observed in our patients

Dream	Behavior
Usual, daily content	
Smoking a cigarette after finishing a roadwork	Fictive smoking in the absence of cigarette (Figure 3)
Eating	Grabs a fictive fork or spoon to the mouth: soup, bread and sandwich can be reconstitued from the behaviors (several examples)
Met an eminent politician, who told him “What, don’t you greet your old friend?” So he greeted him	“Who’s rung the bell? Who has rung?” Later: “Hello” while shaking an invisible hand and sitting on his bed (Figure 5)
Kissing my friend	Mouth giving a kiss
Unknown	Displays pelvic, coitus-like thrusting; masturbating (very rare observations)
Dreaming of urinating in the Senegal river as a child	Urinating in the bed (very rare observation)
A cat was annoying me, so I gave it a kick	He gave a leg kick and hurt himself on the wooden bed
Practicing one’s job	
Teaching (his job when young)	(Professorial) “Can we all return to our seats!What do you do, standing there in the middle? Remove your finger away from the switch! Well, if that’s so, I’ll take the numbers. (Ironic) And… late! (Professorial) Get back to your seats. I’m going to start.”
Inspecting the army (his job when young)	“Raise your hands, raise your hands, raise your hands, I said raise! I didn’t say to pull away! I said: raise your hands!” Here the patient quickly raised his left arm and waved his hand it as if he were showing something. “(Bossy and rhythmic) Halt! (pause) Halt! (pause) Halt!” “(Bossy, like in a military parade) Attention! Gentleman, please, attention! Halt, halt, I said! (pause) Halt, I said!”
Was in a meeting, developing new projects in mining (his former job)	Co-verbal gestures

Sorting buttons in different barrels, for a top designer (Christian Dior, his former job)	Handles out and throws invisible, small objects around the bed
Selling textiles (his present job)	"Someone I know told me (mumbling). Put it six folds, six folds (pause). Listen, 2.30 meters (pause). Give me 500 Euros (pause). Everything is displayed here (two folds, then pause). That's a 110,000 Francs deal (pause). Super quality, eh?" The speech is synchronized with appropriate gestures (points his finger, measures with his hands, flails arms)
Building a staircase (retired carpenter)	He shakes an invisible hand, while he introduces himself as "I am Mr. Do." Later, he seems to draw while whistling, to measure, pull, and then hit something with a fictive hammer.
Shooting several time in the balloon and then scoring a goal (retired professional saucer)	See Figure 3
I am rehearsing a speech for the town council	Declaims political speeches with a loud voice
Leisure dreams	
Picking apples	Grabs her husband's head
Dancing	Says "one, two, three" and twists his hand
Singing under the shower	Sings "It is the most beautiful tango in the world" with appropriate prosody
Attending a show	Claps and gestures thumbs up
Searching for treasure	Removes blanckets out and digs into the mattress
Whistling his dog	Whistles
Bicycling whilebeing chased by a mad dog	Cycling leg movements, while lying on his side
Taking a picture	See Figure 3
Fishing	Throws some fictive worms from his pockets, and then uses his fictive fishing ord
Extraodinary dreams	
I am a police-duck, flying after a pigeon-thief"	His wife observed him squatting on the bed, waving his arms as if flying, shouting 'pin pon' (the two-tone sound of a siren) with a duck's voice
I am flying lying on my back with the feet in front and I am braking with my feet	Extends arms and legs and gives blows all over

Table 4 (Continued)

I fall into a sinkhole, into a black hole	Wakes up shouting
There is a cow in the kitchen, so I try to slowly but firmly push her out	Pushed his wife out of the bed

Is the High Frequency of Violent Dreams in RBD a Recall Bias?

To avoid the recall bias, D'Agostino and coll. prospectively collected dream reports daily over several weeks in patients with RBD (treated with clonazepam) and controls (D'Agostino et al., 2012). They found no between-groups differences in the content of the dreams, suggesting a bias of recall (shifted towards selectively remembering the enacted violent dreams) or a benefit of clonazepam on the abnormal dreaming process itself.

Eventually, Valli and coll. prospectively collected 69 dream reports upon provoked awakening from NREM and REM sleep in 15 patients with Parkinson's disease with (mostly during RBD movements and sometimes during quiet REM sleep) and without RBD (K Valli et al., 2016). The content, action-filledness (actions, environmental events), vividness (cognition, emotions), intensity and threatening elements (including aggression) were not different between groups, although emotions were more negatively toned in those with than without RBD. Further, patients with RBD tended to act out their most intense dreams, and negative dreams may more likely be acted out than positive dreams (K. Valli et al., 2012). Consequently, the acted out dreams are the ones most likely to be remembered afterwards. The retrospective memory bias for intense and aggressive dreams may thus reflect these infrequent tip-of-the-iceberg dreams (and not be a bias after all), although the majority of dreams of patients with RBD are not altered in any way. This interesting study supports the idea that the dream content is similar in patients who enact or not their dreams, presumably suggesting that there is no change in dream content in

Parkinson's disease with RBD but a change restricted to the muscle atonia network.

The dream content was compared in patients with RBD vs. adults with sleepwalking or night terrors (Uguccioni et al., 2013). The reports of the enacted dreams were as long in duration in the sleepwalkers as in subjects with RBD. However, the dreams were more complex and less bizarre in subjects with RBD. Among the 47 enacted dreams obtained in the RBD group, 28 dreams involved a form of threat, including an aggression (57%), a fleeing behavior (18%), a misfortune (14%), an accident (11%), and no failure or disease. The dream occurs alsmot never at home in the room where in the subjects slept. The eyes are usually closed in RBD, as the dream-enactor is attending to the dream environment rather than to the bedside surroundings (in sharp contrast with dreams in NREM parasomnia which often take place in the bedroom).

DREAM-BEHAVIOR ISOMORPHISM IN RBD

Does the Recalled Dream Match the Observed Behavior?

Because patients who are awakened immediately after a behavioral episode during RBD frequently report a dream content that is congruent with the objective behavior observed prior to awakening, these behaviors are believed to represent the acting out of dreams while sound asleep and unaware of one's surrounding. The concordance between the dream actions (reported upon awakening from these behaviors) and the behaviors observed by the bed sharer or by clinicians on the video-monitoring during proven REM sleep in the sleep lab is called isomorphism. The existence of a dreaming-action isomorphism has been supported in numerous case-reports by history and by direct observation in the sleep laboratory. It has been tested in a single controlled study, and debated in the context of the analysis of REM-sleep associated twitches in developing rats.

Figure 4. Dream-behavior isomorphism as illustrated here by an artist (my daugther Cléa Arnulf) from a real medical history. A man with Parkinson's disease and RBD dreamt that he was in a canoe, attacked by caimans, trying to make them flee (left panel). In reality, he was sitting on the bed rowing without paddles, shouting: 'Help, caimans!' getting hold of a heavy oak bedside table and throwing it across the room (right panel).

There have been many incidental home reports of clear dream-action isomorphism in RBD (De Cock et al., 2007; C. Schenck, 2005). As clinicians, we are fond of such stories, because they are important for the diagnosis of the disorder, and unravel all the beauty of the dreaming life, even when dreamt events are rather nightmarish (Figure 4). In a series of patients with Parkinson's disease and RBD, patients and spouses revealed several examples of perfect dream-action isomorphisms (De Cock et al., 2007). Another patient dreamt he was a knight fighting with a sword to save his endangered lady-love; in realty, he was lying in his bed, fighting with an invisible sword, with great agility and shouting 'Manon, Charlemagne!' (a medieval battle cry). In patients with idiopathic RBD followed in Barcelona (Fernández-Arcos et al., 2016), the authros report on a man who roared loudly and woke up, recalling a dream where lions were attacking him. A man dreamt that someone was chasing him to the point that he jumped in a river to escape; instead, he jumped out of his bed. A woman dreamt that someone put a straitjacket on her; in real life, she struggled to take off her pajamas while kicking. A 63 year old man dreamt that a dog was attacking him; asleep in his bed, he said: "a dog wants to eat me" (note the excellent isomorphism here, provided by the concordant sleep talking). In " Paradox

Lost," Carlos Schenck reported on a man who hit his head against the wall, later recalling he was fighting against a mean dog (De Cock et al., 2007; C. Schenck, 2005).

Apart from the medical history, an excellent matching between behaviors and further dream recall has been incidentally observed during video-polysomnography (Oudiette et al., 2009). A retired carpenter with narcolepsy reported after awakening of having drawn and then built a stair in a dream. During the corresponding REM sleep episode, he shook an invisible hand, while he introduced himself as "I am Mr. Do." Later, he seemed to draw while whistling, to measure, to take his tools and then hit something with a fictive hammer for almost 10 min, all behaviors suggestive of the preparation and completion of a handiwork. A patient with idiopathic RBD dreamt of meeting a Minister. The Minister told him: "What? You do not greet an old friend like me." So he shaked his hand. In the corresponding REM sleep episode, he sat in bed and shook an invisible hand while saying "Good day!" (Figure 5A). Only the end and active part of the dream scenario was enacted out in this case. A patient with Parkinson's disease reported a long dream in which he was a knight in the medieval time, riding his horse around France for a whole day. Then he rested in a small inn, sleeping on the hay on the floor, as a Sarrazin fighter entered the room through the window and threatened to kill him with a scimitar. The knight could not grab his sword, but found in the hay a wheat flail, and defended himself with it. In the corresponding video-polysomnography, the patient, lying on his back and sometimes sitting in bed, fought firmly back at an invisible aggressor, and placed on his right hand, using his fist on the positive airway pressure tubing (Figure 5B). All these cases of scenic behaviors illustrate an excellent dream-behavior correspondence, at least for some parts of the dream scenario. However, one may notice that despite the recalled dreams being often long, the observed behaviors were short, possibly acting out only a (final?) part of the dreamer's dream mentation.

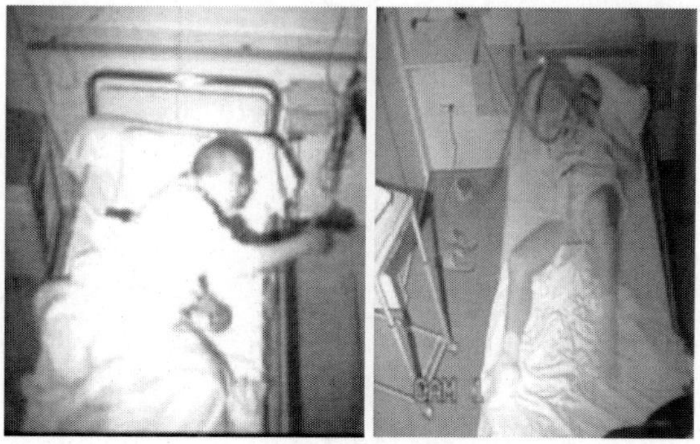

(a) (b)

Figure 5. Dream-enactment during RBD. A Left: this patient with iRBD asleep in REM sleep shakes hand of the Prime Minister saying "Good Day!"; B Right: this patient with Parkinson's disease associated RBD is a knight fighting (with his CPAP tube) a Sarrazin warrior during medieval times.

Formal Testing of the Dream-Behavior Correspondence

Despite the major interest of such studies, only a single study has yet assessed the congruence between the actions performed by RBD patients during REM sleep and the dream content later recalled (K. Valli et al., 2012). Seven blind judges had to match a set of four possible dream contents (collected upon serial awakenings after 10 min of REM sleep in the second and successive episodes, in patients with Parkinson's disease, with and without RBD), of which just one of the dream contents was correct, with motor behaviors videotaped during REM sleep in six patients with RBD. The possibility to match adequately one of these dreams by chance was 25%, as there were four choices. Of the 35 REM sleep awakenings performed, a total of 17 (48.6%) motor-behavioral episodes with recalled dream content were obtained. The average of correctly identified video-dream pairs was 39.5% (range 0-100%), which is significantly above the chance level, but still in less than half of episodes. One may wonder why the concordance did not approach 100%, as in incidental medical histories. However, one should note that this video series contains mainly simple, minor movements, which are difficult to match with any behavior, whereas scenic behaviors are rare, despite being easier to match, due to the complexity of the behaviors and vocalization observed in these cases (K. Valli et al., 2012). An example of dream-behavior pair is, for the motor part on the video: 'Few movements including two whole body jerks and small movements of the right hand. No emotional expression.' And the corresponding dream report was:

> "We drove with a car and there were people running across the street. It was a bit like chaos. And all the people were dressed in dark clothes. I did not recognize the people. It was in the evening. It was after work, on leisure time and they simply went home very fast. I wasn't there in the street in person, I only saw it. Like from nowhere, it was just in front of my eyes. It was a wide street, with telephone poles, and also there was a bush. And there was something behind the bush, I don't know what it was. There were dogs running around. They were frightening. They were running after the people. But they were not aggressive. Big ones. They were darker than shepherds. I can remember no feelings. I felt completely normal, I didn't care about the things I saw. Emotionally neutral."

This is a complex experiment in a small sample, requiring patients with RBD to accept being awakened three times per night and report immediate dream recall. So far, it has not been reproduced.

Dreams Evoking Behaviors or Behaviors Evoking Dreams?

The RBD-associated behaviors are often thought to result from a dysfunction involving atonia-producing neural circuitry in the brainstem, thereby unmasking cortically generated dreams, exactly as if a curtain was placed in front of a theatrical play (normal, atonic condition) or instead raised up (RBD condition). This view may however be too simplistic. It is challenged by two conditions: REM sleep with enhanced chin (postural) muscle tone but without RBD (remembering a raised curtain without any theatrical play behind it) in some patients, and by phasic movements despite preserved chin atonia (a theatrical play made visible through the atonia curtain), presumably indicating that over-activation of the phasic motor system has overwhelmed REM-atonia. If the loss of atonia is an admitted fact, the "dreaming" and cortical origin of behaviors in RBD is still a debate. On the one hand, numerous complex behaviors (including speech and learned behaviors, e.g., smoking, dealing cards, lecturing) can be observed during RBD (I. Arnulf, 2012) . These are not archaic behaviors and could not result from a source other than the motor cortex. On the other hand, Blumberg and coll. (who works on the abundant twiches during REM sleep in neonate rats, and their role in neural development) has suggested another source apart from the motor cortex for the RBD behaviors, as they found for twiches sources (Blumberg & Plumeau, 2016). Early studies in animals show that REM sleep twitches are not driven by the motor cortex, because they persist even when the brain areas located above the peduncle are removed, indicating that the generator of twitches is located between the medulla oblongata and the superior colliculus (Jouvet-Mounier, Astic, & Lacote, 1970). Later, Blumberg and coll. showed that the cortical motor activity increased during REM sleep, but it did so immediately after (and not immediately before) twitches. Because the latencies from twitches to peak neural activation were greater than 100 ms, their conclusion was that the sensory feedback (i.e., re-afference) from twitching limbs was driving activity in motor cortex (Blumberg & Plumeau, 2016). Notably, during wakeful movements, the nervous system drives movements and simultaneously generates a copy of the motor command (the corollary discharge) to inform the sensory cortex of the expected changes. This process helps to distinguish sensations that are self-generated from those that are external. By contrast, during REM sleep the twitches are not accompanied by corollary discharges, and "surprise" the sensory cortex, and also trigger strong activity in the primary motor cortex that is not seen in response to passive movements of the tail when awake (Jouvet-Mounier et al., 1970). These twitches might contribute to an activity-dependent development of the spinal cord, cerebellum, and forebrain and to the construction of internal motor models. Moreover, several authors mentioned that, in animals, the motor cortex is not even necessary to produce complex behavior; for example, chemical and electrical stimulation of some brainstem structures can produce defensive and aggressive behaviors in rats and monkeys that may resemble those reported in human patients with RBD. Accordingly, Mark Blumberg suggested that the brainstem (and possibly the red nucleus) could be one of the sources of the pathological movements and that sensory feedback from moving limbs could secondarily influence the content of dream mentation (Jouvet-Mounier et al., 1970). This reverse hypothesis, which is compatible with the previous, "descending" hypothesis (behaviors are the products of dreaming) suggests that a brainstem motor pattern generator of either simple movements (myoclonic twitches) or patterned behaviors (e.g., defense, attack) would first evoke movements, which in turn, via sensory feedback, would evoke dreams incorporating these stimuli.

RBD as a Model Supporting That Non-dreamers Do Dream

Dreaming may not be a universal experience, as around 0.38% of adults report that they have never ever experienced any type of dreaming, and do not report any dream (or white dream) when awakened at the end of REM sleep periods (Pagel, 2003). Whether these non-dreamers have no dream production or do have recall cannot be tested because there is no reliable marker of dreaming activity to be contrasted with dream recall, except for the recent increased EEG activity within the posterior associative cortex ("the hot zone") after NREM and REM sleep recalls of mentation (Siclari et al., 2017). Because the experiments regarding dream recall postulate that dreams are not directly accessible, the study of dreaming has been restricted to the analysis of recalled sleep mentation after awakenings. However, this limitation can now be circumvented by the discovery of RBD. Indeed, as clinicians, we frequently ask our patients' wives, in presence of their husbands: "Does your husband dream a lot?" If the answer is yes, one may suspect that their husbands have at least some frequent sleep talking and at best some clear-cut RBD.

To determine whether non-dreamers do not produce dreams or do not recall them, we identified subjects with no dream recall but with dreamlike behaviors during RBD (Herlin, Leu-Semenescu, Chaumereuil, & Arnulf, 2015). Among 289 consecutive patients with iRBD or RBD associated with Parkinson's disease who underwent a video-polysomnography and interview on dream recall, eight (2.8%) patients had no dream recall, including four patients who had never-ever recalled dreams, and four patients who had no dream recall for 10-56 years. These non-dreamers did not differ in terms of cognitive, clinical, treatment or sleep measures from dreamers with RBD. All the non-dreamers exhibited, daily or almost nightly, several complex, scenic and dreamlike behaviors and speech, which were also observed during REM sleep on video-polysomnography (e.g., arguing, fighting and speaking). They did not recall a dream following sudden awakenings from REM sleep. The scenic dreamlike behaviors observed during REM sleep in the rare non-dreamers provide strong evidence that non-dreamers produce dreams, but do not recall them.

Therefore, RBD provides a new model to evaluate cognitive processing during dreaming (and its enactment), and subsequent recall. Here is the paradox of RBD: dreams are thought to represent personal experiences; however, in the case of facila expression, speeches and scenic behaviors, an external observer can sometimes know or guess part of the sleep mentation of the dreamers instead of the dreamers themselves (at least when they have forgotten everything following awakening). Of course, the observer cannot see the images or hear the sounds experienced by the dreamer; however, he has privileged visual and auditory access to at least part of the scene played by the dreamers. Thus, RBD-associated behaviors may be considered materialized mental images of which some parts (the motor, facial expression and verbal parts) are made visible to the external observer, while they may not be encoded or recalled by the dreamer. RBD is a unique condition because there are no other conditions in which one may know instead of others what they are thinking and experiencing. This condition may question the very definition of dreams: if dreams are mental contents that occur during sleep and are recalled following awakening, then can RBD behaviors without dream recall be classified as dream-enacting behavior?

RBD as a Model to Determine Whether the Eyes Scans Dream Images

As soon as REM sleep and its association with visual dreams was discovered (Aserinsky & Kleitman, 1953), the question arised whether the eyes scan dream images. Studies have retrospectively compared the direction of eye movements to the dream recall recorded after having awakened the sleeper, with at best one third of directional correspondance, i.e., no better than chance. A main issue is that the eyes and head work in concert to produce gaze (and its target) during wakefulness, and that the neck is paralyzed during normal REM sleep. One way to circumscribe these methodological problems (recall bias, retrograde assessment, neck–eye movement combination) in humans is to study subjects with RBD (in whom the neck is not paralyzed), and to determine directly whether the eyes move in the same directions as the head and limbs (Leclair-Visonneau et al., 2010). In 56 patients with RBD and 17 healthy matched controls, we monitored eye movements by electrooculography in four directions (right, left, up and down), and synchronized with video and sleep monitoring. The RBD-associated behaviors occurred two times more frequently during REM sleep with REMs than without REMs, and more often during or after REMs than before REMs. When REMs accompanied goal-oriented motor behavior during RBD (e.g., grabbing a fictive object, hand greetings, climbing a ladder, sending a kiss with the hand), which happened in 19 sequences, 82% were directed towards the action of the patient (same plane and direction). When restricted to the determinant REMs, the concordance increased to 90%. Rapid eye movements were absent in 40% of behaviors. This directional coherence among limbs, head and eye movements during RBD suggests that, when present, REMs imitate the scanning of the dream scene. Since the REMs are similar in subjects with and without RBD, this concordance can be extended to normal REM sleep. However, these results do not mean that the dreamer actually watches the dream images in RBD. Rather, one common system may simultaneously activate dream images as well as eye and body movements in a coherent fashion [50]. This scenario would support the results from several experiments, including the presence of REMs in the absence of any kind of vision (in neonates, congenitally-blind humans, cats without visual cortex, pontine cats), as well as the temporal association between ponto-geniculo-occipital spikes and REMs in cats.

Conclusion

RBD unmasks part of the dream content, which allows studying the dreaming process in an "online" manner. To a certain degree, what is found within the unmasked material may also apply to normal REM sleep dreaming. The use of RBD to understand the cognitive processes during REM sleep is tightly dependent on the possibility to observe scenic behaviors, which are rare phenomena, compared to the number of simple, jerky movements without any clear purpose (at least for the observer) seen in the sleep lab context. Consequently, we think that, in the future, it could be important to build video banks of all RBD behaviors during REM sleep (and their corresponding dream reports, if available), in order to share them and make progress with this line of research. Scenic RBD is a narrow, but fascinating window upon dreaming. We are convinced that what it may reveal from the normal (and pathologic) dreaming process is just the beginning.

References

American Academy of Sleep Medicine. (2014). *The international Classification of Sleep Disorders, 3rd edition*. Darien, IL: American Academy of Sleep Medicine.

Antelmi, E., Donadio, V., Incensi, A., Plazzi, G., & Liguori, R. (2017). Skin nerve phosphorylated α-synuclein deposits in idiopathic REM sleep behavior disorder. *Neurology, 88*(22), 2128-2131.

Arnaldi, D., Antelmi, E., St Louis, E. K., Postuma, R. B., & Arnulf, I. (2017). Idiopathic REM sleep behavior disorder and neurodegenerative risk: To tell or not to tell to the patient? How to minimize the risk? *Sleep Med Rev, 36*, 82-95. doi:10.1016/j.smrv.2016.11.002.

Arnulf, I. (2012). REM sleep behavior disorder: motor manifestations and pathophysiology. *Mov Disord, 27*(6), 677-689. doi:10.1002/mds.24957

Arnulf, I., Uguccioni, G., Gay, F., Baldayrou, E., Golmard, J.-L., Gayraud, F., & Devevey, A. (2017). What does the sleeping brain say? Syntax and semantics of sleep talking in healthy subjects and in parasomnia patients. *Sleep, 40*, zsx159.

Aserinsky, E., & Kleitman, N. (1953). Regularly occurring periods of eye motility, and concomitant phenomena during sleep. *Science, 118*, 273-274.

Blumberg, M., & Plumeau, A. (2016). A new view of "dream enactment" in REM sleep behavior disorder. *Sleep Med Rev, 30*, 34-42.

Boot, B. P., Boeve, B. F., Roberts, R. O., Ferman, T. J., Geda, Y. E., Pankratz, V. S., Petersen, R. C. (2012). Probable rapid eye movement sleep behavior disorder increases risk for mild cognitive impairment and Parkinson disease: a population-based study. *Ann Neurol, 71*(1), 49-56. doi: 10.1002/ana.22655.

Borek, L. L., Kohn, R., & Friedman, J. H. (2007). Phenomenology of dreams in Parkinson's disease. *Mov Disord, 22*(2), 198-202. doi: 10.1002/mds.21255.

Cle, M., Maranci, J. B., Weyn Banningh, S., Lanfranchi, J., Vidailhet, M., & Arnulf, I. (2019). Smiling asleep: A study of happy emotional expressions during adult sleep. *J Sleep Res, 28*(4), e12814. doi:10.1111/jsr.12814.

Cochen De Cock, V., Debs, R., Oudiette, D., Leu, S., Radji, F., Tiberge, M., Arnulf, I. (2011). The improvement of movement and speech during rapid eye movement sleep behaviour disorder in multiple system atrophy. *Brain, 134*(Pt 3), 856-862.

D'Agostino, A., Manni, R., Limosani, I., Terzaghi, M., Cavallotti, S., & Scarone, S. (2012). Challenging the myth of REM sleep behavior disorder: no evidence of heightened aggressiveness in dreams. *Sleep Med, 13*(6), 714-719.

De Cock, V. C., Vidailhet, M., Leu, S., Texeira, A., Apartis, E., Elbaz, A.,. Arnulf, I. (2007). Restoration of normal motor control in Parkinson's disease during REM sleep. *Brain, 130*(Pt 2), 450-456.

Dodet, P., Chavez, M., Leu-Semenescu, S., Golmard, J., & Arnulf, I. (2015). Lucid dreaming in narcolepsy. *Sleep, 38*, 487-497.

Ehrminger, M., Latimier, A., Pyatigorskaya, N., Garcia-Lorenzo, D., Leu-Semenescu, S., Vidailhet, M., Arnulf, I. (2016). The coeruleus/ subcoeruleus complex in idiopathic rapid eye movement sleep behaviour disorder. *Brain, 139*, 1180-1188.

Fantini, M. L., Corona, A., Clerici, S., & Ferini-Strambi, L. (2005). Aggressive dream content without daytime aggressiveness in REM sleep behavior disorder. *Neurology, 65*(7), 1010-1015.

Ferini-Strambi, L., Oldani, A., Zucconi, M., & Smirne, S. (1996). Cardiac autonomic activity during wakefulness and sleep in REM sleep behavior disorder. *Sleep, 19*(5), 367-369.

Fernández-Arcos, A., Iranzo, A., Serradell, M., Gaig, C., & Santamaria, J. (2016). The clinical phenotype of idiopathic rapid eye movement sleep behavior disorder at presentation: a study in 203 consecutive patients. *Sleep, 39*, 121-132.

Ferri, R., Rundo, F., Silvani, A., Zucconi, M., Bruni, O., Ferini-Strambi, L., Manconi, M. (2017). REM Sleep EEG Instability in REM Sleep Behavior Disorder and Clonazepam Effects. *Sleep, 40*(8). doi: 10.1093/ sleep/zsx080.

Haridi, M., Weyn Banningh, S., Clé, M., Leu-Semenescu, S., Vidailhet, M., & Arnulf, I. (2017). Is there a common motor dysregulation in sleepwalking and REM sleep behaviour disorder? *J Sleep Res, 26*, 614-622.

Herlin, B., Leu-Semenescu, S., Chaumereuil, C., & Arnulf, I. (2015). Evidence that non-dreamers do dream: a REM sleep behaviour disorder model. *J Sleep Res, 24*, 602-609.

Husain, A. M., Miller, P. P., & Carwile, S. T. (2001). Rem sleep behavior disorder: potential relationship to post-traumatic stress disorder. *J Clin Neurophysiol, 18*(2), 148-157.

Jouvet, M. (1962). Recherches sur les structures nerveuses et les mécanismes responsables des différentes phases du sommeil physiologique [Research on the nervous structures and mechanisms responsible for the different phases of physiological sleep]. *Arch Ital Biol, 100*, 125-206.

Jouvet, M. (1979). What does a cat dream about? *TINS*(November).

Jouvet-Mounier, D., Astic, L., & Lacote, D. (1970). Ontogenesis of the states of sleep in rat, cat, and guinea pig during the first postnatal month. *Dev Psychobiol, 2*(4), 216-239. doi: 10.1002/dev.420020407.

Kang, S. H., Yoon, I. Y., Lee, S. D., Han, J. W., Kim, T. H., & Kim, K. W. (2013). REM sleep behavior disorder in the Korean elderly population: prevalence and clinical characteristics. *Sleep, 36*(8), 1147-1152. doi:10.5665/sleep.2874.

Knudsen, K., Fedorova, T. D., Hansen, A. K., Sommerauer, M., Otto, M., Svendsen, K. B., Borghammer, P. (2018). *In-vivo* staging of pathology in REM sleep behaviour disorder: a multimodality imaging case-control study. *Lancet Neurol, 17*(7), 618-628. doi: 10.1016/S1474-4422(18) 30162-5.

Leclair-Visonneau, L., Oudiette, D., Gaymard, B., Leu-Semenescu, S., & Arnulf, I. (2010). Do the eyes scan dream images during rapid eye movement sleep? Evidence from the rapid eye movement sleep behaviour disorder model. *Brain, 133*, 1737-1746.

Li, S. X., Wing, Y. K., Lam, S. P., Zhang, J., Yu, M. W., Ho, C. K., Mok, V. (2010). Validation of a new REM sleep behavior disorder questionnaire (RBDQ-HK). *Sleep Med, 11*(1), 43-48. doi: 10.1016/j. sleep.2009.06.008.

Luppi, P.-H., Clement, O., Sapin, E., Gervasoni, D., Peyron, C., Leger, L., Fort, P. (2011). The neuronal network responsible for paradoxical sleep and its dysfunctions causing narcolepsy and rapid eye movement (REM) behavior disorder. *Sleep Med Rev, 15*(3), 153-163. doi: 10.1016/j. smrv.2010.08.002.

Oudiette, D., De Cock, V. C., Lavault, S., Leu, S., Vidailhet, M., & Arnulf, I. (2009). Nonviolent elaborate behaviors may also occur in REM sleep behavior disorder. *Neurology, 72*(6), 551-557. doi: 72/6/551 [pii] 10.1212/01.wnl.0000341936.78678.3a.

Oudiette, D., Leclair-Visonneau, L., & Arnulf, I. (2010). Snoring, penile erection and loss of reflexive consciousness during REM sleep behavior disorder. *Sleep Med, 11*(9), 953-955.

Oudiette, D., Leu-Semenescu, S., Roze, E., Vidailhet, M., De Cock, V. C., Golmard, J. L., & Arnulf, I. (2012). A motor signature of REM sleep behavior disorder. *Mov Disord, 27*(3), 428-431. doi: 10.1002/mds.24044.

Pagel, J. F. (2003). Non-dreamers. *Sleep Med, 4*(3), 235-241.

Peever, J., Luppi, P. H., & Montplaisir, J. (2014). Breakdown in REM sleep circuitry underlies REM sleep behavior disorder. *Trends Neurosci, 37*(5), 279-288. doi:10.1016/j.tins.2014.02.009.

Postuma, R. B., Arnulf, I., Hogl, B., Iranzo, A., Miyamoto, T., Dauvilliers, Y., Montplaisir, J. Y. (2012). A single-question screen for rapid eye movement sleep behavior disorder: a multicenter validation study. *Mov Disord, 27*(7), 913-916. doi:10.1002/mds.25037.

Postuma, R. B., Iranzo, A., Hu, M., Hogl, B., Boeve, B. F., Manni, R., Pelletier, A. (2019). Risk and predictors of dementia and parkinsonism in idiopathic REM sleep behaviour disorder: a multicentre study. *Brain, 142*(3), 744-759. doi: 10.1093/brain/awz030.

Sastre, J., & Jouvet, M. (1979). Oneiric behavior in cats. *Physiol Behav, 22*, 979-989.

Scaglione, C., Vignatelli, L., Plazzi, G., Marchese, R., Negrotti, A., Rizzo, G., Coccagna, G. (2005). REM sleep behaviour disorder in Parkinson's disease: a questionnaire-based study. *Neurol Sci, 25*(6), 316-321.

Schenck, C. (2005). *Lost Paradox: Extreme-Nights.*

Schenck, C. H., Bundlie, S. R., Ettinger, M. G., & Mahowald, M. W. (1986). Chronic behavioral disorders of human REM sleep: a new category of parasomnia. *Sleep, 9*(2), 293-308.

Schenck, C. H., & Mahowald, M. W. (2002). REM sleep behavior disorder: clinical, developmental, and neuroscience perspectives 16 years after its formal identification in SLEEP. *Sleep, 25*(2), 120-138.

Siclari, F., Baird, B., Perogamvros, L., Bernardi, G., LaRocque, J., Riedner, B., Tononi, G. (2017). The neural correlates of dreaming. *Nature Neurosci, 20*(6), 872-878.

Sorensen, G. L., Mehlsen, J., & Jennum, P. (2013). Reduced sympathetic activity in idiopathic rapid-eye-movement sleep behavior disorder and Parkinson's disease. *Auton Neurosci, 179*(1-2), 138-141. doi: 10.1016/j.autneu.2013.08.067.

Stiasny-Kolster, K., Mayer, G., Schäfer, S., Möller, J., Heinzel-Gutenbrunner, M., & Oertel, W. (2007). The REM sleep behavior disorder screening questionnaire--a new diagnostic instrument. *Mov Disord, 22*, 2386-2393.

Uguccioni, G., Golmard, J. L., de Fontreaux, A., Leu-Semenescu, S., Brion, A., & Arnulf, I. (2013). Fight or flight? Dream content during sleepwalking/sleep terrors vs rapid eye movement sleep behavior disorder. *Sleep Med, 14*(5), 391-398.

Valencia Garcia, S., Libourel, P. A., Lazarus, M., Grassi, D., Luppi, P. H., & Fort, P. (2017). Genetic inactivation of glutamate neurons in the rat sublaterodorsal tegmental nucleus recapitulates REM sleep behaviour disorder. *Brain, 140*(2), 414-428. doi: 10.1093/brain/aww310.

Valli, K., Frauscher, B., Gschliesser, V., Wolf, E., Falkenstetter, T., Schonwald, S. V., Hogl, B. (2012). Can observers link dream content to behaviours in rapid eye movement sleep behaviour disorder? A cross-sectional experimental pilot study. *J Sleep Res, 21*(1), 21-29.

Valli, K., Frauscher, B., Peltomaa, T., Gschliesser, V., Revonsuo, A., & Högl, B. (2016). Dreaming furiously? A sleep laboratory study on the dream content of people with Parkinson's disease and with or without rapid eye movement sleep behavior disorder. *Sleep Med, 16*, 419-427.

Wing, Y. K., Lam, S. P., Mok, V. C., Zhang, J., Tsoh, J. M., & Leung, E. Y. (2018). Early neurodegenerative biomarkers and clinical outcome in psychiatric patients with rapid eye movement sleep behaviour disorder: a prospective study. *Hong Kong Med J, 24 Suppl 3*(3), 24-28.

In: Parasomnia Dreaming
Editor: James Frederic Pagel
ISBN: 978-1-53617-815-9

Chapter 7

DREAMLIKE EXPERIENCES IN ADULT SOMNAMBULISM

Antonio Zadra*

Department of Psychology, Université de Montréal,
Center for Advanced Studies in Sleep Medicine,
Hôpital du Sacré-Cœur de Montréal, Montreal, Quebec, Canada

ABSTRACT

Somnambulism is a common NREM sleep parasomnia affecting between 2%-4% of the general adult population. Behavioral manifestations with ranging degrees of complexity arise from incomplete awakenings, usually from slow-wave sleep. Until recently, it was believed that sleepwalkers were amnesic for their episodes and that complex mental contents were not involved in the behaviors exhibited during somnambulistic episodes. A growing number of clinical and laboratory findings, however, suggest otherwise. This chapter discusses the phenomenology of adult somnambulism, with a focus on dreamlike experiences associated with somnambulistic episodes.

* Corresponding Author's Email: antonio.zadra@umontreal.ca.

Keywords: somnambulism, sleepwalking, dreaming, sleep mentation, NREM sleep, consciousness, perception, sleep forensics, phenomenology

INTRODUCTION

Somnambulism (or sleepwalking) is one of the most frequent and impressive non-rapid eye movement (NREM) sleep parasomnias. The disorder is characterized by behaviors of varying complexity usually initiated during arousals from slow-wave sleep (N3; deep sleep) which culminate in walking around with an altered state of consciousness and impaired judgment (AASM 2014, Zadra et al., 2013). While the symptoms and manifestations of sleepwalking can be considered along a spectrum, most episodes are typically characterized by misperception and relative unresponsiveness to external stimuli, mental confusion, perceived threat or agitation, and variable retrograde amnesia.

PREVALENCE AND CLINICAL CHARACTERISTICS

Sleepwalking is more common in childhood than in adulthood as most children will experience, at least temporarily, one or more of the NREM sleep parasomnias. The prevalence of sleepwalking is around 3% in toddlers (Petit et al., 2007), about 11% at ages 7 and 8 years (Petit et al., 2010), and peaks at about 15% at ages 10 to 12 (Petit et al., 2015). The prevalence of sleepwalking rapidly decreases during adolescence (Klackenberg 1982, Laberge et al., 2000, Petit et al., 2015) to reach about 2-4% in adulthood (Hublin et al., 1997, Ohayon, Guilleminault, and Priest 1999, Ohayon et al., 2012) with no significant gender differences. Most children thus outgrow the disorder during mid to late adolescence, but somnambulism can persist into adulthood in up to 25% of cases (Hublin et al., 1997). Sleepwalking can also appear *de novo* in adults and subsequently increase in severity over time. In children and adults alike, somnambulistic episodes usually occur in the first third of the night when slow-wave sleep is predominant.

A sleepwalker's emotional expression can range from neutral or calm to extremely agitated. Similarly, behavioral manifestations can range from clumsy or simple and isolated actions such as gesturing, fingering bed sheets, and walking about a room to (especially in adults) surprisingly complex and organized behaviors such as rearranging furniture, cooking or eating, getting dressed, and playing a musical instrument. Complex or agitated episodes may begin with simple movements (e.g., sitting up in bed) and culminate in walking, bolting from the room, or potentially injurious behaviors. The sleepwalker's eyes are usually open throughout an episode but they may appear to have glassy stare. Sleepwalkers may spontaneously return to bed or eventually lie down to sleep elsewhere. Given the heterogeneous nature of sleepwalking episodes, their duration can vary from a few seconds to dozens of minutes.

While childhood somnambulism is often transitory and relatively harmless, sleepwalking in adulthood can carry considerable harm potential including placing oneself in dangerous situations (e.g., running into walls and furniture, trying to escape imaginary threats, leaving one's house), destroying property, and having episodes result in serious injuries to the sleeper, bed partner or others (Moldofsky et al., 1995, Rauch and Stern 1986, Schenck et al., 1989, Cramer Bornemann and Mahowald 2017). In fact, somnambulism constitutes a major cause of reported injurious or violent behaviours following an arousal from sleep (Cartwright 2004, Pressman 2007, Siclari et al., 2010) and episodes resulting in injury to the patient or others are more prevalent than generally recognized (Guilleminault et al., 2005, Ohayon, Guilleminault, and Priest 1999, Schenck et al., 1989). Furthermore, reported behaviors in more extreme cases, including driving motor vehicles (Pressman 2011), suspected suicide (Mahowald et al., 2003), and even homicide (Broughton et al., 1994) and attempted homicide have raised fundamental questions as to the medico-forensic implications of these acts (Cartwright 2004, Pressman 2018, Siclari et al., 2010, Cramer Bornemann and Mahowald 2017).

Another, and long neglected question about the nature of somnambulistic episodes, is whether or not sleepwalkers experience some form of sleep mentation or dreamlike activity during their episodes. This question is far from trivial or purely of academic or theoretical interest.

Since somnambulism is typically diagnosed exclusively on the basis of the patient's clinical history, the validity and reliability of the diagnostic criteria are of prime importance. A study (Vignatelli et al., 2005) on the reliability of the diagnosis for various parasomnias criteria found that sleepwalking only received a "fair" interobserver reliability because of disagreement on the "amnesia for the episode" criterion found in standard diagnostic manuals. In other words, when patients presenting with what appears to be a history of sleepwalking disorder report sometimes remembering their episodes or some form of mental content related to their sleepwalking events, some clinicians rule out somnambulism as a possible diagnosis.

Even with the improved diagnostic criteria put forth in the latest edition of the International Classification of Sleep Disorders (American Academy of Sleep Medicine, 2014), disagreements still occur for the "limited or no associated cognition or dream imagery" and "partial or complete amnesia for the episode" criteria (Loddo et al., 2019). In fact, to this day, many clinicians and researchers believe that sleepwalkers are always amnesic for their episodes and that complex mental contents are not involved in the behaviors exhibited during somnambulistic episodes. Although this appears to be true for children, in whom somnambulism may be more likely to express itself in the form of automatic behaviours and complete amnesia, it does not correspond to the clinical presentation seen in many adult sleepwalkers.

Phenomenological Aspects of Somnambulism

That people sometimes engage in complex ambulatory behaviors during sleep has been known for centuries. For example, Shakespeare's Macbeth contains an episode of somnambulism during which a guilt-ridden Lady

Macbeth famously tries to wash imaginary bloodstains from her hands while speaking of the crimes she and her husband have committed. Interestingly, this early 17th century depiction of somnambulism is consistent with the medical view that existed until the mid 1960s which held that sleepwalking represented a dissociative state related to dreaming.

A growing number of findings suggest that Shakespeare's depiction of sleepwalking may not have been that far off the mark. Recent clinical and empirical studies indicate that a significant proportion of adult sleepwalkers recall their episodes upon awakening in the morning and that in a sizable sub-group of sleepwalkers, sleep mentation is not only frequently part of the primary phenomenon, but that it can actually modulate the motor behaviors during an episode.

As detailed in several case reports (Broughton and Shimizu 1995, Pillmann 2009, Schenck and Mahowald 1995), that patients suffering from sleepwalking can and do recall at least portions of their episodes upon awakening has been known for decades. For instance, one paper (Montplaisir and Demers 1983) described how a 20 year-old sleepwalker "woke up" one night and said to the friend sleeping with her: "Look, there is someone in the corner of the room who is coming at us to attack us." Her partner got up, went to the corner, and told her that she had probably dreamt this as there was nobody else in the room. To further reassure the patient, the partner went to take a look in the other rooms of the house, but then heard her scream. The patient had opened the bedroom window and jumped from a height of 7 meters in the hope of escaping the fictional aggressors, breaking both of her legs.

Another detailed report (Schenck and Mahowald 1995) involved a 43-year-old man presenting with a life-long history of somnambulism, including violent and highly complex behaviors (e.g., stabbing at furniture or the air with knives, swinging baseball bats, running out of the house) that repeatedly resulted in injuries to himself and to his wife. While the patient was often amnesic for his episodes, he described events in which his somnambulistic behaviors appeared to result from dreamlike experiences.

When the patient was 25 years-old, he *"jumped out of bed during perceived 'deep sleep with dreaming' in which he believed that someone was in his house and about to attack him ... sill clad in pajamas, he left the house by running through a screen door, entered his automobile and drove 8 kilometers to his parents' home without an accident, and awakened them by pounding on their door."* The patient reported that while driving, he was aware only of driving to his parents' house to escape what he thought was an intruder in his own house.

In another episode, he attempted to strangle his wife while dreaming that he was protecting her. This was how the patient's wife of 15 years described the event: *"He later told me that he was dreaming that someone was trying to strangle me and so he was trying to pry the attacker's hands off me. But actually, his hands were wrapped around my neck, while my hands were around his hands — trying to pry his hands off my neck. It was my screaming that finally woke him up."*

Similarly, laboratory investigations of adult sleepwalkers suggest that when available, patients' phenomenological contents are broadly consistent with the observed behaviors during an episode. For instance, as part of a study on the use of auditory stimuli to experimentally induce somnambulistic events in adults presenting with a history of sleepwalking (Pilon, Montplaisir, and Zadra 2008), one patient experienced an episode during which he quickly removed his pillow and frantically examined the back and sides of his bed. The patient, who had recently become a dad, later reported that that he believed that his newborn had fallen off the bed. In the same study, three patients who experienced induced episodes from the auditory stimuli reported hearing a voice asking them to carry out a precise action that precipitated the behavior. For example, one patient suddenly looked up at the ceiling with a fearful expression, started pointing about with one hand, and then proceeded to remove her electrodes in an agitated manner. She later reported hearing someone tell her to tear away the electrodes or she was going to suffocate because the electrodes were attached to the ceiling.

In a retrospective study of 43 patients with a history of sleepwalking, Oudiette and colleagues (Oudiette et al., 2009) found that 71% of the patients recalled short, unpleasant dreamlike mentations associated with their sleepwalking episodes. Moreover, the patients' reported mentation was congruent with the observed nocturnal behavior, suggesting that sleepwalking may represent the acting out of dreamlike experiences. What's more, one recent investigation (Zadra et al., 2019) of 165 adult sleepwalkers referred to a sleep disorders clinic for chronic somnambulism revealed that the three factors rated by the greatest proportion of sleepwalkers as "often" or "always" precipitating their episodes were stress (71% of all patients), bad dreams and nightmares (47%), and dreams in general (35%).

Finally, as part of an ongoing project on the phenomenology of sleepwalking, dreamlike mentation during somnambulistic episodes was investigated in 94 adult patients (43 men, 51 women, mean age = 32.9 ± 11.7 years) referred to our sleep disorders clinic for chronic sleepwalking. All underwent at least one polysomnographic assessment in the sleep laboratory, were free of any other major sleep disorder, and received a final diagnosis of sleepwalking. Participants also completed a detailed questionnaire assessing various aspects of their sleepwalking, including phenomenological dimensions of their episodes.

Results from the intake questionnaire suggest that a considerable proportion of adult sleepwalkers recall specific elements of their episodes. In fact, fewer than 9% of the patients reported "never" remembering sleep mentation from their episodes upon awakening. In addition, 61% reported remembering specific behaviours that they displayed during their episodes. Similarly, perceptual elements from the sleeper's actual environment during somnambulistic episodes were recalled upon awakening by 75% of patients. Emotions such as fear, anger, frustration, and helplessness were also described by 75% of sleepwalkers as being "often" or "always" experienced during their episodes. Furthermore, the displayed behaviors were construed by most patients as being motivated by an intrinsic sense of urgency or underlying reason (although judgment is often impaired) that accounts for their actions during actual episodes.

Taken as a whole, these descriptive, clinical and empirical findings indicate that invariable and complete amnesia of somnambulistic episodes is not the norm in adult sleepwalkers. Moreover, these results raise important questions about the role that dreamlike mentations play in how somnambulistic episodes are experienced and unfold.

ARE SLEEPWALKERS ASLEEP DURING THEIR EPISODES?

Most sleepwalking arises out of deep N3 sleep. But unlike the dream-enactment behaviors often seen in patients with REM-sleep behavior disorder (see *Chapter 6*), sleepwalkers often appear to be awake during their episodes. For instance, sleepwalkers' eyes are usually open throughout the episode, they often show an awareness of their immediate environment (e.g., negotiating doors, hallways and stairs with relative ease), and can interact with other people nearby. That said, sleepwalkers often misperceive their environment, fail to respond to external stimuli, show signs of mental confusion and impaired judgement, and, when awakened, may have no recall of what they were doing. In other words, they appear to be both awake and asleep at the same time.

In line with an increasingly accepted view that sleep and wakefulness are not mutually exclusive, a concept termed "local sleep" (Nir et al., 2011), there is converging evidence that sleepwalkers are neither fully awake nor fully asleep during their episodes. For instance, studies of somnambulistic episodes recorded in the sleep laboratory reveal EEG patterns indicative of both sleep and wakefulness (Zadra et al., 2004). Similarly, one intracerebral investigation of a sleepwalking episode recorded in an epileptic subject found an EEG pattern of wakefulness in the motor and central cingulate cortices concomitant with increased delta bursts (indicative of sleep) in the frontal and parietal dorsolateral associative cortices (Terzaghi et al., 2009).

Consistent with these results, brain imaging by SPECT during a sleepwalking episode (Bassetti et al., 2000) showed, on one hand, a deactivation of frontoparietal associative cortices which is typical of sleep, and, on the other hand, an activation of the posterior cingulate and anterior cerebellum networks as well as a lack of deactivation of the thalamus which are characteristic of the emotionally-driven behaviour of wakefulness. What's more, one recent study (Desjardins et al., 2017) of EEG functional connectivity prior to sleepwalking showed that this co-existence of deep sleep and wakefulness can be observed up to 20 seconds *before* the episodes actually begin. For all these reasons, sleepwalking may be best conceptualized as a state characterized by a dynamic interplay between sleep and wakefulness and during which dreamlike experiences can arise.

FUTURE RESEARCH DIRECTIONS

Researchers and clinicians are only starting to study and appreciate the role that dreamlike experiences can play in adult somnambulism. The nature of the experiences reported, however, as well as the circumstances under which they occur, raise several interesting questions. Whereas normal REM and NREM dreaming (and even dreaming in REM-sleep behavior disorder) occurs in a virtual, "off-line" world with exceedingly limited awareness of the external world, sleepwalkers usually have their eyes open during their episodes and are typically aware of their immediate physical environment while sleepwalking. In this sense, and as with sleep paralysis (see *Chapter 4*), perceptual experiences during sleepwalking appear to reflect a combination of the real and imaginary. How and why these different elements come together and subsequently coevolve during the episode's duration remains largely unknown.

Another key question concerns the brain mechanisms that can result in the recall of somnambulistic episodes on some occasions, and in complete amnesia for the episode in others. Given that studies have delineated functional differences between high and low dream recallers during wakefulness as well as during sleep (Vallat et al., 2018, Eichenlaub et al., 2014) and that neural correlates of REM and NREM dream recall have been proposed (Siclari et al., 2017, Siclari et al., 2018), it would be interesting to know if similar or analogous patterns may differentiate sleepwalkers with and without frequent episode recall.

Similarly, further work is needed to better understand the nature of the neurophysiological mechanisms underlying the variable encoding and recall of somnambulistic episodes both across and within subjects. For example, it would be fascinating to explore the extent to which such processes may be related to cortical brain oscillations described in relation to episodic memories during wakefulness (Hanslmayr, Staresina, and Bowman 2016,

Griffiths et al., 2019) or to the recall of dreams from NREM sleep (Siclari et al., 2018).

Conclusion

Although sleepwalking is often characterized in terms of its automatic behaviors and retrograde amnesia, ongoing work into the phenomenology of somnambulism indicates that perceptual, cognitive and affective dimensions can play an important role in the subjective experience of adult sleepwalking. Clinicians and researchers may gain valuable information by investigating these dimensions of sleepwalkers' experiences, particularly in episodes characterized by strong emotional reactions and complex behaviors.

References

American Academy of Sleep Medicine. 2014. *The International Classification of Sleep Disorders.* Third Edition. Darien, IL: American Academy of Sleep Medicine.

Bassetti, C., S. Vella, F. Donati, P. Wielepp, and B. Weder. 2000. "SPECT during sleepwalking." *Lancet* 356 (9228):484-5. doi: https://doi.org/10.1016/S0140-6736(00)02561-7.

Broughton, R. J., and T. Shimizu. 1995. "Sleep-related violence: a medical and forensic challenge." *Sleep* 18 (9):727-30.

Broughton, R., R. Billings, R. Cartwright, D. Doucette, J. Edmeads, M. Edwardh, F. Ervin, B. Orchard, R. Hill, and G. Turrell. 1994. "Homicidal somnambulism: a case report." *Sleep* 17 (3):253-64.

Cartwright, R. 2004. "Sleepwalking violence: a sleep disorder, a legal dilemma, and a psychological challenge." *American Journal of Psychiatry* 161 (7):1149-58. doi: https://doi.org/10.1176/appi.ajp.161.7.1149.

Cramer Bornemann, M. A., and M W Mahowald. 2017. "Legal Tpoics in Sleep Medicine." In *Principles and Practice of Sleep Medicine*, edited by T. Roth and W. C. Dement, M. Kryger, 653-660. New York: Elsevier.

Desjardins, M. E, J. Carrier, J. M. Lina, M. Fortin, N. Gosselin, J. Montplaisir, and A. Zadra. 2017. "EEG Functional Connectivity Prior to Sleepwalking: Evidence of Interplay Between Sleep and Wakefulness." *Sleep* 40 (4):01. doi: https://dx.doi.org/10.1093/sleep/zsx024.

Eichenlaub, J. B., O. Bertrand, D. Morlet, and P. Ruby. 2014. "Brain reactivity differentiates subjects with high and low dream recall frequencies during both sleep and wakefulness." *Cerebral Cortex* 24 (5):1206-15. doi: https://doi.org/10.1093/cercor/bhs388.

Griffiths, B. J., G. Parish, F. Roux, S. Michelmann, M. van der Plas, L. D. Kolibius, R. Chelvarajah, D. T. Rollings, V. Sawlani, H. Hamer, S. Gollwitzer, G. Kreiselmeyer, B. Staresina, M. Wimber, and S. Hanslmayr. 2019. "Directional coupling of slow and fast hippocampal gamma with neocortical alpha/beta oscillations in human episodic memory." *Proceedings of the National Academy of Sciences* 116 (43):21834-21842. doi: https://doi.org/10.1073/pnas.1914180116.

Guilleminault, C., C. Kirisoglu, G. Bao, V. Arias, A. Chan, and K. K. Li. 2005. "Adult chronic sleepwalking and its treatment based on polysomnography." *Brain* 128 (Pt 5):1062-9. doi: https://doi.org/10.1093/brain/awh481.

Hanslmayr, S., B. P. Staresina, and H. Bowman. 2016. "Oscillations and Episodic Memory: Addressing the Synchronization/Desynchronization Conundrum." *Trends Neurosci* 39 (1):16-25. doi: https://doi.org/10.1016/j.tins.2015.11.004.

Hublin, C., J. Kaprio, M. Partinen, K. Heikkila, and M. Koskenvuo. 1997. "Prevalence and genetics of sleepwalking: a population-based twin study." *Neurology* 48 (1):177-81. doi: https://doi.org/10.1212/wnl.48.1.177.

Klackenberg, G. 1982. "Somnambulism in childhood - prevalence, course and behavioral correlations. A prospective longitudinal study (6-16 years)." *Acta Paediatrica Scandinavica* 71 (3):495-9. doi: https://doi.org/10.1111/j.1651-2227.1982.tb09458.x.

Laberge, L., R. E. Tremblay, F. Vitaro, and J. Montplaisir. 2000. "Development of parasomnias from childhood to early adolescence." *Pediatrics* 106:67-74. doi: https://doi.org/10.1542/peds.106.1.67.

Loddo, G., L. Vignatelli, C. Zenesini, F. Lusa, L. Sambati, L. Baldelli, V. Favoni, L. R. Pisani, G. Pierangeli, S. Cevoli, and F. Provini. 2019. "Interobserver reliability of ICSD-3 diagnostic criteria for disorders of arousal in adults." *Sleep and Breathing* 23 (4):1309-1314. doi: https://doi.org/10.1007/s11325-019-01937-w.

Mahowald, M. W., C. H. Schenck, M. Goldner, V. Bachelder, and M. Cramer-Bornemann. 2003. "Parasomnia pseudo-suicide." *Journal of Forensic Science* 48 (5):1158-62.

Moldofsky, H., R. Gilbert, F. A. Lue, and A. W. MacLean. 1995. "Sleep-related violence." *Sleep* 18 (9):731-9. doi: https://doi.org/10.1093/sleep/18.9.731.

Montplaisir, J., and L. Demers. 1983. "Le Somnambulism [Somnambulism]." *Union Medical du Canada* 112 (7):619-23.

Nir, Y., R. J. Staba, T. Andrillon, V. V. Vyazovskiy, C. Cirelli, I. Fried, and G. Tononi. 2011. "Regional slow waves and spindles in human sleep." *Neuron* 70 (1):153-69. doi: https://doi.org/10.1016/j.neuron.2011.02.043.

Ohayon, M. M., C. Guilleminault, and R. G. Priest. 1999. "Night terrors, sleepwalking, and confusional arousals in the general population: their frequency and relationship to other sleep and mental disorders." *Journal of Clinical Psychiatry* 60 (4):268-76. doi: https://doi.org/10.4088/jcp.v60n0413.

Ohayon, M. M., M. W. Mahowald, Y. Dauvilliers, A. D. Krystal, and D. Leger. 2012. "Prevalence and comorbidity of nocturnal wandering in the U.S. adult general population." *Neurology* 78 (20):1583-9. doi: https://doi.org/10.1212/WNL.0b013e3182563be5.

Oudiette, D., S. Leu, M. Pottier, M. A. Buzare, A. Brion, and I. Arnulf. 2009. "Dreamlike mentations during sleepwalking and sleep terrors in adults." *Sleep* 32 (12):1621-7. doi: https://doi.org/10.1093/sleep/32.12.1621.

Petit, D., E. Touchette, R. E. Tremblay, M. Boivin, and J. Montplaisir. 2007. "Dyssomnias and parasomnias in early childhood." *Pediatrics* 119 (5):e1016-25. doi: https://doi.org/10.1542/peds.2006-2132.

Petit, D., J. Paquet, Touchette E., and J. Montplaisir. 2010. Sleep: An unrecognized actor in child development. In *Quebec Longitudinal study of child development (QLSCD 1998-2010) - From birth to 8 years of age*, edited by H. Desrosiers and C. Giguère. Montreal: Institut de la Statistique du Québec.

Petit, D., M.-H. Pennestri, J. Paquet, A. Desautels, A. Zadra, F. Vitaro, R. E. Tremblay, M. Boivin, and J. Montplaisir. 2015. "Childhood sleepwalking and sleep terrors: A longitudinal study of prevalence and familial aggregation." *JAMA Pediatrics* 169 (7):653-8. doi: https://doi.org/10.1001/jamapediatrics.2015.127.

Pillmann, F. 2009. "Complex dream-enacting behavior in sleepwalking." *Psychosomatic Medicine* 71 (2):231-4. doi: https://doi.org/10.1097/PSY.0b013e318190772e.

Pilon, M., J. Montplaisir, and A Zadra. 2008. "Precipitating factors of somnambulism: Impact of sleep deprivation and forced arousals." *Neurology* 70 (24):2284-2290. doi: https://doi.org/10.1212/01.wnl.0000304082.49839.86.

Pressman, M. R. 2007. "Factors that predispose, prime and precipitate NREM parasomnias in adults: clinical and forensic implications." *Sleep Medicine Reviews* 11 (1):5-30. doi: https://doi.org/10.1016/j.smrv.2006.06.003.

Pressman, M. R. 2011. "Sleep driving: sleepwalking variant or misuse of z-drugs?" *Sleep Medicine Reviews* 15 (5):285-92. doi https://doi.org/10.1016/j.smrv.2010.12.004.

Pressman, M. R. 2018. *Sleepwalking, criminal bheavior, and relaiable scientific evidence.* Washington: American Psychological Assocation.

Rauch, P. K., and T. A. Stern. 1986. "Life-threatening injuries resulting from sleepwalking and night terrors." *Psychosomatics* 27 (1):62-4. doi: https://doi.org/10.1016/S0033-3182(86)72743-6.

Schenck, C. H., and M. W. Mahowald. 1995. "A polysomnographically documented case of adult somnambulism with long-distance automobile driving and frequent nocturnal violence: parasomnia with continuing danger as a noninsane automatism?" *Sleep* 18 (9):765-72. (p. 766). doi: https://doi.org/10.1093/sleep/18.9.765.

Schenck, C. H., D. M. Milner, T. D. Hurwitz, S. R. Bundlie, and M. W. Mahowald. 1989. "A polysomnographic and clinical report on sleep-related injury in 100 adult patients." *American Journal of Psychiatry* 146 (9):1166-73. doi: https://doi.org/10.1176/ajp.146.9.1166.

Siclari, F., G. Bernardi, J. Cataldi, and G. Tononi. 2018. "Dreaming in NREM sleep: A high-density EEG study of slow waves and spindles." *The Journal of Neuroscience* 38 (43):9175-9185. doi: https://doi.org/10.1523/JNEUROSCI.0855-18.2018.

Siclari, F., R. Khatami, F. Urbaniok, L. Nobili, M. W. Mahowald, C. H. Schenck, M. A. Cramer Bornemann, and C. L. Bassetti. 2010. "Violence in sleep." *Brain* 133 (Pt 12):3494-509. doi: https://doi.org/10.1093/brain/awq296.

Terzaghi, M., I. Sartori, L. Tassi, G. Didato, V. Rustioni, G. LoRusso, R. Manni, and L. Nobili. 2009. "Evidence of dissociated arousal states during NREM parasomnia from an intracerebral neurophysiological study." *Sleep* 32 (3):409-12. doi: https://doi.org/10.1093/sleep/32.3.409.

Vallat, R., J. B. Eichenlaub, A. Nicolas, and P. Ruby. 2018. "Dream recall frequency is associated with medial prefrontal cortex white-matter density." *Frontiers in Psychology* 9:1856. doi: https://doi.org/10.3389/fpsyg.2018.01856.

Vignatelli, L., F. Bisulli, A. Zaniboni, I. Naldi, J. E. Fares, F. Provini, R. Vetrugno, G. Plazzi, P. Tinuper, and P. Montagna. 2005. "Interobserver reliability of ICSD-R minimal diagnostic criteria for the parasomnias." *Journal of Neurology* 252 (6):712-7. doi: https://doi.org/10.1007/s00415-005-0723-1.

Zadra, A., A. Desautels, D. Petit, and J. Montplaisir. 2013. "Somnambulism: clinical aspects and pathophysiological hypotheses." *Lancet Neurology* 12 (3):285-94. doi: https://doi.org/10.1016/S1474-4422(12)70322-8.

Zadra, A., M. Pilon, A. Desautels, and J. Montplaisir. 2019. "Stress, dreams and sleep deprivation as precipitating factors for sleepwalking." *World Sleep* 2019, Vancouver, Canada.

Zadra, A., M. Pilon, S. Joncas, S. Rompre, and J. Montplaisir. 2004. "Analysis of postarousal EEG activity during somnambulistic episodes." *Journal of Sleep Research* 13 (3):279-84. doi: https://doi.org/10.1111/j.1365-2869.2004.00404.x.

In: Parasomnia Dreaming
Editor: James Frederic Pagel
ISBN: 978-1-53617-815-9

Chapter 8

Cognition as a Feature of Consciousness in Sleep Stages: Theoretical Foundations and Research Results

Miloslava Kozmová**, PhD
University of Massachusetts, Boston, Department of Psychology; Beth Israel Deaconess Medical Hospital, teaching hospital of Harvard Medical School, Department of Neurology, Sleep and Inflammatory Systems Laboratory, Boston; and psychotherapist in private practice

Abstract

In the chapter about cognition in sleep stages, the present author addresses, through academic debate and research findings, the question whether during sleep, in dreaming state of consciousness, the individuals are capable of using their mental faculties just as they would do in the wakefulness. In essence, I pose the question: *Could it be that sensory separateness from external world prevents dreamers from thinking their*

* Corresponding Author's Email: kozmova@gmail.com.

own thoughts during dreaming? This inquiry is grounded in two differing viewpoints: (a) The assumed deficiency of cognition due to the dreams considered as hallucinations, and (b) the richness of continuous cognition in specific dreaming situations when the individual dreamer recognizes something personal at stake. The results of this inquiry reflect theorists' and researchers' differing theoretical and research approaches based in phenomenology with focus on dreamers' subjective experiences and neuroscience with neuroimaging techniques with concentration on neural correlates. The present author argues that presumed complementarity of these methodologies, for integrative purposes, will require application of new perspectives about dreamers' actual cognitive capacities in non-lucid dreaming state of consciousness.

> Every man takes the limits of his own field of vision for the limits of the world.
>
> - Arthur Schopenhauer, 1908, p. 69

In 1890, William James (1890/1983) declared that "psychology is the Science of Mental Life, both of its phenomena and of their conditions. The phenomena are such things as we call feelings, desires, cognitions, reasonings, decisions, and the like" (p. 15). W. James's psychology was a psychology of a Thought as a "vehicle of knowing" (p. 345) manifested by agent (person) with mental faculties (e.g., memory, reasoning, volition, imagination, appetite, reminiscences, perceptions, passions, p. 15). Some of these conscious phenomena of persons' mental lives fall under the umbrella of *cognitive processes* (e.g., attention, perceptions, sensations, thoughts, feelings and emotions connected to thinking in contexts, motivations, memories, volition, sense of agency, identity, autonomy, agency, values, beliefs, planning for goal-oriented actions, executive skills). The conditions in which these cognitive elements of consciousness could be studied include—obviously—persons' internal world, but also interpersonal relationships, and external situations and contexts.

William James (1890/1983) also held an opinion that "in dream we forget our waking experiences" (p. 352) in the sense of "the present self and self of yesterday" and asserted that one's continuous personal identity exists only during one's waking life (p. 315). W. James's claim could be understood, as Kahan (2001) proposed, that "dreams do not contain the waking quality of self-consciousness" (p. 336). In other words, W. James assigned the conscious qualities, including cognition—as one aspect of mental life—exclusively to the condition of wakefulness.

The list of authors who in the past (1890-1994, see Kahan 2001, p. 336) and in more recent times agreed with W. James's notion that specific cognitive qualities so abundant during wakefulness are lost during dreaming, is uncommonly long: Freud (1900/1999) considered dreamers' thinking capacities (termed *secondary process* with logical reasoning and decision making) to be suspended in dreaming because the cessation of demands from external environment: With sensory de-linking to the external world, even though Freud (1900/1990) found coherent and reasonable thoughts in dreaming and termed those "complicated feats of thinking" (p. 389), he considered the thinking in dreaming inconsequential for his interpretative technique of dreams. In the dreaming condition, as Freud opined, the individuals' minds regress into "primitive" cognitions of unconsciousness ruled by primary process thinking with emotional ill-logic, magical thinking, condensation, and displacement. He contrasted this state to awake conscious mind which is maintained by rationality, order, and clarity (Freud, 1900/1990, p. ix).

Following Freud's (1900/1999) inflexible "prescription" about determinants of thinking in dreaming state of consciousness, Rechtschaffen (1978), in the Freudian mechanistic viewpoint, declared dreaming consciousness as non-reflective and Kunzendorf (1987-1988) considered dreamers to be without waking life self-consciousness. Foulkes (1990) portrayed the dreamers' experiences' as model of passive awareness (p. 47) with "relinquishment of the ability to exercise voluntary control over the contents of consciousness" (p. 49, cited in Kahan, 2001, p. 336). In the same line of thought regarding capacities for higher-order thinking within sleeping minds of dreamers, Braun et al. (1997) proclaimed that "REM [Rapid Eye Movement] sleep may constitute a state of generalized brain activity with the specific exclusion of executive systems which normally participate in the highest order analysis and integration of neural information" (p. 1190). Similarly, Maquet et al.'s (2005) argued that "volitional control is notoriously decreased in dreams" and that only during wakefulness individual's mental and external "behavior would be usually adapted to the objects and locations internally perceived" (p. 223). This viewpoint continues to be fashionable with Hobson's (2009a) claim that "dreams do not strongly evince the characteristics of secondary consciousness" (p. 803). Hobson defined *secondary consciousness* as "subjective awareness including perception and emotion that is enriched by abstract analysis (thinking) and metacognitive components of consciousness (awareness of awareness)" (p. 803) existing during waking life.

Yet, as could be seen from following illustration, the contents of dreaming consciousness are rich with variety of thinking:

> *I had a destination in mind where I am going. I knew I had to pass J.'s house. I also knew that on the way, either I could pass through the place—the clinic where Dr. P. practiced—or I could avoid it. I decided to go through the place even though I felt a bit of a fear. I thought, What will I do if I meet him in the hallway? Nevertheless, I decided to go through the clinic.* (Kozmová, 2008, p. 4)

In the mode of *prima facie* (at the first sight) consideration, this particular dreamer's higher-order cognitive processes include knowledge, options in decision-making accompanied by volition, evaluation of psychophysiological state with labeling the experienced emotion (fear), and anticipation of future. All of these thought processes are, in this case, part of independent, agentic, autonomous, and goal-oriented collection of thinking within the dreaming consciousness. They appear, firstly, in the dreamer's ability to identify the problem, and secondly, in her subsequent problem-solving attempts. With this example, even without knowing the particular waking life concerns of this dreamer, nevertheless, it seems plausible that "dreams reveal our self-conceptions and our emotional preoccupations" (Domhoff, 1996, p. viii) of lived experiences reflected, with thought processes included, in the experiences in the state of dreaming consciousness. This dream also serves as an illustration, contrary to the claim that lack of external input is responsible for lack of thinking in dreaming (Hobson, Pace-Scott, & Stickgold, 2000a) that dreamers, despite such hypothesis, continue to think their thoughts when needed.

In this critically-oriented chapter, the present author will, with inquiry into theories and research, address the debate about cognitive processes occurring during Non-Rapid Eye Movement (NREM) and Rapid Eye Movement (REM) sleep stages in non-lucid dreaming state of consciousness. The academic discussion is informed by two scientific stances: (a) the viewpoint about thought processes in dreaming considered as hallucinatory and deficient, and conceived as a return to primary consciousness ("simple awareness that includes perception and emotion," Hobson, 2009a, p. 803, not to be confused with "primary awareness," Kozmová & Wolman, 2006, p. 199); (b) the vantage point that privileges looking at the narrative data regarding similarities and differences of dreamers' use of cognitive processes in two states of consciousness—dreaming and waking (e.g., Kahan & LaBerge, 2011). In other words, in this perspective, the researchers' focus is on finding what kind of cognitive processes dreamers are actually capable of using, considering the condition of "when" they are most likely to do so, and characterizing these mental activities.

It is my contention that concentrating exclusively on findings in cognitive deficiencies would follow the line of "relatively limited potential for discoveries" (Schwartz, 2013, p. 121). Instead, by including also demonstrations of dreamers' capacities, this critical chapter might widen the potential circle of discoveries in phenomenological and neuroscientific approach to cognition in dreaming.

The guiding posts or platforms for orientation within this scope of scholarship regarding cognition occurring in dreaming states of consciousness in NREM and REM sleep stages will include the following aspects: (a) summary about sleep stages; (b) main distinctions between lucid and non-lucid dreaming; (c) defining cognition in wakefulness and dreaming states of consciousness; (d) assumptions and postulates of scientific world that continue to prevent recognition of cognitive processes in dreaming and contrasting this stance with delineation of contradictory findings in terms of dreaming persons' cognitive processes; and (e) consideration of research about NREM and REM sleep cognition. Interspersed throughout the chapter will be scientists' assumptions and inferences from neuroimaging findings assumed to be relevant to dreamers' nocturnal mental activities. In conclusion, the present author evaluates the outlined approaches to study of nocturnal cognitive processes in dreaming state of consciousness with view of previously suggested integration (e.g., Nir & Tononi, 2010) between phenomenology and neuroscience.

Sleep Stages

According Carskadon and Dement (2011), human sleep consist of two states: Non-Rapid Eye Movement (NREM, represented in scoring with sign "N") subdivided into N1, N2, N3 stages with N3 designated as slow-wave sleep (SWS) and Rapid Eye Movement (REM, represented in scoring with sign "R"). Sleep starts with NREM sleep, and as night progresses, the sleeping person moves from N1 to N2 to N3 with gradual release in muscle tone and loss of eye movement. Stages are not random, but occur in rhythm as the night progresses. Each cycle of NREM is followed by period of REM sleep in duration 80-100 minutes. As the cycles repeat themselves throughout the night, NREM and REM sleep stages cycle in average of 90 minutes while duration of REM sleep increases and duration of slow wave sleep decreases.

Main Distinctions between Lucid and Non-Lucid Dreaming

One of the most misunderstood comparisons in consciousness studies, dream science, dream psychology, neuroscience, and science of cognitive psychology, is the relationship between lucid dreaming and non-lucid dreaming states of consciousness. Hobson at al. (2000a) consider non-lucid dreaming riddled with "delusional belief that we are awake, the lack of directed thought, the loss of self-reflective awareness, and the lack of insight about illogical and impossible dream experience" (p. 842). Of course, Hobson (2009b) considers non-lucid dreaming a hallucinatory state akin to psychosis and recognition of two states of consciousness--dreaming and waking—as a sure "cure" for this malady (Hobson, 2009b, p. 39). Yet, Nielsen (2011) pointed out that the research of dreams and dreaming consciousness itself is adversely affected by researchers' unfulfilled need for "empirically based consensus on how dreaming as an object of study should be characterized" (p. 595). Nielsen (2011) offered an alternative to hallucinatory conceptualization of dreaming and suggested it to be considered as a form of "imaginative cognition" with "reality simulation" that includes spontaneous imagery (e.g., dreaming and daydreaming fantasy) and instrumental imagery (e.g., waking life autobiographical recall, story creation, and imaginal problem-solving, p. 596).

It could be proposed that the scientific misunderstanding about differences in lucid and non-lucid dreaming lies in the insistence that only when the dreamer is able to recognize in being in two states of consciousness simultaneously, he or she is considered to be rational (Hobson, 2009b) and might be able to use rational thought processes (for contrasting evidence, see Wolman and Kozmová [2007]).

The dreamers' ability to concurrently recognize dreaming and waking states of consciousness is termed "self-reflective awareness" (Fosse, 2000, p. 492). In laboratory conditions, the specific lucid dreamer might be able to demonstrate cognitive/sensory-motoric ability by signalling to experimenters (with pre-determined signals) both the recognition of lucid dreaming and the awareness of external environment at the same time (Dressler et al., 2012).

On an experiential level, at the bare minimum, lucidity occurs when dreamer is "realizing within a dream" that he or she perceives a dream (Waggoner, 2014, p. 194). In the expanded definition, Tart (1988) and Tholey (1988) consider lucidity, in the same way as van Eden (1913), only if the dreamer remembers one's own waking life in full and is able to exercise control over elements of dream and one's own activities. In summary, Hobson (2009b) and Voss, Holzmann, Tuin and Hobson (2009) argue that the superiority of lucid dreaming rests in the dreamers' ability to recognize that one is dreaming (ability to distinguish two states of consciousness) and that only in this condition the dreamer retains his or her rationality of waking state. In contrast, non-lucid dreamers are fully immersed in the uncalled-for situations of the dreams and unless waking out from undesirable or personally significant circumstances, very rarely recognize that they are dreaming (Kozmová, 2012; Kozmová & Wolman, 2006). Non-lucid dreamers are, however, even in this immersion able to think rationally (Wolman & Kozmová, 2007).

Many authors assume that non-lucid dreamers display "deficit" of higher order skills (e.g., Hobson et al., 2000, p. 809), cognitive "defect" and "loss" (Hobson, Pace-Scott, & Stickgold, 2009a, p. 821), "gross cognitive defect, so strongly suggestive of organic brain dysfunction" (Hobson, 2009a, p. 17), or consider "decoupled" prefrontal cortex as reason for dreamers' lack of thinking (Dang-Vu, Schabus, Desseilles, Schwartz, & Maquet, 2007, p. 100). From physiological viewpoint, Hobson (2009a), Voss et al., (2009), and Voss and Voss (2014) presume that the "activation" of dorsolateral prefrontal cortex acts as exclusive prerequisite for lucid dreamers' use of higher order cognition. It could be argued, however, that these deficit-based notions create and maintain very negative profile for research of non-lucid dreaming. As will be seen later, non-lucid dreamers retain, in the absence of lucidity, their mental faculties as well (e.g., Wolman & Kozmová, 2007). In addition, dreamers are able to exercise control over themselves with use of self-organizing and self-regulatory processes, become engaged with their interpersonal relationships and with animate and inanimate environment by using cognitive, kinetic, communicative, emotional, telepathic, and other mental skills (Kozmová, 2008, 2012, 2015, 2017a) even while immersed in non-lucid dreaming state. For this reason, in this chapter, the present author will focus on dream stages in non-lucid dreaming even though there exist many interesting viewpoints about lucid dreaming (e.g., Hurd & Bulkeley, 2014).

Conceptualization of Cognition in Wakefulness

Since the cognitive processes in dreaming seem to be one of the most contested points in dream psychology and consciousness studies (e.g., Hobson, Pace-Scott, & Stickgold, 2000a, 2000b), it seems reasonable to acknowledge first how this phenomenon is conceptualized in both waking and dreaming states of consciousness.

As mentioned earlier, W. James's (1890) consideration of thought processes consisted, among others, of "cognitions, reasonings, decisions" (p. 1). The thought process such as higher order analysis, volition, and adaptation in terms of "a feature of an organism's structure, physiology, or behavior that solves a problem in its life" (Colman, 2001, p. 11) are just few examples of mental capacities of higher-order thinking that individuals use during their wakefulness. Overall, Reber and Reber (2001), defined cognitive processes or *cognition*, as

> a broad (almost unspecifiably so) term which has been traditionally used to refer to such activities as thinking, conceiving and reasoning. Most psychologists have used it to refer to any class of mental 'behaviours' (using the term very loosely) with underlying characteristics of an abstract nature involving symbolizing, insight, expectancy, complex rule use, imagery, belief, intentionality, problem-solving and so forth. While the term is typically applied to mental activities in the general sense, individual thoughts, ideas or pieces of knowledge are also sometimes referred to as cognitions. (p. 128)

Specific examples of cognitions are processes of higher-order cognitions such as meta-cognition (for review, see Kahan, 2001) and executive skills defined as "the ability to plan and coordinate willful action in the face of alternatives, to monitor and update action as necessary and to suppress distracting material by focusing attention to the task at hand" (Jones & Harrison, 2001, p. 464). Executive thought processes are by some authors considered a quintessence of healthy functioning (Workman et al., 2000); they include autonomy or internally directed thinking, internal behavior [self-regulation], and verbal communication (Faden & Beauchamp, 1986). Overall, higher order mental activities consist of motivating oneself, planning, making conscious decisions, organizing internal experiences, reflecting upon results of one's actions, and dealing with situations as these occur in real time by volitionally acting in one's behalf or according to one's self-knowledge (Jenkins, 2001; Rychlak, 1994; Workman et al., 2000). One aspect of higher order cognition is also *agency* which Rychlak (1994) characterized as "the capacity that an organism has to behave or believe in conformance with, in contradiction to, in addition to, or without regard for what is perceived to be environmental or biological determinants" (Rychlak, 1994, p. 308) or "sociocultural and/or biological stimulation" (Rychlak, 1988, p. 509).

Conceptualization of Cognition in Dreaming

In sleep studies, in general terms, cognition is often referenced as *mentation* and is briefly defined as "mental activity during sleep" (Stickgold, 2000, p. 6). In more broader terms, Nielsen (2000) defined *mentation* as "remembrance of any mental activity having occurred just prior waking up" including "static visual images, thinking, reflecting, bodily feeling, or vague and fragmentary impressions . . . orienting, selective attention, sensory discrimination, recognition, rehearsal, memory activation" (p. 853).

Yet, the term *cognition* in research of dreams and dreaming presents a predicament because cognition could be conceptualized, associated with, theorized about, or investigated in variety of different ways that include the following: (a) continuous psychological feature of consciousness that demonstrates dreamers' thinking and other mental activities (e.g., Kahan, 1994; Wolman & Kozmová, 2007); (b) developmental accomplishment (e.g., Domhoff, 2003); (c) dream production (Domhoff, 2003; Mamelak & Hobson, 1989) or "how are dreams generated by mind" (Cavallero & Foulkes, 1993, p. 5); or (d) imaginative cognition (Nielsen, 2011).

Given this cornucopia of meanings about how cognition in dreaming could be conceptualized, for the purposes of this chapter about cognition in sleep stages, I will consider *cognition* as mental processes that are psychologically continuous with the processes of waking consciousness with the exclusion of long-term planning. The focus here is on cognition defined as host of *cognitive abilities* that dreamers might be able to use as their own person (e.g., Wolman & Kozmová, 2007) or as dream characters (Kozmová, 2017b, 2017c, 2018) while being engaged in spontaneously arising or self-created situations or circumstances. These *cognitive abilities* might include capacity for observation, thinking in form of internal deliberation, decision making, planning, goals accomplishments, participation, action, or imagining future scenarios—all accompanied by processes of self-awareness and reflective awareness, volition, internal and external speech (within dreaming environment), focused and directed thinking, and agency.

Research of Cognitive Processes in Dreaming is Rich in History and Scarce in Findings

It seems obvious to suggest that if there is any thinking completed by dreamers in the dreaming state of consciousness, then dreamers would need to be somewhat aware of their own person while thinking, and also be able to report the thoughts. Kleitman's (1963), however, claimed that once individuals are asleep, "the sole portion of . . . existence that 'counts' in any way, sleep [is] appearing as 'time out' from the game of living" (p. 3, cited in Pivik, 2000, p. 687). This viewpoint continued to be propagated for decades: For example, Sparrow (1976/2015) stated that "the dreamers' capacity to respond as an individual is minimal" (p. 12). As an explanation for this scientific stance, Moffitt, Hoffmann, Mullington, Purcell, Pigeau, and Wells (1988), working within dream psychology, upon listing authors who were not considering dreamers' capacities as possible, indicated that this part of science was "operating in the dark" (p. 429). And indeed, in 1990, Foulkes confirmed this statement by claiming that "falling asleep seems to be relinquishment of the ability to exercise voluntary control over the contents of consciousness" (p. 49). Purcell, Moffitt, and Hoffmann (1993) summed up this particular type of scientific approach to study of consciousness in dreaming by stating that "scientific theorizing defines dreaming in terms of cognitive deficiencies" (p. 197) and that "deficiency assumption has carried the day" (p. 203).

When considering how scientists make their decisions about any subject matter of interest, including cognition in dreaming, it seems useful to point out that scientific selection is guided by specific *paradigm* defined as "an internalized model of reality [that] determines one's perception of the world," and "a model, or an organizing principle that governs the overall pursuit in a given field of research" (Sparrow, 1976/2015, p. 57). Since this perception of reality and world could become inflexible, Kuhn (1962) proposed earlier to the community of scientists that it is possible to "see new and different things when looking . . . in places they have looked before" (p. 62). In this viewpoint, the paradigm, as organizing principle of scientific work (Kuhn, 1966), could be challenged and changed. Yet, with the researchers' focus on dreamers' cognitive deficiencies (or "what is not there"), the path of the prevailing paradigm in consciousness studies of cognition follows Hobson's (2007) claim that "science can never conclusively prove a negative" by focusing on research of deficiencies. Even though Hobson countered his own claim by stating that science "can only offer more attractive and concise alternatives" (p. 77), the prevailing paradigm remains "operating in the dark" by focusing on presumed deficiencies ("a negative").

Thus the decades-long predominant theoretical and research focus on demonstration of dreamers' deficits in thought processes (e.g., for adherence to this viewpoint in reviews, see Hobson et al., 2000a) obviously concretized scientists' concerted efforts to avoid focusing on alternatives in the form of existing cognitive processes. For example, Hobson (2009b) reemphasized that dreamers experience "gross cognitive defect, so strongly suggestive of organic brain dysfunction" (p. 17). The uncritical acceptance of deficit theories and deficit-based research explanations (e.g., Braun et al., 1997) could be understood also as a scientific reductionism in terms of renunciation of *self*—the person of the dreamer—from the dream research.

Dreaming Self Is Conceptualized as Static Non-Engaging Entity

Many theorists and researchers working with dreaming consciousness, whether wittingly or unwittingly, let their work be guided by prevalent assumption that cognitive processes in non-lucid dreaming are almost nonexistent due to "deactivated" neural regions which, during waking time, are actively supporting high-order cognition and executive thinking (e.g.,

Pace-Scott, 2011). This epistemological notion has been carried by the idea of mind/brain *isomorphism* (Hobson, 1988), in other words, concomitant, one-to-one correspondence between neural underpinning of dreaming and dreaming persons' mental processes (Hobson, 1992; 1997; Hobson et al., 2000a). Identically, Stickgold (2000) claimed that "dreaming, in both its general form and its specific content, is totally determined by physiological processes" (p. 6). This claim can be traced to one of the most enduring uncritically accepted myths in research of dreaming consciousness: Freud's (1900/1999) mechanistic impersonal viewpoint about trajectory of predetermined processes that drive dreaming mind in the presence of "ego suspension" also known as "suspension of ego control." This surmise of loss of waking life cognitive abilities during dreaming found its home in terminology of deactivated "executive ego function" with its presumed aspect of lack of acting voluntarily (Hobson, 2007, p. 75).

In psychoanalytical terms, the term *ego* is understood as conscious and unconscious part of mind that mediates between one's internal emotional and mental life, demands of external world and its realities, and the "prescriptions" of conscience (Zimmer, Bookstein, Kenny, & Kraeber, 2005, p. 550). Even though the *ego* denotes the "I," or "self" in terms of "conscious, rational aspects of oneself" (Bettelheim, 1982, p. 55), it has been relegated to an impersonal "mental apparatus" designated for "reality testing, thought processes, defensive function" (Zimmer et al., 2005, p. 550).

Freud (1900/1999) assumed that without external input, sleep constitutes time of "suspension of ego control," in other words, time for dreamers' descent into regressive mode of functioning prompting dreamers to "returning to earlier phase of development," (Gabbard, 2004, p. 26) for reason of escaping the waking life "tyranny" of unacceptable thoughts. This assumed nocturnal "suspension" in ego control meant, for Freud, that dreaming was taken over by mechanism of *primary* process thinking characterized by ill-logic, wishes, and magical thinking.

Although Freud (1900/1990) found within dreaming the existence of *secondary process* which he termed "complicated feats of thinking" (p. 389), he decided that these thought processes are unimportant (Foulkes, 1964). Following this dismissal, in scientific community, dreamers and their capacities were considered to be static entities: Earlier theorists assumed dreamers' lack of volition (Schutz, 1962) and lack of self-reflectiveness and intentionality (Globus, 1987, p. 12). In these suppositions, the authors precluded the consideration of actual experiences of dreaming persons from their theorizing. It could be proposed that the unfortunate uncritically accepted "legacy" of notion about diminishment of dreamers' capacity to think for themselves within spontaneously arising dream settings represents continuous propagation of unexamined Freudian thought. For example, Hobson (2009a) in his protoconsciousness theory of REM sleep dreaming, without directly acknowledging Freud's ideas, restated them in the following manner:

> Primary consciousness can be defined as simple awareness that includes perception and emotion. . . . By contrast, secondary consciousness depends on language and includes such features as self-reflective awareness, abstract thinking, volition and metacognition. . . . Dreams that occur during rapid eye movement (REM) sleep have features of primary consciousness but . . . dreams do not strongly evince the characteristics of secondary consciousness. (e.g., Hobson, 2009a, p. 803).

Several earlier authors also worked in this psychoanalytical line of thought: Hartmann (1973) negated the self in dreaming completely and proposed absent sense of self during dreaming (p. 148). Rechtschaffen (1978) also proclaimed that since the non-lucid dreamer is not lucid, he or she is also non-reflective. Hobson (2009b) reemphasized this notion of static suspension by claiming that "most experienced dream reporters say that they decided nothing at all in their non-lucid dreams. Things just happened to them, spontaneously as it were. They did not have the dream as much as the dream had them!" (p. 81). Hobson's personal formulation is based upon mechanistically equating mind with brain (Hobson, Stickgold, & Pace-Scott, 2000a, p. 804); the author proliferates it as valid for all dreamers with claim that "self reflective awareness [about the state of consciousness], background-foreground monitoring, insight and judgment, all dissolve with this loss of my second self, my observing ego" (p. 20). In summary, according to Hobson, not only the hapless static dreamer loses high-order thought (working memory, volition, and self-reflective awareness, p. 80) and is unable to think, but the basic capacity for attention and noticing are lacking as well.

It is true that the sleeping individual is not sensory connected to awake reality (with the exceptions of beforehand arranged laboratory experiments with lucid dreaming, e.g., Dressler et al., [2012] or parents expecting to hear crying child). Thus it can be easily seen how sensory separation from awake world prompts the investigators, as Kahan (2001) highlighted, to think in terms of loss of reality as equaling to the loss of "rationality, order, and clarity" characterizing waking thought (p. 336). Question could be posed, Is *self,* and with it the thought processes, truly disappearing during dreaming?

Self Is Central for Consciousness and Cognition in Dreaming

Kahan and LaBerge (1994) posited that "dreaming consciousness is closer to waking consciousness than usually assumed" and that *self* is "central to what is thought of as human cognition" (p. 259). From developmental perspective of initiation of dreaming process, Domhoff (2003), working with research findings by Foulkes (1999), proposed that as the brain structures which support dreaming *per se* physically mature (in the absence of biological/organic impairment), then in parallel, the personal consciousness (understood as awareness of oneself) also appears. Thus, in an objectively agreeable way, Foulkes (1999) considers research of dreaming as method "to understanding the unfolding of consciousness and personhood" (p. 158). In 1999, the same author pronounced that "dream data . . . remain our best source of information on the development of inner (meaning a conscious) mental life" (p. 158). It stands to reason that this inner life necessarily includes observational cognitions starting with recognitions of at least some elements known to the dreamer from the waking life.

On a continuum, from the earliest age (3-5 years old, Foulkes, 1999), children are able to report their dreams and their dream narratives first consist of recognition of isolated wholly static images: The imagery might include the dreamer yet remains motionless (e.g., at four years and eight months of age, the dreamer recalled *"I was asleep and in the bathtub,"* Foulkes, 1999, p. 159). The earliest reports of dreams, even in the form of static imagery, nevertheless demonstrate, despite Hobson's (2009b) claim of loss of "my second self, my observing ego" (p. 20), the cognitive processes of attention and observation as cognitive achievement in dreaming state of consciousness. Further, remembering (recall) of dreams needs to be considered as cognitive accomplishment in the waking state of consciousness (both perhaps prompted by parents' appreciation for this kind of experience).

As children's neural structures become more mature (Domhoff, 2003), the narratives demonstrate more complexity of thought with descriptions of internal environments similar to the ones existing in the external world: Seven years old children could report dreams with "active characters and simple story lines" (Foulkes, 2017, p. 7). The following illustration demonstrates that the individual young dreamer (aged eight years, eight months) could also report feelings if prompted to do so by researchers: *"I could not get them down, and I was scared when I couldn't get down from the balloons,"* Foulkes, 1999, p. 163). Later on, children are able to use basic comparative/evaluative thought processes (e.g., Foulkes [1999]: *"It looked like it was in a school,"* in a report of a girl, ten years and five months old, in Foulkes, 1999, p. 163).

As can be seen from these examples, the start of cognitive developments of "dreaming person in his dreams" (Tholey, 1989, p. 567) is sequentially well documented: First, within the spontaneously appearing, uncalled for, and unanticipated content of dream, gradually, a child as an observer, secures its presence in the dreaming state of consciousness by noticing images and scenery and then becomes involved or engaged in dream circumstances. He or she includes oneself into the dreaming experiences by observing, experiencing, becoming aware of feelings, and using evaluative cognition when participating in behaviors and actions. The young dreamer is also able to have communicative exchanges with dream characters (e.g., *"And then I went home and asked my mom if I could buy it,"* in Foulkes, 1999, p. 163: the dreamer is eight years and eight months old; see also Kozmová, 2018a, b). The older dreamers' thoughts and action could even reflect a moral stance based on one's convictions about sanctity of life (self-preservation) or moral viewpoint on behalf of self or others (dream characters) in distress (Kozmová, 2015, 2017b, 2017c, 2018).

This presence of oneself in dreams deserves explication of Tholey's (1989) concept of "dreaming *person* in his dream" (p. 567) and subsequently, an individual dreamer's *self-awareness*. These terms will be explained first from the viewpoint of waking state of consciousness since most likely that is where cognition starts to develop. The abovementioned concepts also bear relevance to the scientific debate in studies of consciousness involving cognitive processes and their continuity or discontinuity in non-lucid dreaming.

Who Is the "Person" in Dreaming?

Regarding one's existence in times of wakefulness, without taking into consideration dreaming state of consciousness, Kohut and Wolf (1978) characterized an individual as a *person* in terms of "independent recipient of perceptions and a center of initiative" (p. 414). Atwood and Storolow (1984) usefully expanded this concept of *person* as "an experiencing subject and agent who initiates action" (p. 34). Gabbard (2004) also characterized *person* as an "agent or author of his or her own life" (p.154) which implies, in some way, of going about living one's life and one's involvement in it (and not just being hapless observer and receiver). Gusnard (2005) determined that organismic persons—the individual *selves in action*—are able to use a "capacity for intelligently strategizing their movements relative to their particular [internal and external] environment" (p. 680). In this case, the term *organismic* denotes "an individual constituted to carry on activities of life by means of organs separate in function but mutually dependent" (Merriam-Webster, 1988, p. 831).

It is worth mentioning that, most likely unknown to Gusnard (2005), Storolow, Brandschaft, and Atwood (1987) conceptualized *self*, in the experience distant terms, as "psychological structure" and in more experience near terms, as an aspect of "organization of [one's] experience."

It could be proposed that when making sense of ongoing experiences with regard to goal achievements, the mental activity includes some form of evaluation as part of executive skills defined as "the ability to plan and coordinate willful action in the face of alternatives, to monitor and update action as necessary and to suppress distracting material by focusing attention to the task at hand" (Jones & Harrison, 2001, p. 464). In this sense, organizing one's experiences implies and includes various degrees of self-awareness consisting of sensory and emotional awareness, use of one's thinking capacities (in no particular order) in terms of describing, reflecting, interpreting in terms of previous experiences or seeking to think in alternate ways in spite of it, evaluating, putting it in perspective, seeking meaning, and maybe even transcending experiences and transcending one's self. This whole organizing process includes creating a story in terms "where," "when," "what," and "who" (Kahan & LaBerge, 1996, p. 237).

In experiential terms, organization of one's experiences also implies putting one's subjectivity in language words and labeling one's internal states and communicating, if needed, with external environment. Thus in far reaching consequences, this internal work could be considered motivational self-regulating state which serves as a prerequisite for volitional strategizing including planning and decision making (unless under duress) in which one' exercises capacity for executive skills. The question could be posed, could the individual dreamer during dreaming continue to be this *cognitively intact self* as known from waking life?

Experiencing during Dreaming and Self-Awareness in Primary and Reflective Awareness

In the science of dreaming consciousness, the dream reports (narratives) and interviews (e.g., Kozmová & Wolman, 2006; Wolman & Kozmová, 2007) about dreaming subjective experiences represent an essential venue for explorations of and potentially making predictions about unobservable, invisible, and imperceptible internal (mental) processes of individual dreamers. *Self-awareness* could be considered one of such cognitive processes.

There are several conceptual precursors to Kozmová and Wolman's (2006) investigation of *self-awareness* in non-lucid dreaming. From waking life psychology, as Freud (1965) explained, internally, one could become conscious of one's own person experiencing in subjective manner as an "I" and consider this experience as a person who thinks about "me." Farthing (1992) divided this combination of cognitive processes into (a) *primary consciousness"* as a mixture of "percepts and feelings, and thoughts and memories rising in direct response to them" with these elements of consciousness occurring in real time; and (b) *reflective consciousness* with focus on "thinking *about* one's direct subjective experience" (p. 13). This *reflective* type of thought brings forward "the self in an act of knowing" (Damasio, 1999, p. 130).

In the studies of dreaming consciousness, precursor to investigation of *self-awareness* (Kozmová & Wolman, 2006) became Kahan's (1994) analogy to primary and reflective consciousness, namely dreamers' "unreflected" and "reflected" experiences (Kahan, 1994, p. 177).

In combination with the abovementioned ideas about types of thinking, Kozmová and Wolman's (2006) study of *self-awareness* in dreaming tested May's (1967) waking life postulate that "the human dilemma is that which arises out of a man's capacity to experience himself as both subject and object at the same time. The important point is a process of oscillation between the two" (pp. 8-9). Based upon this conclusion about human condition, Kozmová and Wolman hypothesized that dreamers are aware of themselves not only through their own recognizing, feeling, thinking, and acting person, but also as a thinking reflecting persons in a manner similar to self-awareness during waking state of consciousness despite the fact that they, as dreamers, remain separated from the external awake world.

Kozmová and Wolman's (2006) goal in their research of *self-awareness* in non-lucid dreams with retrospectively-based interviews with 91 volunteers (M=49, age range 19-69 years; F=42, age range 20-80 years) was to map out existence of *self-awareness* and its varieties. The authors conceptualized processes of *primary awareness* as dreamers' "attention focused on internal psychophysiological states during which an individual recognizes, through memory, cognition, perception or feeling, his or her own person acting or experiencing" (p. 199). In this phenomenological awareness in terms of actual experience, the authors then inferred qualitative aspects of primary consciousness (similarly to Schwartz's [2000] definition of phenomenology as systematic inquiry into qualitative elements of dreamers' mental experiences, p. 58). By researcher (MK) asking questions about interviewed individual's existence ("Are you in your dreams?") and ways of recognizing oneself ("How do you know it is you?," p. 204), the following exact host of cognitive processes by which dreamers recognize themselves and their personal existence in non-lucid dreaming state of consciousness was brought forward by now awake dreamers (p. 208): Perceptual (visual experience and visual self-representation, auditory, kinesthetic); experiential (being self-intuitive, experiencing emotions, feelings, events); cognitive (use of inductive reasoning, being aware of one's own thinking, maintaining waking perspective, recognition of self in transformation); and by memory-based recognition (recognition of one's own characteristic either mental, physical or auditory; and recognition of locality and people).

In terms of functional awareness (the manner in which the dreamer exists in the dream based on level of activity and engagement), the participants reported their own absence, or ability to observe the scenery (which "may or may not include the dreamer"), or being active and interacting with dream characters (Wolman & Kozmová, 2006, p. 208).

Kozmová and Wolman (2006) further defined *reflective awareness* (known as second part of self-awareness in waking life and similar to secondary consciousness with abstract thinking [Hobson, 2009a] that contains within it processes of higher-order cognition [Farthing, 1992]) as "instances in dreaming or waking in which an individual verbalizes assessive or evaluative thoughts about ongoing internal events (perceptions, feelings, memories, images, hallucinations, and such) and/or external events (actions, behaviors, situations, environments)" (p. 200). Kozmová and Wolman assumed and found support for notion that within the dreams, the basis of the oscillatory change from primary to reflective awareness could be initiated by dreamers' experiences of positive or negative emotions and their verbal assessment or evaluation (e.g., *"I said, this is fantastic, I am really having fun,"* p. 205) or by noticing personally meaningful oddities (e.g., in participant's dream, her parents overtook the world, forced their children to shave their heads, and made them work. The following internal commentary: *"Where is my world? Get me out of here!"* p. 209, illustrates internal evaluation).

In Kozmová and Wolman's (2006) study, 8.2% of male participants (out of N=49 males) and 9.5% of female participants (out of N=42 females) reported occurrence of verbalizing assessive or evaluative thoughts based upon experiencing positive emotions (e.g., *"I remember thinking it was nice to have a new talent");* 32.6% of males and 16.% of females noticed negative feelings (*"I kept telling myself, this is ridiculous, I work so much during a day and now I have two jobs,"* p. 209); and 12% of males and 19% of females reported noticing of personally meaningful oddities (*"Mother, how can you speak to me, you are dead!,"* p. 209) relevant to their waking life or their sense of self. In total, 53% of males and 45% of females reported some variation of instances where experiences of feelings and noticing of oddities evoked evaluative thought.

Based upon Kozmová and Wolman's (2006) results, it could be proposed that for many dreaming individuals the *sense of self* brought about by experiencing of feelings and personally-defined oddities remains intact despite sensory disconnection from external world. Furthermore, it is noteworthy that this oscillation between direct experience and thinking about the experience in terms of its evaluation could occurs as early as age seven (Foulkes, 1999) and even in the nightmares (Kozmová, 2019).

When looking closely at the processes of *self-awareness* associated with oscillating *reflective awareness,* one can notice the difference between Kahan's (1994) and Kozmová and Wolman's (2006) distinct conceptualizations of *reflective awareness* and highlight a potentially major impact on understanding the dreamers' existing or developing/emerging capacities in non-lucid dreaming. While Kahan (1994)in her definition of *reflective awareness* focused on any "ongoing thoughts, feelings, or actions" (p. 177), Kozmová and Wolman, in addition to any thoughts, feelings, or actions about ongoing situation emphasized volitionally-based "assessive" and "evaluative" thought processes evoked by experience of positive and negative feelings and noticing of personally meaningful oddities. It could be proposed that this *evaluative/assessing* skill implies and takes into consideration that dreamers remain aware of their personal beliefs and values and that their *self, sense of self, sense of historical and personal time,* and *memory* remain intact during non-lucid dreaming. This ability to be oneself during dreaming implies reassurance of continuity of *sense of self* (which includes remembering individual identities and roles [e.g., *"There is silence in the room. Then I scream. I realize that the red wine is Christ's blood, but I am a Jew. I scream "I am a Jewess!"* Kozmová, 2008, p. 129]).

Kozmová and Wolman (2006) concluded that "the exact same mental processes used to determine the experience of the self in waking are also used during dreaming," and that these processes are isomorphic (p. 213). The authors further argue that dreaming self remains intact:

> It is our contention that one's sense of self persists and is maintained in both the waking and dreaming states. This notion is reassuring insofar as we do not have to think of our "selves" dissolving or disappearing when we sleep only to reappear or reassemble upon waking. The variety of thought and perceptual activity during sleep, subject to extensive combinations and permutations, does not require the dissolution of the self, which is a constructed reality, built proprioceptively, developmentally, and experientially during life. We further contend that self-awareness in dreaming appears to obey the same procedural requirements as self-awareness in waking, namely, the use of perceptual, experiential, cognitive, and memory-based data to verify, reinforce, and preserve the subjective sense of the existence of self (pp. 211-212).

It is noteworthy to mention then it is also possible, from therapeutic standpoint, to notice in dreams times of dreamers' periods of greatest vulnerabilities and even shattered sense of self (Kohut, 1977) and therapeutically reinstate the processes of organizing the self and ownership of self-awareness in times of wakefulness. In this case, the cognitive elements of consciousness in dreaming would directly and fruitfully, in cross-state fashion, inform organizing of one's experience in wakefulness.

Theoretical Underpinnings and Research of NREM Sleep Cognition

The investigations of mental activities in NREM sleep has been from beginning of its distinction from REM sleep periods influenced by Freud's

(1956) theory of dream formation. Freud proposed that dreaming is an unconscious state in which there is no mental activity one could be aware of and later report to someone till the latter parts of night when dreamer is closer to awakening (p. 281). In his opinion, only then the consciousness could become connected to dreaming content (p. 576).

As a valid reason for this cognitive "unconsciousness" in dreams from earlier parts of night, Freud (1956) considered dreaming to be affective in nature and motivated by drives of sex and hostility (aggression) based upon earliest experiences of one's parental relationships. In his drive theory, in dreams, these now unacceptable associated affects and wishes were mechanistically directed toward satisfaction regardless of the adult's reality occurring outside of the sleep. He assumed, because of need for concealment of these objectionable thoughts from oneself during times of wakefulness, that unconscious developmentally immature *primary process thinking* based in mechanisms of condensation, displacement, and symbolization takes over the dreamers' minds. Freud thought that when dreamer regresses into *primary process* during dreaming with its distortion, disguise, and repudiation of personally "inappropriate" ideas from consciousness, the ego control over one's thinking becomes "suspended." In this state, during the majority of the time of night, dreamer could get the desired fulfillment of wishes with immediate sensory/affective satisfaction without any monitoring from one's conscience or from external world.

For dreamer, as Freud (1900/1990) assumed, *displacement* acts as means to minimize the conflict and tension associated with feelings of anxiousness and being overwhelmed by specific thoughts. As explained by McCullough, Vaillant, and Vaillant (1986, p. 69), the adaptive mechanism of *displacement*, with or without conscious awareness, serves to redirect and express "affect, impulse, or action" toward less emotionally-arousing substitution that nevertheless is similar to original object [e.g., person, idea, etc.]. The mechanism of *condensation*, according to Gabbard (2004), works with combining one aspect of one's life (wish, feeling, impulse, idea) into "one manifest dream image" (symbol) with unconscious goal to disguise feelings associated with several different individuals assembled into that one image (p. 118). As can be seen, these mechanisms are impersonal without dreamers' input. Under these circumstances, as Freud (1900/1990) asserted, the *secondary process* (conscious logical reasoning and decision-making redirecting these supposed wishes toward more acceptable forms of rational mental functioning) is "suspended": Within dreaming, with lack of external and internal monitoring, the feelings that could not be during day become expressed in a motoric manner (through inappropriate physical or communicative behavior), became expressed in a perceptual-hallucinatory manner during dreaming.

Many of earlier investigators of NREM mentation (e.g., Foulkes, 1962) have been influenced by Freud's (1956) claims about unconscious processes occurring during earliest phases of night and his quest for origins of dream formation. Generally, participants in earlier studies in which researchers proposed isomorphic relationship between REM sleep and its recorded physiology (e.g., Asserinsky & Kleitman, 1955) were able to recall dreams from REM sleep periods. Thus the investigative focus shifted to NREM sleep periods to see whether there could be any mental activity going on during NREM sleep and thus confirm or refute Freud's (1956) claim about the lack of mental activity in earlier part of the night.

Foulkes's (1960, 1962) research about dream reports from different stages of sleep could be considered an important milestone in research of cognition in NREM sleep stages since it clearly demonstrated how pervasive Freud's lack of appreciation for dreamers' conscious cognitive activities during dreaming has become.

Foulkes (1962) tested hypothesis of isomorphic correspondence of ocular, EEG, and mental conscious activities in NREM and REM stages of sleep (p. 24) with goal to find out "formation" of dreams occurring in the first REM sleep period of the night. His eight male participants, ages 17-27, either employed or in school, spent in laboratory total of 57 nights and were woken up with loud bell and experimenter calling dreamer's name to make sure participant is awake (except one of the participants woken up only by experimenter's voice). Experimenter asked each of them if "anything was 'going through his mind'" because Foulkes (1960) realized that participants themselves do not label all cognitive and perceptual occurrences (upon asking) as "dreams" (Foulkes, 1962, p. 15). The participants reported "dream" or "thought" or anything else that was going through their minds before they heard the bell. Upon completion of report, experimenter asked questions about settings and their familiarity in dream, presence and involvement of other people (familiar or not), dreamer's presence and his own feelings and emotions and emotions of others, whether the now-awake person was dreaming or "what was happening before the bell rang," whether the dream was visual, any conversation and dreamer himself conversing, scenes' shifts, awareness of dreaming, clarity of imagery, coherence of dream at the time of its happening, time of its occurrence, colors in dream, estimate the length of dreaming, existence of other dreams or "anything else" even if dreamer could not recall it, and dreamer's and others' engagement in physical activity (p. 16). Participants reported the dreams or "thoughts" and were questioned about them in detail. In the morning, participants again were asked to recall all dreams and filled out rating form for each dream (*Dream Information Survey*) that focused, without separating dream life and wake life, on activity in dream, control, emotions, anxiety, clarity of content, distortion, relation of content to participant personal life, pleasantness-unpleasantness, frightening and dramatic quality, activity-passivity of oneself, and degree of sense-making out of dream (p. 16).

Foulkes (1962) classified reports with visual, auditory, or kinesthetic imagery as "dreams"; in addition, if the dreamer reported he experienced another identity as himself or found himself in another place than laboratory, the reports were also termed dreams. Foulkes categorized as "thoughts" occurrences reported without imagery or with identity different then dreamers' own, or occurring within laboratory settings. From REM sleep, from 94 reports, 89 of them Foulkes considered dreams, and 5 of them thinking reports; from NREM sleep, from 100 reports, 73 he considered dreams, and 27 of them thinking reports. Qualitatively, if extrapolating from his reports of qualitative characteristic of REM and NREM dreams (p. 19, Table 4) in terms of mental processes, one could considered ratio of "formal differences" in dreams that Foulkes found between REM reports and NREM reports in the following terms: Reports of emotions dreamer experienced (in REM, N=88 and in NREM, N=95), physical movement of oneself (N=73 and N=57), and of others (N=66 and N=51).

Interestingly, even though Foulkes (1962) collected data about actual cognitive processes—conversations by dreamer or dream characters—which is considered an evidence of dreamers' ability to think (Wolman and Kozmová, 2007, Kozmová, 2017, 2018), Foulkes did not report this data.

Foulkes (1962) also differentiated between "organismic involvement" of dreamer in dreams; he understood it as reports of emotions of self and others, anxiety, visual scenery, physical activity, and reports of violence-hostility (p. 21). The least "organismic involvement"—in his perspective, the dreamers' *thinking*—he found to be "almost exclusively outside REM" sleep periods (p. 20). In his words, reports from REM sleep demonstrated more "organismic involvement in affective, visual, and muscular dimensions and were more highly elaborated than NREM reports" (p. 24). Foulkes concluded that his research did not confirm earlier prediction about "suspension of consciousness" during sleep as Kleitman and Asserinsky (1957) suggested at that time.

One unfortunate occurrence stemming from Foulkes (1962) research is that later researchers continue to call NREM sleep dreams as more "thought-like" (for review, see Nielsen, 2000) with unreported caveat that frequent awakenings through the night in laboratory conditions evoke in sleeping persons more thought-like occurrences which would be, in different circumstances, dreams, p. 863). Participants in others studies, however, continuously reported dreams from NREM sleep stages that are, from researchers' standpoint, undistinguishable from REM sleep reports (e.g., Strauch & Meier, 1996).

In later report from the same research (Foulkes, 1962), Foulkes (1964) stated that participants produced "reports of mental phenomena from NREM sleep almost as often they did from REM sleep" (p. 240). Since dreamers themselves labeled NREM reports more often as "thoughts" then those reported from REM sleep, and content of NREM reports was focused on "recent and everydayish activities," Foulkes considered them less distorted (p. 241). This finding contradicted Freud's (1956) idea of dreams from the first part of the night as disguise. Foulkes (1964) considered REM sleep to be filled with more distortion and secondary elaboration (revisions, during wakefulness, as attempts to make out of a dream a story). This notion of *secondary elaboration/revision* occurring while awake, however, needs to be distinguished from dreamers' *thinking* during dreaming as higher-order cognition (e.g., Kozmová, 2012) and from Hobson's (2009a) *secondary consciousness* defined as cognitive processes which include logic (p. 21). Hobson (2009b) broadened *secondary consciousness* as "subjective awareness including perception and emotion that is enriched by abstract analysis (thinking) and metacognitive components of consciousness" (p. 803). It needs to be pointed out, however, that despite Hobson's (2009a, 2009b) frequent objections about lack of thought in dreaming, non-lucid dreamers are capable of using processes of secondary consciousness in both *descriptive* or *problem-solving dreams* (Kozmová, 2012, 2015, 2017a; Wolman & Kozmová, 2007).

Foulkes (1964) also openly highlighted that even though Freud (1956) freely admitted the existence of dreams with mentation and without condensation, distortion, dramatization, and wish fulfillment, he stopped thinking about this type of dreams (p. 331). As Foulkes (1964) critically pointed out, this theoretical position that only toward morning the consciousness "attracts" developed dream and thus dreamer can become conscious only in REM sleep dreams, remains "misleading characterization of sleep consciousness" (Foulkes, 1964, p. 243). It could be concluded that mentation in the sense of dreamer having his own thoughts and thinking capacities continuous with awake life and thus preserved in both sleep stages will require, from scientific community, a direct acknowledgment that the consciousness in dreams is of a very personal quality: It includes dreamers' sense of self; his or her capacities to think for oneself; to display moral stance, beliefs, and values; and use of his or her thinking capacities when it matters or when the dreamer's high-order cognition and executive skills are emerging or are already developed.

Arkin, Antrobus, Toth, Baker, and Jackler's (1972) study is an example of how thinking within the scientific community slowly changed from mechanistic approach to dreaming mind that highlighted processes of condensation and displacement to considering dreamer as maintaining a personal sense of self and cognitive abilities.

Given that the main pull of research of NREM dream reports started by trying to find contradictory support for Freud's (1956) hypothesis that consciousness is not part of sleep in the first NREM dreams after falling asleep, the researchers' work within the area of cognition (e.g., Arkin et al., 1972) was later on focused on further evidence that dreams with individuals' awareness are occurring with NREM sleep.

Arkin et al. (1972) wanted to know, from analysis of 74 pairs of dream reports (one from vocalization, one silent) from 23 sleep talkers, whether there are differences in mental content of NREM dream reports if participants are woken up within the NREM without vocalization ("silent NREM") and also from NREM with participants' sleep utterances. The authors considered sleep utterances as representing "stream of ongoing NREM mentation" (p. 434) which some dreamers might vocalize while asleep.

Arkin et al. (1972) hypothesized that the spontaneous utterances would be "objective indicator of the presence and *content* of NREM mentation" (p. 427) and that these utterances do not reflect merely "transitory arousal states *unrelated* to previous sleep mentation" (p. 428). The authors thus speculated that this vocalization of dreaming experiences would then support the notion that NREM dreams do indeed contain mentation. Each report was scored for one of the following items: Participant recalled his vocalizing during sleep experience; or recalled someone else vocalizing; or recalled oneself or someone else anticipating vocalization; or recalled mentation of references to written or printed words, or to sound; or no vocalization, words, or sounds (p. 431). Authors concluded that there is a "strong degree of association between sleep speech and mentation reports in which the subject [in the current scientific consciousness he would be called participant] remembers himself as vocalizing in an imagined sleep experience" (p. 432). Arkin et al. also stated that both reports ("silent" NREM and vocalizing NREM) contain recalled mentation content and concluded that "experimenter-observed vocalization is associated with the subject having an imagined experience of vocalizing during sleep" (p. 433). The authors' important assertions are that for varieties of speech to occur, three conditions—two personal and one mechanistic—need to be satisfied:

1) the need to assert oneself or enhance the sense of self-existence and/or the presence of an emotion of proper intensity, and

2) a breach of motor outflow thresholds for vocalization due to diminution of threshold and/or increase of motor discharge. (p. 434)

Arkin et al. (1972) further assert existence of "imagined vocalization" as an "inner dialogue" which they assumed serves the purpose (function) of "preservation of the subjective sense of self" as well as preparation for readiness for the awakening (p. 435). It could be proposed that Arkin et al. acknowledged the dreamers as experiencing some form of personal concerns in the actual dream in contrast to Freud's (1956) notion on dream as a disguise.

In the research in which Fosse, Stickgold, and Hobson (2004) investigated assumed changes in thinking within NREM and REM sleep, the authors hypothesized that NREM sleep dreams will have more thinking and REM sleep dream more imagery throughout the night. In this research, for 14 nights, 16 undergraduate participants 19-26 years old (eight males and eight females), slept at home and experienced instrumental awakenings from Nightcap sleep monitoring system by beeping in all sleep stages. The 14-day research protocol included selected nights for awakenings and days with paging for wake mentation reports. Participants were instructed to "remember what was going on in your mind prior to your being beeped, i.e., anywhere up to fifteen mines before the beep" (p. 299). The questions they were answering upon awakening or beeping were about location, presence of others, activities, seeing, thinking, feeling, and happenings around (p. 229).

Fosse et al. (2004) guided their research by division of reports into "thinking" and "hallucinating." Thinking in a form of *directed thinking* was conceptualized as "any continued mental effort or occupation, including contemplating, reflecting, and evaluation, as well as attempts to decide, figure out, grasp, and plan" (p. 229). Hallucinations were defined as "internally generated (endogenous) sensations in any sensomotoric modality, most commonly a visual, auditory, or kinesthetic" (p. 299). Participants reported 229 REM and 165 NREM narratives from 88% of

REM awakenings and 66% of NREM awakenings. Judges scored reports as directed, conceptual thoughts; directed thoughts; images; images/hallucinations; and hallucinations (p. 299). Inter-rater reliability was achieved at 66%-75%.

Fosse et al. (2004) reported that incidence of hallucinations was in REM reports 82% and in NREM reports 60%. Directed thoughts were described in 38% of NREM and in 22% of REM reports. The authors reported no changes in directed thinking during subsequent REM sleep reports through the night and that with increasing nighttime, NREM reports of hallucinations increased but thoughts decreased.

With regard to Fosse's et al. (2004) research, on a conceptual level, it could be argued that not every dream reports shows a content that requires or inspires directed thinking: For example, *descriptive dream* is conceptualized as dream that "contain the descriptions of scenery, the situations, the actions of the dreamer or other characters, and his or her observations or experiences within that dream" (Kozmová, 2008, p. 72). In this type of dream, the dreamer either does not have yet developed thinking capacities for use in dreams or is not inspired to think in direct mode. Also, not in every dream the dreamer is experiencing immediate problems or concerns which might also feel to dreamer that some thinking effort is necessary. But the main objection regarding Fosse's et al. research could be toward characterizing dreams as "hallucinatory" because they do not contain directed thinking; this point needs to be emphasized especially if the content of dream (including subjective experience of dreamer) does not require thought processes or the dreamer is not yet able to use them.

Theoretical Underpinnings and Research of REM Sleep Cognition

Despite Snyder's (1970) research of dream narratives in which he found that dreamers are able to use "inferential thinking, remembering, deciding" (p. 139), theorizing about dreamers' cognitive processes in REM sleep dreaming has been consumed by efforts to demonstrate lack of dreamers' thought processes. The supposition about dreamers' specific changes in mental capacities has been perpetuated by dedicated efforts to prove that physiology of dreaming is *isomorphic* or concomitant (in one-to-one relationship) with its formal psychological aspects, in this case, thinking (Hobson, 1992; 1997).

The theories that came about cognitive processes in dreaming highlighted correspondence between mental activities and brain processes (Hobson & McCarley, 1977, p. 1335) and downplayed the possibilities of dreamers direct intellectual engagement with dreaming experience. For example, for Hobson (1992), regarding cognition, the amount of an input source—external sensory data—was directly relevant to the level of cognition that is possible in a dreaming state. As I already mentioned, in dreaming, there exists a predominant sensory disengagement with external reality; and again, Hobson et al. (2000a) equaled the lack of contact with external sources during sleep to lack of cognition in non-lucid dreams and left cognitive abilities to occur and be continuous with wakefulness only during lucid dreaming (Hobson & Schredl, 2011).

Kahn and Hobson (1994) concurred with the same deficiency theme in their *self-organizing theory of dreaming* in the claim about dreamers' "inability to direct either . . . thoughts or actions in dreaming" (p. 2). Regarding executive skills, the authors argued that dreamers are without choices within the course of the dream because the "system itself 'decides' which of many possible behaviors it will adopt" (p. 11). Obviously, in this situation, the consciousness was relegated to impersonal system and left the person of the dreamer without recourse. Kahn, Combs, and Krippner (2002) further declared that brain in the state of dreaming exists with "weakened volition, reduced logic, and diminished self-reflection" (p. 311) and that the mind in sleep is not able to reach for "planning and volitional actions" (p. 315). Again, the person of the dreamer is not considered in this viewpoint: The authors claim that the individual dreamer experiences "loss of a continuous objective sense of self" (p. 318).

One of the continuous reasons for this negative stance toward dreamers' thinking in non-lucid dreaming are the earliest neuroimaging studies (e.g., Braun et al., 1997, Maquet et al., 1996) showing dorsolateral prefrontal cortex (DLPFC), as seat of executive skills in wakefulness, in REM sleep dreaming not activated to the level of wakefulness. In that case, Hartman (2000) proposed that dreamers lose self-reflection and stated that substitution for it is "total therenesss" (p. 184); he meant that immersion within the dream precludes dreamers' from "thinking, planning, and organizing" (p. 185). Stickgold (2000) claimed that dreaming content needs to be considered "totally determined by physiological processes" (p. 6). By inference, theorists decided that deactivation of DLPFC (e.g., Braun et al., 1997) as s crucial neural structure important for executive thought in waking would mean "lack of directed thought" (Hobson et al., 2000a, p. 842) and thus "deactivated ego function" with its aspect of acting voluntarily (Hobson, 2007, p. 75).

Interestingly, Ioannides, Kostopolous, Liu, and Fenwick (2009) found, by using magnetoencephalography, that in REM sleep there is activated left DLPFC. The authors also highlighted the methodological challenges of earlier PET studies conducted, for example, by Braun et al. (1997): Ioannides et al. state that the crucial "dorsal areas, like DMPFC . . . were outside of field of view of early PET scanners" (p. 465). Yet, the early methodologically insufficient neuroimaging findings remain unquestioned and are predominantly referred to, and taken as a gold standard for theoretical predictions and explanations about cognition including executive functions in REM sleep non-lucid dreaming (e.g., Domhoff, 2018; Maquet et al., 2005). Even though Dawson and Conduit (2011) emphasized that if any neuroimaging findings as data are interpreted retrospectively, the theorists and researchers need to make explicitly known "a network of assumptions" including suppositions, explications, and clarifications about neuroimaging methodology (p. 150), this assertion is not heeded (e.g., Domhoff, 2018).

Self-Reflectiveness

Purcell et al. (1993) in their *Dream Control Scale Categories in Abbreviated Form* in the process level represented what kind of role(s) is dreamer capable of playing in the dream content. It needs to be emphasized that dream content, for the most part, unless dreamer is personally engaged in responding, is uncalled for and unsolicited even though the dreamer might incubate a dream (e.g., White & Taytroe, 2003); the content of incubated dreams is possible to use during wakefulness.

The following levels from Purcell et al.'s (1993) *Dream Control Scale Categories* are directly important for investigation of thought processes in dreaming: Dreamer observes and reports but does not participate; the dreamer is present and observes, and expresses personal response to situations or events or even may think about the events; dreamer is engaged, but ineffectively; the dreamer is impactful and more successful in the efforts; the dreamers is not displaying the full capacities to deal with situation; the dreamer is in full control of his or her capacities and the dream situation, and might be emotionally present and expressive (has a control "in" the situation); the lucid dreamer has a control "over" part of whole dream and is able to manipulate dream mechanics and state explicitly the awareness of dreaming; and dreamer is able to use dream lucidity to "exerts or attempts to exert control over dreamed events or outcomes" (p. 219). Categories in Purcell et al. scale outline the multiplicity of ways a dreamer might be able to become engaged in dreaming.

At the same time, Kahan (1994) introduced similar approach to study of consciousness in dreaming, specifically cognition. Instead of focusing on deficiency of cognition including high-order cognitive skills or correspondence of mental activities with presumed physiological underpinnings, she centered on evidence that *reflective awareness*, defined as "the awareness of one's thoughts and action and the deliberate direction of them" (p. 177), exists in dreaming (Purcell, Mullington, Moffitt, Hoffman, and Pigeau, 1986).

Moffit et al. (1988) considered *reflective awareness* as one aspect of self-reflectiveness characterized as "self-organizing and self-regulating processes [that] occur within dreaming as well as waking" (p. 431). In the hindsight, it could be proposed, that this set of processes conceptualized by Moffitt et al. and elaborated in Purcell et al. (1993) could be considered precursors to research of *executive skills* later conducted by Kozmová (2008, 2012, 2015, 2017a, b, 2018; for all currently known terminology of high-order cognitive skills in dreaming, see also Table 1 in Kahan & LaBerge, 2011, p. 495).

Kahan (1994) utilized two different approaches to study of self-reflectiveness: (a) Evidence with *Dream Self-Reflectiveness Scale* (SR, Purcell et al., 1986) with a third person measure; and (b) participants' own "rating of phenomenal qualities of their dream experiences" (Kahan, 1994, p. 179). SR contains, in addition to different points of view (observer, participant, noticing of oddities, control, intentionality, lucidity) also cognition in terms of "dreamer thinks over an idea" (p. 179).

In Kahan's (1994) research, undergraduate participants, 18 females and 12 males, age range 18-24 years, wrote for duration of three weeks dream journal five times a week; total of dreams was 239 reports and these were scored with SR scale by method of third person scoring and differently by participants using *Dream Rating Scale* in which they evaluated their experience with focus on control (others' action, object characteristics, actions, own thoughts and feelings, dream environment, and other, p. 183). The area of cognition in terms of "dreamer thinks over an idea or has definite communication with someone" (p. 181) became one of the most frequently occurring scores (43%). With this result in high level of self-reflectiveness in terms of awareness and control of own thoughts and feelings, Kahan concluded that early hypothesized "single-mindedness" in dreaming (Rechtschaffen, 1978) so highly prized in theorizing about mental activities in dreaming, underestimates frequency and types of cognition in dreams. Kahan also suggested that "reflective awareness and behavior regulation is separable dimension of dream cognition" (p. 190).

In the research subsequent to investigation of *reflective awareness* in dreaming (Kahan, 1994), Kahan and LaBerge (1996) compared the incidence of cognition and metacognition in dreaming and waking again with the method of comparison of first and third person ratings. In this study with 43 participants, 22 males and 21 females, undergraduates, ages 18-37, who recalled dreams 3-5 times a week, 6% indicated remembering lucid dreaming in past six months. With *Metacognitive, Affective, and Cognitive Experience* (MACE) questionnaire developed prior to the study (Kahan, LaBerge, Levitan, & Zimbardo, 1997), the researchers measured incidence of "choice, internal commentary, sudden attention, focused attention, public self-consciousness, emotions, self-reflectiveness, and thwarted attention" (p. 239).

It could be proposed that MACE questions (Kahan & LaBerge, 1996) point to possible presence of several cognitive processes dreamers might have used, for example: "At any time did you choose between alternative action after confirmation of options?"; "Did you internally comment on any event, or wonder about anything?"; "Did you focus for a period of time on accomplishing a particular objective?"; or "Did you think about your own thoughts, feelings, attitudes, motivations, or behavior?" (pp. 248-249). Participants used the questionnaire for analysis of incidence of particular elements in both dream reports and sampling waking experience. Narrative reports were also scored by independent scorers with removed identification of dreaming or waking reports. In finding particular incident relevant to questions, scorers marked it as "yes" with inter-rater reliability of 86%. Interestingly, the highest rated processes were incidences of internal commentary equally often occurring in both dreaming and waking and the lower incidence of choice in dreaming in comparison to waking reports. Self-reflection occurred commonly in both states of consciousness. Based upon their findings, the authors proposed that "dreaming and waking cognitive and metacognitive processes are more similar than different" (p. 244).

In Kahan and LaBerge's (1996) research, the interesting metacognitive processes could be considered "the ability to divide one's attention (e.g., between ongoing events) and one's reaction to them; to focus one's attention on task or goal, and to be self-reflective," p. 246). The authors concluded that both dreaming and waking states of consciousness are characterized by the same range of cognition and metacognition. In other words, it could be emphasized that dreamers do retain and use some their waking life cognitive abilities.

REM Sleep Thought Processes Are Rational

Precursor to later research of executive thought processes (Kozmová, 2008) was Wolman and Kozmova's (2007) research inspired by their findings that self-aware dreamers are able to recall their thoughts from dreams (Kozmová & Wolman, 2006). Despite earlier predictions that dreamers are not able to sustain *directed thought* (Hobson et al., 2000a, p. 842; *directed thought* later detailed as "any continued mental effort or occupation, including contemplating, reflecting, and evaluating, as well as attempts to decide, figure out, grasp, and plan," Fosse, Stickgold, & Hobson, 2004, p. 299), Wolman and Kozmová (2007) focused precisely on the quality of thinking processes. The authors decided to investigate hypothesized variety of rational thoughts in dreaming. Since in their study of five males (age range 22-58 years) and five females (age range 25-49 years), there were no prior arrangements made for eliciting dreams (for 14 consecutive days) from any particular stage, as Meier and Strauch (1996) concluded, dreams spontaneously recalled at home are considered to originate from the latter REM sleep periods.

Wolman and Kozmová (2007) operationalized *rational thought* as "a mental process that utilizes an individual's internal logic and is based on an idiosyncratic belief system. This mental process intervenes between sensory perception and the creation of meaning, and leads to a conclusion or to taking acting" (p. 841). The authors explained that the rational process as part of consciousness begins when an individual whether in waking life or in dreams "registers sensory perceptions in the widest sense of the term" (both internal and external, memory, and any other percepts that an individual registers through awareness, p. 841). Then, as the authors continued their explanation of rationality, the individual uses his or her own experiences again in the widest sense of term (e.g., available beliefs whether personal or societal, memory, self-knowledge, recognition of emotions and feelings, habitual thought, one's social role) and combines it with reasoning rules (e.g., "associations, connections, compare and contrast reasoning, combinations; conditional probability and exclusivity; evaluations, inferences; analogies; interpretations; extrapolations; intuitions; and induction and deductions" (p. 841; based in part on Harman, 1995; Sternberg & Smith, 1988). The rational process ends when the person or dreamer arrives at satisfactory or unsatisfactory and reasonable or unreasonable conclusion which may or may not lead the person to action taking (p. 841).

Based upon operationalized *thought unit* ("any part of a reported dreaming narrative that is coherent, and analyses, explains, or elaborates the descriptive dreaming experience," p. 844) Wolman and Kozmová (2007) randomly selected five though units form each dream suitable for scoring from total of randomly selected 40 dreams (with 454 thought units). Thus from each participant (10 total) they selected four dreams from which they randomly selected five scoring units. Two independent judges scored 200 delimited thought units with inferences about "predominant internal organization of thought" (p. 844, see also Smith, 2000) based upon 232 agreed-upon codes from *Code Book of Rational Thought Processes* (Kozmová & Wolman, 2005) with 91% of agreements between independent scorers.

The results of the inferential analysis in Wolman and Kozmová (2007) demonstrated that dreamers are capable of using eight different categories of thought processes: Analytical, executive, subjective, memory and time awareness, perceptual, affective, intuitive/projective, and operational (for descriptions of thought processes, see Table 1, p. 845).

Examples of these rational thought processes from Wolman and Kozmová (2007) research are: "*I express amazement as I had expected a simple civil proceedings and have no knowledge of a criminal case* (analytical); *I saw that everyone in the room was crouched down on the floor"* (perceptual-visual); *"I see a map of Europe sketched on it"* and *"When September 1 came, the woman said: 'This is the end of summer"* (memory and time awareness, p. 846); *"I feel scared, sad, mad, and quite confused by this carnage"* (affective); *"I was attempting to learn the stroke with them"* (executive); *"My friend, the nurse I work with, is already at the dance"* (subjective); *"It seems that he is not going to leave soon"* (intuitive); *"the hole is about 40 cm in diameter"* and *"I read about it in newspaper"* (operational with illustration of measuring and reading abilities, p. 847). The authors expressed appreciation for complexity of dreamers thinking and concluded with assertion that thought processes are directed and rational and not hallucinatory or bizarre as earlier authors (e.g., Fosse et al., 2004) claimed.

Executive Processes Are Attenuated but Not Suspended in Dreaming

Kahan and LaBerge (2011) revisited similarities and differences in cognition between dreaming and waking states of consciousness; this time they focused on high-order cognition. The authors espoused "alternative view that dreaming is continuous with waking" (p. 3). They divided continuities in terms of content (what") and cognitive and perceptual processes (the "how" of subjective experience) meaning that the same cognitive processes of waking are operative in dreaming. They agreed with Cavallero and Foulkes (1993) that "generating of dream experience involves the same cognitive and perceptual systems operating during waking" (p. 133).

In this replication of their study from 1996, Kahan and LaBerge (2011) added experience samples from late-night REM sleep and waking (see protocol in Kahan, 2001) with 6 men and 10 women, age range 20-47 years. Participants reported 5-6 dreams weekly. With MACE questionnaire, they assessed pervasiveness of seven metacognitive activities previously established regarding control (Kahan & LaBerge, 1996) by responding to the question: "During how much time did you;" (engage in each activity)" (p. 7) in extent form 0-4 (none, some, all). In addition, for estimation of incidence of cognitive processes, participants used *Subjective Experiences Rating Scale* (SERS) that contained (among sensory and affective qualities, and reality orientation relative to everyday life) the following cognitive processes: Thinking, planning, imagining, evaluation, remembering, feeling, talking, listening, attending to the outer world, attention to the inner world with thoughts, feelings, memories (p. 18). Participants scored higher pervasiveness in waking experience high on choice and reflective awareness; the rest of the results did not differ on the remaining five questions (attention captured, suddenly focused attention, public self-consciousness, reflective awareness, own thoughts or feelings, and reflective awareness external events (p. 10).

Regarding incidence of metacognitive activities, any answer other then 0 (not at all) represented presence of that activity. Kahan and LaBerge (2011) reported similarities to previous findings in choice, sudden and focused attention, reflective awareness of own behavior and external events and reported higher differences in waking regarding reflective awareness of own thoughts or feelings (52%) than in dreaming (35%). In terms of pervasiveness, the SERS scoring of nine cognitive activities, the authors reported no statistical significance.

In conclusion, Kahan and LaBerge (2011) highlighted that high-order skills of choice, planning and focused attention are the same in dreaming and waking experiences, and that claims of "presumed suspension of executive processes in REM dreaming" (e.g., Braun et al, 1997; Hobson, 2009[b]; Hobson et al., 2003b; Maquet et al., 1996) are now challenged by the results of the present study" (p. 12). They concluded that in dreaming, "the executive functions of working memory are attenuated in dreaming, but not suspended" (p. 15).

Mapping and Characterizing Executive Thought Processes

With the understanding that eight basic though processes non-lucid dreamers are using were classified as rational (Wolman & Kozmova, 2007), Kozmová (2008) decided to investigate *executive thought processes* defined as cognition that

> represents the dreamer's agency (authorship of intended or completed action), which is demonstrated by using a variety of higher-order cognitive processes. Executive thought process consists of thinking about taking an action and/or its initiation and completion. It implies intentionality; opportunity; planning; choice; judgment; decision-making; problem-solving; and creativity. Using these processes then contributes to a greater cognitive and behavioral flexibility and solidifies the psychological agency of a person. (Original definition by Kozmová & Wolman, 2005, expanded by M. Kozmová and R. Wolman, personal communication, March 11, 2007)

While previously investigating rational thought in non-lucid dreaming, Wolman and Kozmová (2007) came across, in the executive thought process defined as "higher order cognitive processes. Decision making, problem solving, planning, and agency" the existence of *problem-solving* efforts that dreamers made (e.g., *"I made a decision to go further,"* p. 845). When using the *subjective thought*, dreamers retained their memory of waking lives' in collection of "personal history, characteristics, appearance, beliefs, needs and desires, skills and goals," (p. 845). Thus it became clear *if* there would be a way to map out the problem-solving category of executive skills, in other words, the mental activities that dreamers, in their role of self (the dreaming person in one's dream) as the individual dreamers are capable of using or which could be emerging if not used before, *then* the dreamers would need to find themselves in the specific condition (situations, circumstances) that called for use of some form of thinking especially problem-solving thinking (Kozmová, 2008).

Kozmová (2008) hypothesized that because the content of the non-lucid dream is unpredictable, unsolicited, and dreamers find themselves, with few exceptions upon wanting to wake up, experiencing it and being immersed in it, it seemed that if the dreamers find themselves in difficult, problematic, challenging, threatening, or curious situations, they might start, with conscious effort, to rectify the situation to find a solution or to seek a way out. With this hypothesis, the author worked in direct contrast to isomorphic assertion that "dreaming, in both its general form and its specific content, is totally determined by physiological processes" (Stickgold, 2000, p. 6) and she believed that, for individual dreamer, the specific problem-solving condition will give rise to particular mental strategies, tactics, or maneuvers.

Kozmová (2008) conducted the research with 1,298 cross-cultural dreams (dreams from 100 males and 100 females from Argentina, Brazil, England, Japan, two republics of former Soviet Union- Ukraine and Russia, and 32 females and 66 males from USA), ages 20-70 years, from participants in dream seminars conducted in years 1990-2004 who contributed one self-selected dream under the protection of Institutional Review Board at Saybrook Graduate School.

Kozmová (2008) justified the selection of cross-cultural dreams by argument that dreamers from the same national origin might be culturally conditioned to use the same problems-solving strategies. Thus, in order the map out the scope of the problem-solving strategies as part of executive skills that are actually possible in non-lucid dreaming, the author used dreams from males and females from individualist and collectivistic countries. With the same groups of dreams, the researcher later expanded investigations of executive skills with direct and indirect modalities of problem-solving (Kozmová, 2012, 2015, 2017a, b, 2018).

Kozmová (2008) original work was guided by the research question "What is the scope of cognitive problem-solving strategies that dreamers are capable of employing for resolving situations encountered during dreaming?" (p. ii) with three assumptions that contradicted the previous notion of dreamers as being static and without volition (Hobson et al., 2000a). The investigator assumed that dreamers will use their perceptions of encountering unexpected situations and realize the existence of clearly definable or undefinable problem, dilemma, threat, difficulty, or curiosity and they will follow the need to take some form of mental action (p. 52). Further, if the dreamers make active efforts, it will be with different mental tactics and movements; and these mental activities will be noticeable in the dream reports (p. 52).

From the pilot study (Kozmová, 2007), it became obvious that not every dream is suitable as a discernible display of dreamer's mental efforts. Kozmová (2008) divided the dreams into *descriptive* and *problem-solving.* The *descriptive* dream was operationalized as dream that "contains the descriptions of scenery, the situations, the actions of the dreamer or other characters, and his or her observations or experiences within that dream" (p. 72). The *problem-solving dream* was characterized as dream in which

> an independent reader would be able to discern that the dreamer was propelled or prompted to make an active choice in the presence of a situation or event, one which he or she might have recognized, defined, or identified, and explicitly described within the dream. Alternatively, in the report it could be observable that during the dream the dreamer might have implicitly hinted at and not describe, yet acted upon, some precipitating event. The discernible active choices as responses to internally generated, perceived, and later described situations or images could be visible and construed as reactions to an emotionally realistic immediacy of the dream or to some other unidentifiable cause. The overt or covert events, to which dreamers reacted, could be characterized by an independent reader as realistic or imagined vicissitudes of life in general (e.g., ambiguities,

puzzlements, distresses, dilemmas, conflicts, concerns, difficulties, worries, threats, or dangers) (pp. 72-73).

Example of descriptive dream:

I have an invitation to visit a house. It is shaped like a heart. I can tell that a very loving family lives there. I enter the house and I am given a red valentine. It is also shaped like a heart, just as the house has a similar shape. The feelings are loving and kindly. (Dream 34, male, Russia, 1996)

Example of problem-solving dream:

I am sitting in a room with my husband. Because I have lost my power of speech, he has to guess what my thoughts are. It is a frustrating and time-consuming task. Eventually he gives up and I am left trying to make a conversation using "Scrabble" game letter pieces. (Dream 2, female, England, 1994)

From 1,288 archival cross-cultural non-lucid dreams (10 lucid dreams were not part of research), 979 dreams (67%) were of a *problem-solving* type. The investigator analyzed these dreams with the method of grounded theory (Glaser & Strauss, 1967). The purpose of method is to generate, from observed data, a description of the object of research by method of constant comparative analysis of instances of participants' reactions to investigated phenomenon. From these instances will emerge coded categories, their dimensions, and properties (often described by *in vivo language*—in the language of participants) (Glaser & Strauss, 1967) thus the theory emerges from research data (p. 46). In this case, since the dreaming experiences are not directly observable, the researcher intuited that there could exist *a nocturnal problem-solving phenomenon* (Kozmová, 2008). She was looking for dreamers' attempts to resolve or find solutions within the immediacy of their dreams; she used constant comparative analysis with specific questions designed for this project (e.g., "(a) *How does this problem-solving activity of the dreamer differ from the problem-solving activity of a different dreamer?* (b) *How does the dreamer proceed to deal with this situation?* (c)

How does the dreamer react?" Kozmová, 2008, p. 75). From this analysis, the researcher, who was "theoretically sensitive" (p. 46), formed the concepts and formulated the theory. The detailed work with method with examples is described in Kozmová (2012, 2015, 2017).

Kozmová (2008) reported that the analysis of 979 problems-solving dreams by method of grounded theory solidified intuited existence of *phenomenon of nocturnal cognitive problem-solving* strategies with multilayered description. From this description then the researcher posited a substantive *theory of dynamic strategizing engagements* (pp. 155-157): The theory denotes the scope of nocturnal mental activities that dreamers use, with their own authority, as problem-solving tactics, maneuvers, and reactions to situations; it includes dream characters' involvement in problem-solving and also dreams' spontaneous, without dreamers' effort, solving contributions to problematic or curious situations dreamers encounter or create in their dreams.

The actual *phenomenon of nocturnal problem-solving strategies* (Kozmová, 2008) consists of problem-solving in different contexts: Self-preservation, intrapersonal, interpersonal, and inter-object relationships. The phenomenon's three distinct modalities are: (a) *direct* modality when dreamers deliberately act and behave, think, and use emotions in self-regulatory way and in action-prompting or action-preventing manner; (b) *self-monitoring activities (problem-solving by proxy)* when dreamers become reflecting observers and commentators who are aware of problems dream chracters' experience; and (c) *indirect* modality includes strategies independent of dreamers' volition such as solutions offered by dream characters or the dream itself (Kozmová, 2008, p. iii).

Direct Modality of Using Behaviors-Actions to Problem Solve

When directly problem-solving (Kozmová, 2008) their own self-preservation, dreamers used 20 different strategies. Example is an escape from place of harm: "*The house caught fire because of the salute. My sister and I jump out the window, and run away with mom and my boyfriend to survive";* self-defense: "*. . . very malevolent man there, a fakir from India. He looks into my eyes and explains his intention. He wants to harm me in* *some way. I find a sword and I cut off his head"* (p. 115); communication: *"We begin to make turns and somersaults, and he keeps flying faster without my assent. I ask my friend to stop. I shout"* (p. 116).

In problem-solving with personal meaning, dreamers used five different strategies. Examples are goal-oriented behavior: *"I drive the truck to safety and get it to a repair station. The repairman fixes the truck and I go on my way"* (Kozmová, 2008, p. 118). In the context of interpersonal relationships, ten different behavior-action oriented strategies included saving other character's life: *"But suddenly she began to sink, and a lot of people gathered around the pool and I rushed to save her"* (p. 119).

In the secondary analysis of actions and behaviors as dreamers' problem-solving efforts (Kozmová, 2017), the author analyzed separately, again with the method of grounded theory (Strauss and Glaser, 1967), the dreamers' volitional kinetic and communicative cognitive strategies accompanied by evaluative thought (e.g., *I saw, in the reflection of the water, my head with spots on my face. . . . I went inside the water . . . When I did this, the spots on my face were no longer there,"* p. 39). Collectively, in problems that consisted of threats to dreamers' lives; dreamers used 12 strategies, in solving threat to their bodily integrity one strategy; in threat to psychological integrity, equilibrium, self-esteem, or loss of external support seven strategies.

In problems with personal significance or meaning, dreamers used 12 differing strategies such as using intentional behavior in order to reach a goal by head movement: *"I was in a kitchen area in the back, helping a couple of men and women prepare for the break time. I was eager to hear the lecture so I frequently stuck my ear out the serving window"* (Kozmová, 2008, p. 41). Within interpersonal relationships, dreamers used 21 different communicative or kinetic strategies followed by evaluative thought (e.g., as an advice seeking with future goal-orientation: *"I asked him what to do in order to learn from my dreams and put the knowledge to use when I am awake"* (p. 42).

Direct Modality of Thinking

In life threatening situations, in intrapersonal, interpersonal, and inter-object problematic situations, dreamers used total of 35 different thinking strategies (e.g., memory based personal knowledge: *"I am with my wife. I know she has been dead for several years,"* Kozmová, 2008, p. 125); comparison of inter-personal and inter-object situation: *"I have a spider on my pajamas, and I am afraid it will bite me. Then my father has his arm around me, even though he died three years ago. We walk around the house where I grew up and down the street. An old grade school friend appears, and he hugs me. It was a pleasant moment in contrast to the terrifying spider,"* (p. 125).

In the secondary analysis (Kozmová, 2012), with the question of "What are the scope and purpose of thinking/strategizing efforts as one of three previously identified direct mental problem-solving modalities?" (p. 47), from the analysis with method of grounded theory emerged core aggregate of executive function that consists of eight types of executive thought processes: Analytical, decision-making, defense mechanisms, evaluative, goal-oriented/goal directed, interpretative, motivational, and self-determinative.

Analytical thought processes consisted of 18 strategies, e.g., comparison between mental state and physical state:

> *I cannot swim, nevertheless, I move effortlessly toward the shore. Close to the shore, I put my feet down and find myself submerged in the water without panic. I note with mild curiosity that I am relaxed in a luminous pool, breathing quite naturally.* (p. 55)

Decision-making thought process had four types of strategies, e.g., emotion driven: *"I don't trust the officers and use every trick in the book to get out* (Kozmová, 2012, p. 57).

Defense mechanisms included 11 strategies, e.g., anticipation of the outcome of imagined situations:

> *I start to go through and get stuck. I fear someone may steal my wallet but I know I will have to take off my clothes to make it through. Even then, I will be bruised and cut. I am just about to make it through when I wake up.* (p. 58)

Evaluative thought processes consisted of 12 strategies, e.g., evaluation of emotional consequences of plan: *"It is dangerous to cross the embankment but it is possible"* (p. 59).

Goal-oriented/goal directed thought process included eight strategies, e.g., mental rehearsal and visualization of future action:

> *I dream that a band of bats is flying around me. It is night. I know that to get out of trouble, I will have to fight them, and break the wings of some of these bats. But I am terrified of the bats and thought of having to fight with them.* (p. 60)

Interpretative thought processes consisted of seven strategies, e.g., sensing: *"I sense they want me to undress"* (p. 60).

Motivational thought processes included seven strategies, e.g., contemplating personal preferences: *"I would rather join my three companions and not return to work"* (p. 61).

Self-determinative processes had eight strategies, e.g., making decisions about one's own destiny: *"I decide to look for someplace to die,"* (p. 62).

Direct Modality of Emotions as Elements of Problem-Solving

Dreamers used emotions for strategizing and problem-solving in a self-regulatory way and in an action-prompting or action-preventing manner with 33 different cognitive components of problem-solving (Kozmová, 2008). Example of self-regulatory way in which emotions play role is the dreamer's own evaluation: *"I did not feel fear; I was very impressed"* (p.140). Example of emotion as a motivator for action is the following dream: *"So they are leaving me. I have been very cold and indifferent to them, so they have given up on me. I was scared and sad, crying and trying to stop them"* (p. 139). In action-preventing role, the dreamer makes a decision in order to secure different future: *"I overcome my fear and leave by the window, which is now a door, and climb without weapons downstairs, leaving the residence because there is nothing there for me"* (p. 140).

In the secondary analysis (Kozmová, 2015) of 29 dreams with different emotional qualities emerged 86 different components of cognitive and psychological processes accompanied by dreamers' emotional awareness in emotional signaling, stimulating, self-regulating, or action prompting roles in five different contexts: Interpersonal, intrapersonal, group, animate environmental, and inanimate environment.

Kozmová (2015) proposed that the awareness of emotions as a part of dreamers' problem-solving repertoire allows dreamers to shift from a passive role—passive immersion—in *primary consciousness* (defined as "simple awareness that includes perception and emotion," Hobson, 2009, p. 803) into active and participatory mode of *secondary consciousness* defined as "subjective awareness including perception and emotion that is enriched by abstract analysis (thinking)" (Hobson, 2009, p. 803). In this case, the dreamers are able to use enhanced thought processes of high-order cognition in the form of reflective and intellectual abstract analyses. An example of dreamer using emotions as part of cognitive/psychological processes of problem-solving when in engagement with personal tasks, undertakings, and activities, would be emotions serving as "a signal, acknowledgement, and understanding cause of one's own internal state": *"I am working very hard in my office I am in pain because the work is so hard"* (p. 9). In interpersonal context, emotion serves as "a signal of quality of interpersonal interaction": *"I open my eyes and notice that the coordinator is watching me with a smile that is both satisfied and sarcastic. I hold back my anger"* (p. 10). Example of the experiencing, participating and using emotions within group context would be "emotion as a silent evaluator of possible threat": *"We walk slowly through the place, looking at the men and women who are working there. At the front of another door is a very large arcade. Three suspicious-looking men enter"* (p. 11). Example of using emotion in animate environmental context as "an indicator of physiological disturbance" in a problematic situation is a dream: *"I see it is a large insect. It bites me in the leg and I can feel the pain. My leg turns red where it bit me"* (p. 12). The example of using emotion in inanimate environmental context in a role of an "evaluating aspect of possible solution to a problem" is the following dream:

> *A small mechanism in the embankment's center picks up nearby rocks and grinds them into dust. The movement of its blade and the rock mashing is unmovable. The mechanism picks up huge rocks and throws them toward our side. The rocks fly in disorder. I feel_danger but it is useless to run away*. (p. 13)

The Indirect Problem-Solving Modality

Kozmová (2008) described the indirect problem-solving modality as instances when the dreamer is not directly involved but uses self-monitoring mental activities and actually solves the problem by proxy. It represents situation when the dreamer is not directly involved in problem, but is capable of describing it through one's own eyes, uses metacognitive thinking, is able to comment on characters' mental states, and involves himself in "reasonings, feelings, and outlook toward the situation" (p. 97). Further, the dreamer is aware of problems other characters' experience and is able to describe how they deal with situations. Dreamer could become concerned and emotionally stimulated. Example of observational, cognitive, and metacognitive abilities which include emotional acknowledgment is the following dream in which dreamer recognizes a problem, notices characters' emotional states, yet remains removed from it:

> *There is blood all over the street. Some people are dead. Others are wounded. There is anguish, fear, and grief that the survivors are expressing. I think that I am a king or member of the royal family. I am indifferent to all of this. My body does not register any emotion. The soldiers use sharp weapons to cut the people up. I think the royal family sent the soldiers out to kill the townspeople. The survivors are very sad.* (p. 97)

In the indirect problem-solving, while working with the 979 cross-cultural dreams (Kozmová, 2008) in attempts to map out range of dreamers' problems-solving abilities that become exemplification of their executive skills, it became obvious that sometimes dream characters become involved not only in problem-solving when dreamers experience difficulties, but they also instigated problems for the dreamers. The range of the dream characters' cognitive problem-solving abilities in benevolent way towards the dreamers and characters' negative approach and instigation of difficulties in malevolent manner is described in Kozmová (2017b, 2017c, 2018). When taking variety of dream characters' positive responses to dreamers' difficulties and dream characters' negative instigation of problems for dreamers, the author arrived at the conclusion that cognitive capacities dream characters displayed could be considered processes of dreamers' own thinking repertoire.

While contemplating the research about REM sleep dreams and dreamers' cognitive abilities, it could be concluded that there are now in existence several descriptions of what kinds of cognitive capacities dreamers in non-lucid dreaming state are actually capable of using. So far, the discovered range of dreamers' cognitive abilities points to dreamers' active involvement in the continuity of thinking. The conditions when the dreamers might be capable to do so include the following: (a) the individual dreamer had previously, during waking time, developed high-order thought and executive skills; (b) the high-order thought and executive skills could be emerging during dreaming; and (c) within the dream, the dreamer has reasons—experiences of conflicts, problems, difficulties, or curiosities—that affect the him or her in a personal manner enough to make attempts to seek solution or decide not to use one's own abilities.

CONCLUSION

In this critically-focused chapter, the present author aimed to demonstrate, within an academic debate about sleep stages and cognition in dreaming, how by theorists' and researchers' selected scientific notions, hypotheses, and frameworks have a potential to shape the research findings. For example, *if* in the research of cognitions of NREM and REM dreaming the goal was to find "the negative," or in psychoanalytical thought, *the suspension of ego control* (Freud, 1956) and in the mind/brain isomorphic paradigm, the "deactivated" DLPFC (e.g., Braun et al., 1997), *then* that is what researchers will find (e.g., Fosse et al., 2004). Hobson, Pace-Schott, and Stigckold (2003) articulated this *negatively-focused* aspect of research again in their statement that "neglect of . . . psychological features" in dreaming cannot be compensated by technological sophistication of neuroimaging (p. 231).

It could be proposed that when Freud (1900/1990) became aware of complex thought in dreaming which he called "complicated feats of thinking" (p. 389) and consciously bypassed it, he thus left legacy of "neglect" regarding cognitive processes. Kahan and Claudatos (2016) summarized this continuous state of affairs in dream research when they openly state that

> some important dream features may not be targeted because of the investigator's theoretical orientation or current research interests. For example, if an investigator assumes that dreaming is deficient or lacking in high-order cognitive processes (choice, planning, reflective awareness) (Hobson, 1988; Hobson, Pace-Schott, & Stickgold, 2003; Rechtschaffen, 1978), then a measure designed to assess dream cognition would likely not target these mental processes. (p. 161)

It could be argued that the alternative approach to scientific neglect of features of consciousness in research, specifically cognition, is to continue to look at "what exists" in terms of what kind of mental activities dreamers become engaged in particular circumstances during their experiences in dreaming (e.g., Kozmová, 2012, 2015, 2017a). As an example of this approach is Kahan's (1994) and Kahan and LaBerge's (1996, 2011) closer look at the existence of dreamers' control level (which could be considered as presence of volitional process, Kozmová, 2017a) investigated with MACE questionnaire (Kahan et al, 1997).

Hobson et al. (2000) proposed that in future, all mental aspects of dreaming will be explained by "distinctive physiology of REM sleep" (p. 799). For scientists (e.g., Kussé et al. 2010; Nir & Tononi, 2010; Schwartz & Maquet, 2002) hoping for integration of two kinds of approaches to study of dreaming—neuroimaging for investigations of physiology and phenomenology for investigations of dreamers' subjective experiences—the scientific challenge of including demonstrated cognitive processes is not a simple one. In the past, as Montangero (2018) explained in historically-based review of dreaming and research of REM sleep, the scientists, without careful examination of origins and premises, relied on "received view" which proved to be erroneous (p. 32). Similarly, regarding cognitive processes in dreaming, it could very well be that relying on isomorphic one-to-one correspondence between neural correlates and volitional mental events might not be a fruitful way.

Previously, Gusnard and Raichle (2001) highlighted that "baseline or resting state of brain function involving specific set of mental operations" and increases and decreases in neural activation are difficult to interpret specifically regarding goal-oriented tasks (p. 685). Even though it is cheallenging to estimate the prevalence of problem-solving dreams, given that "dreams reveal our self-conceptions and our emotional preoccupations" (Domhoff, 1996, p. viii), it stands to reason that problem-solving might be a regular feature of dreaming consciousness. Smith et al. (2004) predicted that in REM sleep dreaming, the dreamers could "experience a discrepancy between an existing state and a preferred, or more valued state" and then engage in "motivated behavior" (p. 504). Indisputably, the dreamers' attempts to problem-solve require motivation toward goal-orientation; both of these processes are part of executive skills (Kozmová, 2008; 2012) and thus part of mental activities belonging to individuals.

The detailed classification of dreamers' volitional mental means or problem-solving strategies in form of actions-behaviors, thinking, or using emotions (Kozmová, 2008, 2012, 2015, 2017a, 2017b, 2018) demonstrates that dreamers remember who they are. Individually, with few exceptions which might be developmentally based or exist under stressful conditions (e.g., Kozmová & Wolman, 2006; Kozmová, 2018) the dreamer remains to be his or her own person in the dream. Specifically, in problem-solving dreams, when individual dreamers encounter something personal at stake, they start to be engaged and involved "to make the wrong right" with already developed or emerging thought processes; they could also make a decision not to be involved (Kozmová, 2008, 2012, 2015, 2017a, 2017b, 2018). Kozmová found the following perturbing situations that upset and disturb the dreamers' equilibrium: Self-preservation; violations of personal idiosyncratic values, preferences, and beliefs; experience of unpleasant feelings and health concerns; focus on maintenance of identity and cherished roles or on goals of gaining new knowledge; interpersonal problems; or generally taking care of things that matter to dreamers in the immediacy of dreaming situations. Experiences of more positive feelings or emotions also lead dreamers to pondering personally-defined oddities or to evoking curiosity and subsequent creative problem-solving. All of above-mentioned situations within dreams evoke in dreamers need to use executive skills whether it is by the individual dreamer directly or by dream characters who are being proactive and helpful to the dreamer (Kozmová, 2008, 2017b, 2017c, 2018). Conversely, the dreamer, in a form of dream characters, might be working, again, with the use of mental activities, against oneself (Kozmová, 2017b, 2017c, 2018). To sum up, the sophisticated richness of thought processes, as a baseline in terms of dreamers' mental events and activities, is already mapped out.

Previously, Schredl and Erlacher (2011) questioned not only "how direct" but also whether there could even be a one-to-one relationship or correlation between "all dreamed activities" (dream content) and objectively measured brain activation (p. 101). There are at least two specific types of inquiries that could be contemplated with regard to this questioned neurobiological and psychological correlation: *If the decades- old neuroimaging studies demonstrate deactivated prefrontal cortex deemed by neuroscientists necessary for high-order cognition, then why and how, under the assumption of isomorphism, the dreamers remain capable of thinking in executive mode?* Or, *Could it be that dreamers' capacities of using executive processes, as part of existing feature of consciousness in the state of dreaming, are independent of neural correlates?*

It could be argued that unresolved difficulties in area of scientific epistemology will continue to prevent understanding of relationship between non-lucid dreamers' existing cognitive capacities and the nocturnal brain events. For example, the tendency to consider that biology of brain is responsible for dreamers' mental capacities is a not an axiom, but a questionable assumption: Non-lucid dreamers are able to notice and experience problematic situations and to think about them in problem-solving manner even in the absence of activated DLPFC. For this reason, there might be, in the spirit of open-minded inquiry, other applicable explanations for the phenomenon of nocturnal cognitive problem-solving strategies which include identified, categorized, and characterized mental events (Kozmová, 2008).

The fact remains that even though in the published scholarship the results of phenomenological studies about cognition are not openly regarded (e.g., Domhoff, 2018, Pace-Scott, 2011; Schwartz, 2013), the dreamers continue, while separated on a sensory level from awake world, to demonstrate multiplicity of cognitive processes in situations that personally matter to them. From this viewpoint, it could be proposed that during problem-solving subjective experiences, the dreamers' use of executive skills and higher-order cognition in non-lucid dreaming state of consciousness might serve *transcending function*: In condition of being sensory-separated from awake world, the assumed biologically decoupled nocturnal neural regions considered to be determinants of dreamers' deficient cognitions are continuously surpassed by volitional activities of mentally engaged dreamers with their intact autonomy and agency. In other words, dreamers' cognitive capacities which transcend the constraints of "deactivated" neural correlates deemed necessary for cognition indicate existence of sets of cognitive processes independent of brain biology.

REFERENCES

[1] Arkin, A. M., Antrobus, J. S., Toth, M. F., Baker, J., & Jackler, F. (9172). A comparison of the content of mentation reports elicited after nonrapid eye movement (NREM) associated with sleep utterances and MREM "silent" sleep. *The Journal of Nervous and Mental Diseases, 155, 6,* 427-435.

[2] Asserinsky, E., & Kleitman, N. (1955). Two types of ocular motility occurring in sleep. *Journal of Applied Physiology, 8,* 1-10.

[3] Atwood, G., & Stolorow, R. (1984). *Structures of subjectivity: Explorations in psychoanalytic phenomenology.* Hillsdale, NJ: The Analytic Press.

[4] Bettelheim, B. (1982). *Freud and man's soul.* New York, NY: Vintage Books.

[5] Braun, A. R., Balkin, T. J., Wesenten, N. J., Carson, R. E., Varga, M., Baldwin, P. et al. (1997). Regional cerebral blood flow throughout the sleep-wake cycle—an (H2O)–O-15 PET study. *Brain, 120,* 1173-1197. doi: http://dx.doi. org/10.1093/brain/120.7.1173.

[6] Carskadon, M.A., & Dement, W.C. (2011). Monitoring and staging human sleep. In M. H. Kryger, T. Roth, & W. C. Dement (Eds.), *Principles and practice of sleep medicine*, 5th edition, (pp. 16-26). St. Louis: Elsevier Saunders.

[7] Cavallero, C., & Foulkes, D. (1993). Memory and dreaming. In C. Cavallero, & D. Foulkes (Eds.), *Dreaming as cognition* (pp. 1-17). New York, NY: Harvester Wheatsheaf.

[8] Colman, A. M. (2001). *A dictionary of psychology.* New York: Oxford University Press.

[9] Damasio, A. R. (1999). *The feeling of what happens: Body and emotion in the making of consciousness.* New York, NY: Harcourt Brace.

[10] Dang-Vu, T., Schabus, M., Desseilles, M., Schwartz, S., & Maquet, P. (2007). Neuroimaging of REM sleep and dreaming. In D. Barrett & P. McNamara (Eds.), *The new science of dreaming: Biological aspects* (Vol. 1, pp. 95-113). Westport, CT: Praeger/Greenwood.

[11] Dawson, J. L., & Conduit, R. (2011). The substrate that dreams are made on: An evaluation of current neurobiological theories of dreaming. In D. Cvetkovic, & I. Cosic (Eds.), *States of consciousness: Experimental insights into meditation, waking, sleep and dreams* (pp. 133-156). Heidelberg, Germany: Springer-Verlag.

[12] Dement, W., & Kleitman, N. (1957). The relation of eye movements during sleep to dream activity: An objective method for the study of dreaming. *Journal of Experimental Psychology, 53,* 339-346.

[13] Domhoff, G. W. (1996). *Finding meaning in dreams: A quantitative approach.* New York: Plenum.

[14] Domhoff, G. W. (2003). *Scientific study of dreams: Neural networks, cognitive development, and content analysis.* Washington, DC: American Psychological Association.

[15] Domhoff, G. W. (2018). *The emergence of dreaming: Mind-wandering, embodied simulation, and the default network.* New York, NY: Oxford.

[16] Dresler, M. (2012). Neural correlates of dream lucidity obtained form contrasting lucid versus non-lucid REM sleep: A combined EEG/fMRI case study. *Sleep, 35*(7), 117-120. http://dx.doi.org/10.5665/sleep.1974.

[17] Faden, R. R., & Beauchamp, T. L. (1986). *A history and theory of informed consent.* New York, NY: Oxford University Press.

[18] Farthing, G. W. (1992). *The psychology of consciousness.* Englewood Cliffs, NJ: Prentice Hall.

[19] Fosse, R. (2000). REM mentation in narcoleptics and normals: An empirical test of two neurocognitive theories. *Consciousness and Cognition, 9,* 488-509.

[20] Fosse, R. Stickgold, R., & Hobson, J. A. (2004). Thinking and hallucinating: Reciprocal changes in sleep. *Psychophysiology, 41,* 298-305.

[21] Foulkes, D. (1960). *Dream reports from different stages of sleep.* Unpublished doctoral dissertation, University of Chicago, Chicago, Illinois.

[22] Foulkes, D. (1962). Dream reports from different stages of sleep. *Journal of Abnormal and Social Psychology,* 65(1), 14-25.

[23] Foulkes, D. (1964). Theories of dream formation and recent studies of sleep consciousness. *Psychological Bulletin, 62*(4), 236-247.

[24] Foulkes, D. (1990). Dreaming and consciousness. *European Journal of Cognitive Psychology, 2*(1), 39-55.

[25] Foulkes, D. (1999). *Children's dreaming and the development of consciousness.* Cambridge, MA: Harvard University Press.

[26] Foulkes, D. (2017). Dreaming, reflective consciousness, and feelings in the preschool child. *Dreaming,* 27(1), 1-13. http://dx.doi.org.ezp-rod1.hul.harvard.edu/10.1037/drm0000040.

[27] Freud, S. (1956). *Interpretation of dreams.* New York, NY: Basic Books.

[28] Freud, S. (1965). Lecture XXXI: The dissection of the psychical personality. In D. C. Funder & D. J. Ozer, (Eds.), *Pieces of the personality puzzle: Readings in theory and research,* pp. 227-235. New York, NY: W.W. Norton. From *Introductory lectures on psycho-analysis,* by Sigmund Freud, in *The Standard Edition of the Complete Works of Sigmund Freud,* translated by James Strachey, pp. 48-72. Translation © 1965, 1964, 1963 by James Strachey.

[29] Freud, S. (1999). *The interpretation of dreams.* (J. Crick, Trans.). New York: Oxford University Press. (Original work published 1900).

[30] Gabbard, G. O. (2004). *Long-term psychodynamic psychotherapy: A basic text.* Washington, DC: American Psychiatric Publishing.

[31] Glaser, B. G., & Strauss, A. L. (1967). *The discovery of grounded theory: Strategies for qualitative research.* Chicago, IL: Aldine.

[32] Globus, G. (1987). *Dream life, wake life.* Albany, NY: State University of New York Press.

[33] Gusnard, D. A. (2005). Being a self: Considerations from functional imaging. *Consciousness and Cognition, 14*(4), 679-697, doi: 10.1016/j.concog.2005.04.004.

[34] Harman, G. (1995). Rationality. (2nd ed.). In E. E. Smith & D. N. Osherson (Eds.), *Thinking: An invitation to cognitive sciences.* Vol. 3, (pp. 211–275). Cambridge, MA: MIT Press.

[35] Hartmann, ER. (1973). *The functions of sleep.* New Haven, CT: Yale University Press.

[36] Hartmann, E. (2000). We do not dream of 3 R's: Implications for the nature of dreaming mentation. *Dreaming, 10,* 103-111.

[37] Hobson, J. A. (1988). *The dreaming brain: How the brain creates both the sense and nonsense of the dreams.* New York, NY: Basic Books.

[38] Hobson, J. A. (1992). A new model of brain-mind state: Activation level, input source, and mode of processing (AIM). In J. S. Antrobus & M. Bertini (Eds.), *The neuropsychology of sleep and dreaming* (pp. 227-245). Hillsdale, NJ: Erlbaum.

[39] Hobson, J. A. (1997). Consciousness as a state-dependent phenomenon. In J. D. Cohen & J. W. Schooler (Eds.), *Scientific approaches to consciousness* (pp. 379-396). Hillsdale, NJ: Erlbaum.

[40] Hobson, J. A. (2007). Current understanding of cellular models of REM expression. In D. Barrett & P. McNamara (Eds.), *The new science of dreaming.* Volume 1. Biological aspects (pp. 71-83.) Westport, CT: Praeger.

[41] Hobson, J. A. (2009a). *Dream consciousness.* The William James Lectures at Roehampton University. London, England: Roehampton University.

[42] Hobson, J. A. (2009b). REM sleep and dreaming: Towards a theory of protoconsciousness. *Nature Reviews: Neuroscience, 10(*11), 803-814. doi:10.1038/nrn2716.

[43] Hobson, J. A., & McCarley, R. W. (1977). The brain as a dream state generator: An activation-synthesis hypothesis of the dream process. *American Journal of Psychiatry, 134,* 1335-1348.

[44] Hobson, J. A., Pace-Schott, E. F., & Stickgold, R. (2000a). Dreaming and the brain: Toward a cognitive neuroscience of conscious states. *Behavioral and Brain Sciences, 23,* 793-842.

[45] Hobson, J. A., Pace-Schott, E. F., & Stickgold, R. (2000b). Consciousness: Its vicissitudes in waking and sleep. In M. Gazzaniga (Ed.), *The new cognitive neurosciences* (2nd ed.; pp. 1341-1354). Cambridge, MA: A Bradford Book.

[46] Hobson, J. A., Pace-Schott, E. F., & Stickgold, R. (2003). Dream science 2000: A response to commentaries on dreaming and the brain (authors' responses). In E. F. Pace-Schott, M. Solms, M. Blagrove, & S. Harnad (Eds.), *Sleep and dreaming: Scientific advances and*

reconsiderations (pp. 231-247). Cambridge, United Kingdom: Cambridge University Press.

[47] Hobson, J. A., & Schredl, M. (2011). The continuity and discontinuity between waking and dreaming: A dialogue between Michael Schredl and Allan Hobson concerning the adequacy and completeness of these notions. *International Journal of Dream Research, 4*(1), 3-7.

[48] Hurd, R., & Bulkeley, K. (2014). *Lucid dreaming: New Perspectives on consciousness in dreaming.* Vol. 1: Science, psychology, and education. Vol. 2. Religion, creativity, and culture. Santa Barbara, CA: Praeger.

[49] Ioannidess, A. A., Kostopoulos, G. K., Liu, L., & Fenwick, P. B. C. (2009). *MEG identifies dorsal medial brain activations during sleep. Neuroimaging*, 44, 455-468. doi: 10.1016/k.neuroimage.2008.09.30.

[50] James, W. (1890/1933). *Principles of psychology.* New York, NY: Henry Holt.

[51] Jenkins, A. H. (2001). Individuality in cultural context: The case for psychological agency. *Theory and Psychology, 11,* 347-362. doi: 10. 1177/0959354301113004.

[52] Jones, K., & Harrison, Y. (2001). Frontal lobe function, sleep loss and fragmented sleep. *Sleep Medicine Reviews, 5*(6), 436-475. doi: 10. 1053/2mrv.2001/0203.

[53] Kahn, D., & Hobson, J. A. (1994). *Self-organization theory of dreaming*. Retrieved August 12, 2004, from http://www.asdreams. org/jopurnal/artciles/33_kahn_hobson.htm.

[54] Kahan, T. L. (1994). Measuring dream self-reflectiveness: A comparison of two approaches. *Dreaming, 4,* 177-193. https://doi.org/ 10.1037/h0094411.

[55] Kahan, T. & LaBerge, S. (1994). Lucid dreaming as metacognition: Implications for congitve science. *Consciousness and Cognition, 3,* 246-264. https://doi.org/10.1006/ccog.1994.1014.

[56] Kahan, T. L. (2001). Consciousness in dreaming: Metacognitive approach. In K. Bulkeley (Ed.), *Dreams: A reader on religious, cultural, and psychological dimensions of dreaming,* (pp. 333-360). New York, NY: Palgrave.

[57] Kahan, T. L., & Claudatos, S. (2016). Phenomenological features of dreams: Results from dream log studies using the Subjective Experience Rating Scale (SERS). *Consciousness and Cognition, 41, 159*-176. http://dx.doi. org/10.1016/j.concog.2016.02.007.

[58] Kahan, T. L., & LaBerge, S. (1996). Cognition and metacognition in dreaming and waking: Comparison of first and third person ratings. *Dreaming, 6,* 235-249. doi.org/10.1037/ h0094459.

[59] Kahan, T. L., & LaBerge, S. (2011). Dreaming and waking: Similarities and differences revisited. *Conscious ness and Cognition, 20*(3), 494-514. doi: 10.1016/j. concog.2010.09.002.

[60] Kahan, T., LaBerge, S., Levitan, L., & Zimbardo, P. (1997). Similarities and differences between dreaming and waking cognition: An Exploratory study. *Consciousness and Cognition, 6,* 132-147. https://doi.org/10.1006/ccog.1996.0274.

[61] Kahn, D., Combs, A., & Krippner, S. (2002). Dreaming as a function of chaos-like stochastic processes in the self-organizing brain. *Nonlinear Dynamics, Psychology, and Life Sciences, 6,* 311-322.

[62] Kahn, D., & Hobson, J. A. (1994). *Self-organization theory of dreaming*. Retrieved August 12, 2004, from http://www.asdreams. org/jopurnal/artciles/33_kahn_hobson.htm.

[63] Kleitman, N. (1963). *Sleep and wakefulness* (2nd ed.). Chicago, IL: University of Chicago Press.

[64] Kohut, H. (1977). *The restoration of the self.* New York, NY: International Universities Press.

[65] Kohut, H., & Wolf, E. S. (1978). The disorders of self and their treatment: An outline. *International Journal of Psychoanalysis*, 59, 413-425.

[66] Kozmová, M. (2007). *Dual phenomenon of problem and problem-solving strategies in manifest content of females from England: Testing a suitability of grounded theory.* Unpublished manuscript, Saybrook Graduate School and Research Center, San Francisco.

[67] Kozmová, M. (2008). *The investigation of nocturnal cognitive problem-solving using cross-cultural dreams.* (Unpublished doctoral dissertation). Saybrook University, San Francisco, CA.

[68] Kozmová, M. (2012). Dreamers as agents making strategizing efforts exemplify core aggregate of executive function in non-lucid dreaming. *International Journal of Dream Research,* 5(1), 47-67. doi: http://dx.doi.org/10.11588/ ijodr.2012.1.9159.

[69] Kozmová, M. (2015). Emotions during non-lucid problem- solving dreams as evidence of secondary conscioussness. *Comprehensive Psychology, 4,* Article 6. doi: 10.2466/09.CP.4.6.

[70] Kozmová, M. (2017a). Non-lucid dreamers actualize volition as ego executive capacity by engaging in problem-solving. *International Journal of Dream Research, 10*(1), 30–62.

[71] Kozmová, M. (2017b). Review of varieties of dream characters' presence in non-lucid dreams. *Activitas Nervosa Superior, 59*(2), 37–42. doi: 10.1007/s41470-017-0009-9.

[72] Kozmová, M. (2017c). Dream characters' types of involvement and range of relational capacities in non-lucid problem-solving dreams (Part 1). *Activitas Nervosa Superior, 59*(2), 91–105. https://doi.org/ 10.1007/s41470-017-0010-3.

[73] Kozmová, M. (2018). Dream characters' types of involvement and range of relational capacities in non-lucid problem-solving dreams (Part 2). *Activitas Nervosa Superior, 60,* 48-68. https://doi.org/10. 1007/s41470-018-0020-9.

[74] Kozmová, M. (2019). *Unpublished data.*

[75] Kozmová, M., & Wolman, R. N. (2006). Self-awareness in dreaming. *Dreaming, 16*(3), 196-214. doi: 10.1016/j.concog.2006.09.009.

[76] Kuhn, T. (1962). *The structure of scientific revolutions.* Chicago, IL: Chicago University Press.

[77] Kuhn, T. (1966). *The structures of scientific revolutions.* Chicago, IL: University of Chicago Press.

[78] Kunzendorf, R. (1987-88). Self-consciousness as the monitoring of cognitive states: A theoretical perspective. *Imagination, Cognition, and Personality, 7,* 3-22. https://doi.org/10.2190/TW3U-0886-Q5TA-XHM4.

[79] Kussé, C., Muto, V., Mascett, L., Matarazzo, L., Foret, A., Shaffill-LaBourdiec, A., & Maquet, P. (2010). *International Review of Neurobiology, 92,* 87-99. doi: 10.106/S0074- 7742(10)92005-9.

[80] Mamelak, A. N., & Hobson, J. A. (1989). Dream bizarreness as the cognitive correlate of altered neuronal behavior in REM sleep. *Journal of Cognitive Neuroscience, 1,* 201-222.

[81] Maquet, P., Peters, J. M., Aerts, J., Delfiore, G., Degueldre, C., Luxen, A. et al. (1996). Functional neuroanatomy of human rapid-eye-movement sleep and dreaming. *Nature, 383,* 163-166.

[82] Maquet, P., Ruby, P., Maudox, A., Albouy, G., Strpenich, V., Dang-Vu, T., Desseiles, M., . . . Laureys, S. (2005). Hu- man cognition during REM sleep and the activity profile within frontal and parietal cortices: A reappraisal of func- tional neuroimaging data. *Progress in Brain Research, 150,* 219-227. doi: 10.106/S0079-6123(05)50016.

[83] May, R. (1967). *Psychology and the human dilemma.* Princeton, NJ: D. Van Nostrand.

[84] McCullough, L., Vaillant, C., & Vaillant, G. E. (1986). Toward reliability in identifying ego defenses through verbal behavior. In G. E. Vaillant (Ed.), *Empirical studies of ego mechanism of defense,* (pp. 61-072). Washington, DC: American Psychiatric Press.

[85] Merriam-Webster. (1988). *Webster's ninth new collegiate dictionary.* Springfield, MA: Author.

[86] Moffitt, A., Hoffmann, R., Purcell, S., Piegeau, R., & Well, R. (1988). Dream psychology: Operating in the dark. In J. Gackenbach & S. LaBerge (Eds.), *Conscious mind, sleeping brain: Perspectives on lucid dreaming* (pp. 429-4329). New York, NY: Plenum.

[87] Montangero, J. (2018). Dreaming and REM sleep: History of scientific denial who disappearance entailed a reconciliation of the neuroscience and the cognitive psychological approaches to dreaming. *International Journal of Dream Research,* 30-45.

[88] Nielsen, T. A. (2000). Covert REM sleep effects on REM mentation: Further methodological considerations and supporting evidence. *Behavioral and Brain Sciences, 23,* 1040-1056. https://doi.org/10. 1017/S0140525X0000399X.

[89] Nielsen, T. (2011). Dream analysis and classification: The reality simulation perspective. Chapter 51. In M. H. Kryger, T. Roth, & W. C. Dement (Eds.), *Principles and practice of sleep medicine. 5th* ed., (pp. 595-603). S. Louis, MI: Elsevier Saunders.

[90] Nir, Y., & Tononi, G. (2010). Dreaming and the brain: From phenomenology to neurophysiology. *Trends in Cognitive Sciences, 14*(2), 88-100. doi: 10.1016/j.tics.2009.12.001.

[91] Pivik, R. T. (2000). Sleep and dreaming. In J. T. Cacioppo, L. G. Tassinary, & G. T. Berntson (Eds.), *Handbook pf psychophysiology,* 2nd ed., (pp. 687-716). Cambridge, MA: University Press.

[92] Pace-Schott, E. F. (2011). Neurobiology in dreaming. In M. H. Kryger, T. Roth, & W. C. Dement (Eds.), *Principles and practice of sleep medicine*, 5th ed., (pp. 563-575). St. Louis: Elsevier Saunders.

[93] Purcell, S., Mullington, J., Moffitt, A., Hoffman, R., & Pigeau, R. (1986). Dream selfreflectiveness as a learned cognitive skill. *Sleep, 9,* 432-437. https://doi.org/10.1093/sleep/9.3.423.

[94] Purcell, S., Moffitt, A., & Hoffmann, R. (1993). Waking, dreaming, and self-regulation. In A. Moffitt, M. Kramer, & R. Hoffmann (Eds.), *The functions of dreaming* (pp. 197-260). Albany, NY: State University of New York Press.

[95] Reber, A. S., & Reber, E. S. (2001). *The Penguin dictionary of psychology* (3rd ed.). London: England: Penguin Books.

[96] Rechtschaffen, A. (1978). The single-mindedness and isolation of dreams. *Sleep, 1,* 97109.

[97] Rychlak, J. F. (1994). *Logical learning theory: A human teleology and its empirical support.* Lincoln: University of Nebraska Press.

[98] Schopenhauer, A. (1908). *Studies in pessimism: A series of essays.* London: Swan Sonnenschein.

[99] Schutz, A. (1962). On multiple realities. In M. Natanson (Ed.), *The problem of social reality: Collected papers of Alfred Schutz,* vol. 1. (pp. 207,-259). Hague, Netherlands: Martinus Nijhoff.

[100] Schwartz, S. (2013). Complementarity of dream research and neuroimaging of sleep. In E. Nofzinger, P. Maquet, & M. J. Thorpy

(Eds.), *Neuroimaging and sleep and sleep disorders,* (pp. 121-128). Cambridge, England: Cambridge University Press.

[101] Schwartz, S. (2000). A historical loop of one hundred years: Similarities between 19th century and contemporary dream research. *Dreaming, 10,* 55-66.

[102] Schwartz, S., & Maquet, P. (2002). Sleep imaging and the neuropsychological assessment of dreams. *Trends in Cognitive Sciences,* 6(1), 23-30. https://doi.org/10.1016/S1364-6613(00)01818-0.

[103] Snyder, F. (1970). The phenomenology of dreaming. In L. Madlow & L. H. Snow (Eds.), *The psychodynamic implications of the physiological studies on dreams,* (pp. 124-151). Springfield, IL: Charles C. Thomas.

[104] Sparrow, G S. (1976/2015). *Lucid dreaming: Dawning of the clear light.* Virginia Beach, VA: A. R. Press.

[105] Sternberg, R. J., & Smith, E. E. (1988). *The psychology of human thought.* New York, NY: Cambridge University Press.

[106] Stickgold, R. (2000). The questions of dream research. *Sleep and Hypnosis, 3,* 4-8.

[107] Stickgold, R., Scott, L., Fosse, R., & Hobson, J. A. (2001). Brain-mind states: I. Longitudinal field study of wake-sleep factors influencing mentation report length. *Sleep, 24*(2), 171-179.

[108] Stolorow, R., Brandschaft, B., & Atwood, G. (1987). *Psychoanalytical treatment: An intersubjective approach.* Hillsdale, NJ: The Analytic Press.

[109] Strauch, I., & Meier, B. (1996). *In search of dreams: Results of experimental dream research.* Albany: State University of New York Press.

[110] Tart, C. (1988). From spontaneous event to lucidity: A review of attempts to consciously control nocturnal dreaming. In J. I. Gackenbach and S. LaBerge (Eds.), *Conscious mind, sleeping brain: Perspectives on lucid dreaming,* (pp. 67-100). New York, NY: Plenum Press.

[111] Tholey, P. (1988). Model for lucidity training as a means for self-healing and psychological growth. In J. Gackenbach, & S. LaBerge (Eds.), *Conscious mind, sleeping brain,* (pp. 263-285). New York, NY: Plenum Press.

[112] Tholey, P. (1989). Consciousness and abilities of dream characters observed during lucid dreaming. *Perceptual and Motor Skills, 68*(2), 567–578.

[113] Van Eeden, F. (1913). A study of dreams. *Proceedings of the Society for Psychical Research*, *26,* 431-461.

[114] Voss, U., Holzmann, R., Tuin, I., & Hobson, J. A. (2009). Lucid dreaming: A state of consciousness with features of both waking and non-lucid dreaming. *Sleep, 32,* 1191- 1200. https://doi.org/10.1093/sleep/32.9.1191.

[115] Voss, U. & Voss, G. (2014). A neurobiological model of lucid dreaming. In R. Hurd, & K. Bulkeley (Eds.), *Lucid dreaming: New perspectives on consciousness in sleep.* Vol. 1: Science, psychology, and education (pp. 23-36). Santa Barbara, CA: Praeger.

[116] Waggoner, R. (2014). Learning the depth of lucid dreaming. In R. Hurd, & K. Bulkeley (Eds.), *Lucid dreaming: New perspectives on consciousness in sleep.* Vol. 1: Science, psychology, and education, (pp. 193-212). Santa Barbara, CA: Praeger.

[117] White, G. L., & Taytroe, L. (2003). Personal problem-solving using dream incubation: Dreaming, relaxation, or waking cognition? *Dreaming, 13,* 193-209.

[118] Wolman, R. N. & Kozmová M. (2007). Last night I had the strangest dream: Varieties of rational thought processes in dream reports. *Consciousness and Cognition, 18,* 838-849. https://doi.org/10.1016/j.concog.2006.09.009.

[119] Workman, R. H., McCullough, L. B., Molinari, V., Kunik, M. E., Orengo, C., Khalsa, D. K. et al. (2000). Clinical and ethical implications of impaired executive control functions for patient autonomy. *Psychiatric Services, 51,* 359363.

[120] Zimmer, R., Bookstein, P. M., Kenny, E., & Kraeber, A. K. (2005). Glossary. In E. S. Person, A. M. Cooper, & G. O. Gabbard (Eds.), *The*

American Psychiatric Publishing textbook of psychoanalysis (pp. 547-562). Arlington, VA: American Psychiatric Publishing.

In: Parasomnia Dreaming
Editor: James Frederic Pagel

ISBN: 978-1-53617-815-9

Chapter 9

New Neuroimaging Technologies and Online Methods for Dream Content Analysis Make It Possible to Study Dreaming in Non-Disruptive and Low-Budget Ways in Sleep Medicine Clinics

G. William Domhoff**, PhD
Distinguished Professor Emeritus of Psychology and Research Professor, Department of Psychology, University of California, Santa Cruz, US

Abstract

After presenting an overview of information on the neural basis for dreaming, the gradual cognitive development of dreaming in children, and the quantitative findings on dream content collected in laboratory and non-

---

* Corresponding Author's Email: domhoff@ucsc.edu.

laboratory settings, this chapter suggests ways in which sleep medicine clinics could engage in low-cost, completely objective, mostly automated dream research without interfering with normal clinical routines. The process would begin by adding a few simple questions about the patient's personal dream history to the intake questionnaire, which could be answered in writing, or by use of a voice recorder in a private space. The intake questionnaire first of all might lead to studies of the way in which lesions or major traumas effect the neurocognitive basis of dreaming. The intake questionnaire also might ask patients to write down the Most Recent Dream they can recall, a method that leads to samples similar to representative samples collected through two-week dream diaries. The Most Recent Dream method has been used successfully with pre-adolescents and adolescents in middle schools in four different countries, and with older adolescents in high schools, as well as with adults. Similarly, Most Recent Dreams could be collected at each follow-up visit, and would be useful in comparing diagnostic groups or patients put on new medications. Finally, the intake questionnaire might ask new patients to indicate if they have ever kept a dream journal containing 50 or more dream reports for a period of 30 or more days—and if so, when and for how long. Dream journals provide an excellent context for pre-and-post studies of medications as well as for general content analyses. Once the dream reports collected by means of the Most Recent Dream method, or from a dream journal, are digitized, the use of the keyword search program on dreambank.net, which includes numerous word strings of proven value, would solve the need for fast and reliable content analyses. The coding system and spreadsheet available on dreamresearch.net also could be used to supplement the results from keyword searches. Within this framework, a wide range of studies would be possible in addition to those mentioned earlier, such as studies of patients with extremely high or extremely low dream recall, or in studies of unique individual cases of potential theoretical significance in specifying the neural network for dreaming, including those rare but invaluable patients who say they have lost dreaming or have never dreamt.

Introduction: The Where, When, How, and What of Dreaming

The numerous discoveries about dreaming and dream content that followed in the 35 years after the discovery of the REM/NREM sleep cycle in the 1950s, which often have been replicated, put dream research on a solid foundation for the first time. In addition to revealing the generally realistic nature of dream content, its consistency over time, and its continuity with waking person concerns, it revealed more cross-cultural similarities in dream content than cultural differences (e.g., Domhoff, 1996, 2003, 2018; Foulkes, 1985; Snyder, 1970). The new discoveries included the unexpected laboratory finding that dreaming is a gradual cognitive achievement that develops slowly between the ages of 5 and 12 (Foulkes, 1982, 1999; Foulkes, Hollifield, Bradley, Terry, & Sullivan, 1991). In the 1990s, these findings gradually were encompassed within the context of findings on the neural substrate that subserves dreaming, based on converging evidence from neuroimaging and lesion studies, which revealed that the neural substrate that subserves dreaming is far more circumscribed than previously imagined (Braun et al., 1997; Domhoff, 1996; Foulkes, 1999; Maquet et al., 1996; Nofzinger, Mintun, Wiseman, Kupfer, & Moore, 1997; Solms, 1997).

In the second decade of the twenty-first century, this new understanding of dreaming was extended by the realization that the neural substrate that supports dreaming is a portion of the waking default network, which supports imagination, daydreaming, creativity, mind-wandering and other forms of self-generated, spontaneous thought during waking life. More specifically, dreaming is based on highly activated neural substrates located in *portions* of the default network, including its dorsomedial prefrontal subsystem, its medial temporal subsystem, the lingual gyrus, and the caudate nucleus (Domhoff, 2011a; Domhoff & Fox, 2015; Eichenlaub, Bertrand, & Ruby, 2014; Fox, Nijeboer, Solomonova, Domhoff, & Christoff, 2013).

The two subsystems referred to in the previous paragraph, the dorsal medial prefrontal cortex and the posterior cingulate cortex, carry out slightly different imaginative tasks in waking life while at the same time working together in a harmoniously fashion. The dorsal medial prefrontal cortex system, which includes the dorsal medial prefrontal cortex, the temporoparietal junction, the lateral temporal cortex, and the temporal pole of the temporal lobe, is preferentially activated by instructions to think about the person's present situation or present mental state ("present self"). The medial temporal lobe system, which includes the ventral medial prefrontal cortex, posterior inferior parietal lobule, retrosplenial cortex, parahippocampal cortex, and hippocampal formation, is preferentially activated by thinking about personal situations and decisions in the future ("future self") (Andrews-Hanna, Reidler, Sepulcre, Poulin, & Buckner, 2010, pp. 554, 559). These results were supported and augmented in a later meta-analytic study (Andrews-Hanna, Smallwood, & Spreng, 2014).

Although the default network is normally constrained during waking by the frontoparietal control network, the dorsal attentional network, and the salience network, those three networks are significantly deactivated when dreaming occurs. Thus, dreaming is enabled by a relatively unconstrained and highly activated portion of the default network. Within this context, it is also now known that only brain lesions within the default network can lead to the loss or alteration of dreaming, and that only electrical brain stimulation in the medial temporal lobe, which is part of the default network, can lead to a sense of dreaming, or of recalling a dream (Bischof & Basset, 2004; Curot et al., 2018; Domhoff, 2018; Poza & Marti-Masso, 2006; Solms, 1997).

The inclusion of networks for mental imagery within the default network is crucial because vivid mental imagery is one of the striking characteristics of dreams (Moulton & Kosslyn, 2011, p. 98). In addition, there are overlaps between the default network and the social brain, with one comprehensive synthesis concluding that parts of the social brain are the "social-affective part" of the default network (Amft, Bzdok, Laird, Fox, & Schilbach, 2015). Then, too, the wide range of thoughts related to the past, present, and future that can arise during drifting waking thought and dreaming may be possible because there is a "striking overlap between the cortical network that mediates contextual associative thinking" and the default network (Bar, 2011, p. 17, including Figure 2.3). In addition, the inclusion of language areas in the temporal lobes within the default network is consistent with the frequency and specificity of language use in dreams (Foulkes et al., 1993; Meier, 1993). Finally, the fact that the medial prefrontal cortex, a key hub in the self-system, is part of the default network may help account for the highly personal focus of most dream content (e.g., Amft et al., 2015; D'Argembeau et al., 2012; Di & Biswal, 2014; Heatherton, 2010; Jenkins & Mitchell, 2011). One such study found that focal lesions to the medial prefrontal cortex reduce self-referential processing of stimuli, which indicates that this brain region is "important for self-referential processing and the neural representation of self" (Philippi, Duff, Denburg, Tranel, & Rudrauf, 2012, p. 475).

At the cognitive level, dreaming appears to be based on the process of simulation, a form of off-line thinking in which people imagine themselves as being in a hypothetical scenario and exploring possible outcomes (Buckner, Andrews-Hanna, & Schacter, 2008; Schacter, Addis, & Buckner, 2008). Moreover, dreaming is an enhanced form of simulation, called "embodied simulation," in which people experience themselves as being in hypothetical scenarios that include a vivid sensory environment, interpersonal interactions, and emotions (Domhoff, 2019; Gibbs, 2006)}. The defining feature of dreaming is therefore the sense of being a participant in (or observer of) an ongoing real event, which sometimes is thought of as a real experience for several seconds or minutes after awakening.

As a result of the realistic feelings that often accompany dreaming, dreams have parallels with theatrical plays because the dreamer and the other characters are usually engaged in one or another activity or social interaction within a setting, or series of settings, and sometimes express thoughts or emotions. In a study based on a normative non-lab sample containing 991 dream reports from young women and men, 86.9% of the dream reports included a social interaction or shared social activity, 6.7% included the dreamer seeing, hearing, or thinking about another dream character, 2.2% included only the dreamer and at least one animal, and 4.3% included only the dreamer engaging in an activity (Domhoff & Schneider, 2018, p. 10 Table 3, for a summary of the findings).

Dreaming may therefore be best defined as a simulated subjective experience in which dreamers experience themselves as embodied participants in (or embodied observers of) an event, almost always involving characters, activities, and/or social interactions. This sense of experiential involvement in an event is what distinguishes dreaming from other forms of thinking during sleep, such as isolated imagery or repetitive thoughts, as well as distinguishing dreaming from mind-wandering, whether the mind-wandering occurs during NREM 2 or as part of waking thought when not engaged in a focused task. This point is reinforced by the fact that dreams can last as long as 15-30 minutes, are experienced as "real" while they are happening, and are often remembered upon awakening, at least temporarily, as an actual experience.

In addition, it can be stated that dreaming occurs spontaneously if and only if the following six specific conditions exist. There is:

1) a mature neural substrate that can support the cognitive process of dreaming, a qualification that allows for neural development in childhood;
2) an intact neural substrate that can support the cognitive process of dreaming, which takes into account the loss of dreaming due to specific brain lesions;
3) an adequate level of cortical activation, which is provided by subcortical ascending pathways and crucial regions in the hypothalamus;
4) an occlusion of external stimuli, which likely involves the posterior thalamus;
5) a cognitively mature imagination system, which allows for childhood cognitive development, a necessity that is indicated by the virtual lack of dreaming in preschoolers, by the low frequency, complexity, and emotionality until ages 9-11, and its lack of adultlike content until ages 11-13;
6) a loss of conscious self-control, which may be neurologically mediated by the decoupling of the frontoparietal, dorsal attentional, and salience networks from the functional subsystems of the default network that remain activated during dreaming.

Based on these understandings, it is possible to envision a neurocognitive framework for thinking about dreams that could be tested and augmented as part of the routine practices with a sleep medicine clinic, with a focus on both group studies and unique individual cases of the kind that sometimes have greatly advanced the understanding of waking thought (e.g., Phineas Gage and H.M.).

If such a view were to be supported, it could lead to a neurocognitive approach to dreaming that includes all three of the "distinct but interdependent levels" that are the hallmarks of all theorizing in cognitive neuropsychology: the underlying neural substrate, the cognitive processes it supports, and the output or behavioral level, which means "verbal reports" in the case of dreaming (Ochsner & Kosslyn, 2014, p. 2).

The examples of possible dream studies in a sleep medicine clinic in the next two sections of this chapter assume that the primary mission of sleep medicine clinics is to help people with sleep problems, and that there can be no changes in normal clinical routines. The examples also assume that the studies have to be unobtrusive and low in cost. Moreover, these examples are not exhaustive, nor necessarily relevant to the interests of practitioners in specific sleep clinics. They are meant to stimulate ideas.

Neurocognitive Studies of Dreams in a Sleep Medicine Clinic

Based on both tantalizing findings in past studies and new possibilities that have arisen with the increase in waking neuroimaging findings that may provide new leads for the study of dreaming, this section provides examples of how participants in such studies could be located among patients that arrive to sleep medicine clinics, as well as technologies that might make these studies feasible.

The process of finding patients that might be usefully studied could begin with questions that concern any changes in their dreaming over the past year, with the goal of finding patients who may have started to have nightmares during the sleep-onset process, started to have nightmares during sleep, started to dream excessively, stopped dreaming, or who no longer see visual imagery during dreaming. In studying such patients, there are predictions, based on past lesion findings, that could be made as to the specific neural changes that might be found through neuropsychological examinations and neuroimaging studies (Bischof & Basset, 2004; Domhoff,

2018, Chapter 5, for a summary; Poza & Marti-Masso, 2006; Solms, 1997, for all past studies through the mid-1990s as well as his own original findings).

Questions concerning changes in dreaming also might be useful in locating patients who started having frequent nightmares in relation to an extreme trauma. There is some evidence, which needs further development, that patients with frequent nightmares, including those who are victims of PTSD, have abnormally activated brains during sleep; at least one of these studies implicates the threat-detection system that has key hubs in one region of the amygdala (e.g., Ebdlahad et al., 2013; Germain, Jeffrey, Salvatore, Herringa, & Mammen, 2013; Marquis, Paquette, Blanchette-Carrière, Dumel, & Nielsen, 2017; Mysliwiec et al., 2014; Simor, Horváth, Ujma, Gombos, & Bódizs, 2013). It seems likely that the normal neural substrate that supports dreaming is expanded in nightmare sufferers.

Finally, it would be very worthwhile to have a question that could find the rare few people who say they never have dreamt. A complete study of a few non-dreamers might prove to be very valuable in understanding dreaming. The potential for such studies is seen in an earlier study in a sleep medicine clinic, which began with a question included on the intake questionnaire over a five-year period (Pagel, 2003). The researcher found 16 patients that reported they had never dreamt; they repeated this claim in a follow-up telephone interview, which led to the estimate that 0.38 percent of the total patient population consisted of people who never had dreamt. (Pagel, 2003, p. 238). They were then studied during one-night visits to the clinic, in which none of them reported a dream from a total of 36 REM sleep awakenings. There were no differences between these non-dreaming patients and a control group of rare dreamers, who did recall a few dreams from REM sleep awakenings, on standard polysomnographic, clinical, and demographic variables.

However, there could be no thought at that time of carrying out neuroimaging studies because the default network had just been discovered and was barely known outside the small group that studied it in the late 1990s, according to two first-hand accounts (Buckner, 2012; Raichle, 2009). The first major publications on it occurred just after the turn of the century

(Gusnard, Akbudak, Shulman, & Raichle, 2001; Raichle et al., 2001). In addition, the possible basis of dreaming in portions of the default network was not discussed until 10 years later (Domhoff, 2011b; Pace-Schott, 2011).

Without the information that a highly active default network during waking was linked to mind-wandering and daydreaming, it was not possible to refer these patients for neuroimaging studies. Nor would it have seemed feasible at that time to ask these non-dreamers about their daytime mental imagery, which might have helped to link dreaming to a lack of waking visual spatial skills, or by an inability to generate any mental imagery. Such a possibility in the case of visual mental imagery had been noticed on the basis of very low dream recall from REM awakenings in two average-IQ preadolescent boys in a longitudinal study of children and adolescents ages 3-15 that included waking tests of visual-spatial skills (Foulkes, 1982, pp. 181, 225-226). Similar findings also were found for a few of the respondents in a study focused on adult non-dreamers, who were recruited through newspaper advertisements; however, the sample was too small to be more than intriguing and was soon lost from sight (Butler & Watson, 1985).

The range of studies of the cognitive process of dreaming and the neural substrate that supports dreaming that are possible might be widened if a sleep medicine clinic has a working relationship with units for the treatment of specific disorders, such as OCD, panic disorders, or PTSD. Perhaps just as valuable, if not more so, although not necessary, it might be useful for some clinics if they could refer sleep patients for neuropsychological assessments that might reveal focal brain injuries that would make it possible to shed new light on dream disorders, such as the loss of visual dreaming, excessive dreaming, a sudden increase in nightmares, or the complete loss of dreaming.

Neuroimaging studies of dreaming may become even more feasible with the increasing availability of the lower-cost neuroimaging technology called functional Near Infrared Spectroscopy (fNIRS), which in its portable form involves a device \ about the size of a large microwave. It readily used by easily placing a "swimming cap" on the patient's head, which is similar to an EEG cap. Not only is fNIRS less expensive to purchase ($100-200,000), it does not require the presence of a technological expert (Cui, Bray, Bryant,

Glover, & Reiss, 2011). Neuroimaging studies based on the use of fNIRS could become even less costly if clinics were in larger units that already have fNIRS machines that are only used during the day.

Beyond the cost advantages, fNIRS is in many ways superior to fMRI, even though it does not reach into subcortical regions. It is portable, tolerates motion well, and can be utilized at the person's home (e.g., Balconi & Molteni, 2016; Bulgarelli et al., 2018; Cui et al., 2011; McKendrick, Parasuraman, & Ayaz, 2015). It is used by some neurocognitive researchers to study adults while they are walking and to study the prefrontal cortex in infants while sitting in their mother's laps (Grossmann, 2013; Hamaker & Wichers, 2017). Equally important, fNIRS already has been used to detect the default network studies not focused on dreams (Durantin, Dehais, & Delorme, 2015; Harrivel, Weissman, Noll, & Peltier, 2013; Sasai et al., 2012). It has been used in studying sleep in at least one instance, but the participants were taking an afternoon nap while sitting in a chair, so if it remains to be seen how well it might work in a bedroom at night (Nguyen et al., 2018). Finally, fNIRS works readily with other measuring technologies, including high-density EEG (Pan, Borragán, & Peigneux, 2019).

Circling back to the issue of the rare few people who are non-dreamers (Pagel, 2003), it might be that an ideal study of them would involve the placement of a portable fNIRS device in their bedrooms at home, which could be further supplemented by adding a program to an iPhone that would signal researchers to call the sleepers and have them report what was going through their minds at the time of the call.

Studies of Dream Content Collected in a Sleep Medicine Clinic

With or without studies focused on dreaming as a neurocognitive process, it would be possible to do excellent studies of dream content in a sleep medicine clinic. First of all, such studies could focus on large samples of dream reports, which could be collected through the Most Recent Dream Method. In previous studies, the Most Recent Dream method (hereafter called the MRD method) involved asking students in classrooms and adults at meetings or conferences if they would voluntarily and anonymously write down the most recent dream they remember. The MRD form emphasizes the dream may have been recalled earlier that day, a few days before the request, or in the previous week or month, but it should be the most recent one they recall. To prime for recency, potential participants were asked to write down the date of the dream before they wrote out the dream report, which usually takes anywhere from five to 15 minutes with adults. Potential participants were told that they could choose not to participate; those who choose not to do so were told on the form that they could write what they wished to on the form and then not turn it in, or simply sit quietly and read. (An example of the form that is used appears in Appendix X, and a pdf of the form can be download from DreamResearch.net at https://dreams.ucsc.edu/Forms/most_recent_dreams.html).

The MRD method has been used to study normative samples of college students in Iran, Japan, South Africa, the United States, and Wales, as well as the elderly in Switzerland and animal rights activists in the United States (Domhoff, 1996, p. 67; Domhoff, Nishigawa, & Brubaker, 2004; Lewis, 2008; Malcolm-Smith & Solms, 2004; Malcolm-Smith, Solms, Turnbull, & Tredoux, 2008; Mazandarani, Aguilar-Vafaie, & Domhoff, 2013; Strauch, 2003; Tartz & Krippner, 2017). It also has proven to be useful with preadolescents as young as the fifth grade (ages 10-11) in Spain, Italy, and Greece as well as the United States (Avila-White, Schneider, & Domhoff, 1999; Crugnola, Maggiolini, Caprin, Martini, & Giudici, 2008; Karagianni, Papadopoulou, Kallini, Dadatsi, & Abatzoglou, 2013; Oberst, Charles, & Chamarro, 2005; Saline, 1999).

In addition to collecting a MRD each time a patient arrives to the clinic, it would be useful to ask patients on the intake questionnaire or during an intake interview if they have ever kept, or are keeping, a dream journal for a period of 30 or more days that contains 50 or more reports-- and if so, when and for how long. Such dream journals, rare though they indeed are, could be invaluable in pre-and-post studies of individuals who are suffered a brain injury or a psychological trauma, and for studies of the effects of new medications on dream content. (Based on surveys in classrooms and informal surveys of audiences at talks, this author estimates that about 1-3% of people from ages 16 through adulthood have kept a dream journal at least once for 30 days or more).

Dream journals are kept by a diverse set of individuals in many parts of the world for their own widely separate reasons, which very rarely have anything to do with psychotherapy or an interest in any dream theory (e.g., Bell & Hall, 1971; Domhoff, 2018, Chapter 4; Hall & Lind, 1970; Smith & Hall, 1964). In the case of dream research in a sleep medicine clinic, dream journals are ideal as baselines for pre-post studies with unique patients or patients who are about to be put on new medication. (A pilot study that provides a good example of the possibilities for such studies is summarized at the end of this section.) Even without the possibility of studying the effects of medications with a person who has kept a dream journal, there is much to be learned from any dream journal that is identified and offered for study.

The dream series contained in dream journals are a form of unobtrusive, nonreactive archival data, uninfluenced by the demand characteristics, expectancy effects, or social desirability factors that can arise as subtle confounds in experimental settings using either humans or other animals (e.g., Kihlstrom, 2002; Rosenthal & Ambady, 1995). #2113; Harari, 2016 #2618; Webb, 2000 #2639}. They are not case studies, at least in the usual clinical sense of that term, because their quantitative nature makes them directly comparable, and they can be compared with norms. In those regards, the quantitative studies of dream journals are more akin to personality tests such as the MMPI or the items used in the test associated with the Five-Factor Model.

Quantitative studies of about two dozen dream series using the HVdC (1966) coding categories have revealed that the main characters, avocations, and personal preoccupations expressed in the dream reports are consistent over months, years or decades (Domhoff, 1996, Chapter 7; 2018, Chapters 3-4). Moreover, dream series provide a good opportunity for studies examining the degree to which the conceptions and concerns found in dream reports through quantitative content analyses also might be present in waking thought. This search for continuities between dreaming and waking thought can be studied by asking participants to confirm or reject a series of inferences based on blind quantitative analyses of a dream series, in which the dream researcher knows little or nothing about the dreamer at the outset. These inferences are based on comparisons of the results with normative samples to find high or low frequencies, percentages, and ratios. This strategy is possible because studies of both group data and dream journals have shown that the frequency of the appearance of an element in dream reports is related to the intensity of a waking personal concern (Domhoff, 1996, Chapter 8; 2018, Chapters 3-4; Hall & Van de Castle, 1966, pp. 13-14).

Studies of roughly a dozen dream series based on this strategy have shown that there is usually considerable continuity between the frequency with which specific people and avocations appear in a dream series and the intensity of waking personal concerns about those people and avocations, thereby replicating the general finding that frequencies in dream series reveal the intensity in concerns in waking life (Domhoff, 1996, Chapter 8; 2018, Chapters 3-4, for summaries).

As to how to analyze the dream reports, whether they are collected through the MRD method or from individual dream journals. there are two methods. The first and oldest one is based on a time-tested and frequently used coding system, which contains 10 general categories and numerous subcategories, rests on the nominal (categorical) level of measurement, has high reliability, includes normative baseline findings for both women and men, and is available to all dream researchers on DreamResearch.net (Hall & Van de Castle, 1966). It includes a spreadsheet (DreamSAT) for analyzing the codings, and provides p values and effect sizes, and a correction formula for multiple tests of the same pair of samples. The results provided by DreamSAT are displayed in either tables or bar graphs called h-profiles. The one drawback for use in low-budget studies in an applied context is that it is labor-intensive.

The second method, and the one more likely to be used in a clinical setting, at least for preliminary studies, is based on word, phrase, and word-string searches that are facilitated by an online platform for dream researchers that is accessible at DreamBank.net (www.dreambank.net). It can be used to analyze any set of digitized dream reports, which can be stored in a password-protected private space, which is accessible only to the researchers who collected the dream reports. The findings from the searches, which appear within a space of several seconds, are automatically analyzed using programs built into the website (see Domhoff & Schneider, 2008; 2020, for accounts of the capabilities built into DreamBank.net).

DreamBank.net also contains 25,000+ dream reports in English and 5,000+ in German, which were collected in a variety of lab and non-lab settings (but never from the Internet). Specific sets or series in this large collection can be used as normative baselines for purposes of some studies. Using DreamBank.net as a starting point or model, it also might be feasible for a sleep medicine clinic, or several sleep medicine clinics, to develop a common data-analysis unit specific to their own needs and goals. Relevant software could be borrowed from DreamBank.net, and the expert who created DereamBank.net, Adam Schneider, could be consulted (http://adamschneider.net/adam.html).

The creation of word strings, whether they are generic, and can be used with any set of dream reports, or individually tailored for the study of individual dream journals, carry out all four steps of a quantitative content analysis in a matter of seconds or minutes. The word strings entered into the search program define the category's breadth and limits exactly. Since the same results are guaranteed each time the sample is searched, the categories have perfect reliability. The search engine also provides instant frequency counts. It also creates analyses in the form of percentages. The fourth step of any content analysis, a comparison with a normative or control group, can be carried out by applying any given word string to the dream reports from the women or men that were used in creating the Hall and Van de Castle (1966) normative findings, or by making use of a relevant dream series that is available on DreamBank.net. (Hereafter the Hall and Van de Castle (1966) coding system will be called HVdC).

Generic word strings concern topics that are likely to be relevant in the lives of most individuals and at the same time are described in a relatively circumscribed number of words that are commonly used by most people.

Examples include word strings for sensory terms, pet animals (e.g., dogs and cats), and religion. For the purposes of research in a sleep medicine clinic, the generic emotion word strings are far more relevant. They are comprehensive, are based on words that reflect the five emotion categories in the HVdC system, and provide reasonably similar results to those provided by a labor-intensive coding for emotions using the HVdC categories. They have been used in several studies (Domhoff, 2018, Chapters 3-4, for summaries).

A good example of the usefulness of an individually tailored word string can be found in a study of 2,022 dream reports from a college student (Domhoff, 2018, pp. 42-45). This dream series also was studied using the generic word string for emotions mentioned in the previous paragraph as well as with selected HVdC coding categories. The convergent results are summarized briefly in a chapter on dream series (Domhoff, 2018, pp. 93-95). The dream series itself can be found on DreamBank.net under the pseudonym "Kenneth," along with most of the other dream series that have been studied in the past.

More recently, a study using individually tailored word strings, created separately for two different dream journals from young women, one containing 4,329 dream reports, the other containing 664 dream reports, showed that there is both consistency and continuity in dream content from ages 13-14 to the mid-20s (Domhoff & Schneider, 2020). (These two dream series are available on DreamBank.net under the pseudonyms "Izzy" and "Jasmine").

A Content Analysis Based on Group Data from Psychiatric Patients

Two types of studies, as briefly mentioned earlier, would be possible using MRD's First, it would be possible to do studies of patients with specific diagnoses, such as panic disorder, chronic anxiety disorders, OCD, sleep-onset disorders, or PTSD. Second, sleep medicine clinics that provide medications to patients who visit regularly for check-ups or prescription renewals would be good candidates for well-controlled studies. In either case, or in the case of studies that might attempt to determine which kinds of medications work with which kinds of patients, the patients would be asked to provide an MRD upon their initial arrival to the clinic, and then to provide and MRD on each subsequent visit. If 125 patients eventually provided at least original in-take and last-visit dream reports, but hopefully with reports from intermediate visits as well, it would be possible to carry out studies of changes in dream content over time to determine the effects of the medication on dream content, and perhaps to determine if any changes are correlated with waking tests or patient ratings on a well-being scale. In particular, it might be expected that the medications would reduce aggressive interactions, increase friendly interactions, and lead to a decrease in negative emotions in dream content.

There have, in fact, been few solid studies of the dreams of patients with one or another diagnosis, as numerous review articles have shown down through the decades (Domhoff, 1996, pp. 184-189; Kramer, 2010; Skancke, Holsen, & Schredl, 2014). Nor have there been any studies leading to consensus on the effects of medications on dream content, a topic that is reviewed further in the context of the subsection on the analysis of individual dream journals.

One of the few rigorous studies of differences among various types of psychiatric patients, published 54 years ago, demonstrates what might be possible with group data collected via MRDs. It showed that the various patient groups all included less characters who were named as friends and less friendly interactions. (These two categories differ because friends do not always have friendly interactions with the dreamer, and friendly interactions can happen with family members and strangers as well as friends.) The study was based on 211 dream reports collected from 50 male patients, who were grouped into four diagnostic categories: five patients who were both schizophrenic and alcoholic, 20 patients who were schizophrenic, 15 patients who were alcoholics, and 10 patients with a variety of other diagnoses.

Using the HVdC coding system, the four sets of dream reports were compared with each other and with the HVdC male norms for characters, social interactions, success and failure, misfortune and good fortune, and eating and drinking. Surprisingly, there were very few differences among the groups or with the male norms (Hall, 1966). However, there was one potentially useful difference between the dream reports of the 50 patients as a whole and the male norms: lower percentages and ratios having to do with friends and friendly interactions. Specifically, the patients' friends percent (all friends divided by all characters), F/C index (the total number of friendly interactions divided by the total number of characters, and percentage of dreams with at least one friendly interaction were below the male norms. Since most of these differences are slightly larger for the 105 dreams reports from schizophrenics, Table 1 compares the findings from their dreams with the male norms on the six indicators related to aggression and friendliness.

A Drug Study Based on a Dream Journal

As in the case of past studies of people with different psychiatric diagnoses, which have added up to very little, so, too, the past studies of the effects of medications on dream content have been mixed at best (Armitage, Rochlen, Fitch, Trivedi, & Rush, 1995; Roth, Kramer, & Salis, 1979). One of the few systematic studies of the effects of a medication that detected any differences involved a relatively short series of dream reports from a young college woman suffering from a variety of symptoms. Since the study involves only one participant and has fewer than the 125 dream reports for each condition that would be ideal in terms of sample size, it is best viewed as a pilot study.

The student had written down 33 dream reports in her first year of college as part of the therapy she was receiving from the campus counseling center for anxiety and mild depression. After abandoning the therapy because she did not feel it was helping her, during her junior year she saw a psychiatrist and was placed on the SSRI sertraline (Zoloft), with her dosage gradually increasing from 25 to 100 milligrams before her light-headedness and panic attacks subsided. A year later, she recorded 40 dreams while still taking Zoloft for possible use in a research project in a psychology course. Based on the pre-medication and on-medication dream series, Kirschner (1999) discovered that the second set of dream reports greatly differed from the first set on several HVdC content indicators. In particular, she moved further above the norms on the friends percent and the F/C index, and closer to the norms on the A/F% (all aggressions divided by all aggressions + friendliness), the A/C index (all aggressions divided by the total number of characters), and the percentage of dreams with at least one aggressive interaction (Domhoff, 2015, pp. 75-72 and Table 6.2 for a summary of the main results; Kirschner, 1999, for the original study).

Although this particular study, and most previous studies of dream journals, have been carried out using the HVdC coding system, it is also possible to do studies using the generic work strings that work with any sample of dream reports, as discussed at the outset of this section.

Conclusion

As the overview of the scientific literature at the outset of this chapter suggests, there is now a solid foundation for future dream research because much is known about the where, when, how, and what of dreaming. However, there is even more that is not known at the same time as the number of dream researchers is dwindling, and only three or four of them have access to sleep-dream labs or neuroimaging technology. Ironically, dream researchers may have worked themselves out of government and foundation financial support by showing that dreaming is more frequent, regular, and mundane than had been imagined in the past, and therefore likely to be of major help in solving the problems of psychosis, as often claimed in the past, or in providing deep insights into people's darkest secrets, as also often asserted in the past.

Not only are there fewer dream researchers, but it may be that many future advances might be more possible in clinical settings, including sleep medicine clinics. More needs to be known about PTSD nightmares and about nightmare sufferers in general, which in both cases may involve expansions or disruptions of the neural substrate that normally supports the relatively mundane process of dreaming, with 96-99% of dreams never remembered by most people, and the few that are remembered soon forgotten.

Similarly, there are people who have trouble falling asleep, and more needs to be known about the role of the default network and/or overly vivid dreaming in creating this disorder in one or more stages of the eight-stage sleep onset process. The role of overly activated default networks and/or dreams may be a factor in one or more of the types of insomnia, a topic that has not been discussed in this chapter.

Over and beyond the role of dreaming and nightmares in various parasomnias, there is also the possibility of new general discoveries via very atypical patients, such as those who turn out of have previously undetected brain lesions, or a lack of visual imagery, or developmental irregularities at the level of cognition. And once again, there are the elusive non-dreamers, and the lure of what might be revealed about the basis of dreaming, or about its adaptive function, by studying them with comprehensive cognitive, neuropsychological, and neuroimaging assessments.

The search for the Phineas Gage or H.M. of dreaming is just beginning.

References

Amft, M., Bzdok, D., Laird, A., Fox, P., & Schilbach, L. (2015). Definition and characterization of an extended social-affective default network. *Brain Structure & Function, 220*, 1031-1049.

Andrews-Hanna, J., Reidler, J., Sepulcre, J., Poulin, R., & Buckner, R. (2010). Functional-anatomic fractionation of the brain's default network *Neuron, 65*, 550-562.

Andrews-Hanna, J., Smallwood, J., & Spreng, R. (2014). The default network and self-generated thought: component processes, dynamic control, and clinical relevance. *Annals of the New York Academy of Science, 1316*, 29-52.

Armitage, R., Rochlen, A., Fitch, T., Trivedi, M., & Rush, A. (1995). Dream recall and major depression: A preliminary report. *Dreaming, 5*, 189-198.

Avila-White, D., Schneider, A., & Domhoff, G. W. (1999). The most recent dreams of 12-13 year-old boys and girls: A methodological contribution to the study of dream content in teenagers. *Dreaming, 9*, 163-171.

Balconi, M., & Molteni, E. (2016). Past and future of near-infrared spectroscopy in studies of emotion and social neuroscience. *Journal of Cognitive Psychology, 28*, 129-146.

Bar, M. (2011). The proactive brain. In M. Bar (Ed.), *Predictions in the brain: Using our past to generate a future* (pp. 13-26). New York: Oxford University Press.

Bell, A., & Hall, C. (1971). *The personality of a child molester: An analysis of dreams*. Chicago: Aldine.

Bischof, M., & Basset, C. (2004). Total dream loss: A distinct neuropsychological dysfunction after bilateral PCA stroke. *Annals of Neurology, 56*, 583-586.

Braun, A., Balkin, T., Wesensten, N., Carson, R., Varga, M., Baldwin, P., . . . Herscovitch, P. (1997). Regional cerebral blood flow throughout the sleep-wake cycle: An (H2O)-O-15 PET study. *Brain, 120*, 1173-1197.

Buckner, R. (2012). The serendipitous discovery of the brain's default network *Neuroimage, 62*, 1137–1145.

Buckner, R., Andrews-Hanna, J., & Schacter, D. (2008). The brain's default network: Anatomy, function, and relevance to disease. *Annals of the New York Academy of Sciences, 1124*, 1-38. doi:10.1196/annals. 1440.011

Bulgarelli, C., Blasi, A., Arridge, S., Powell, S., de Klerk, C., & al., e. (2018). Dynamic causal modelling on infant fNIRS data: A validation study on a simultaneously recorded fNIRS-fMRI dataset. *Neuroimage, 175*, 413-424.

Butler, S., & Watson, R. (1985). Individual differences in memory for dreams: The role of cognitive skills. *Perceptual and Motor Skills, 53*, 841-864.

Crugnola, C., Maggiolini, A., Caprin, C., Martini, C., & Giudici, F. (2008). Dream content of 10- to 11-year-old preadolescent boys and girls. *Dreaming, 18*, 201-218.

Cui, X., Bray, S., Bryant, D., Glover, G., & Reiss, A. (2011). A quantitative comparison of NIRS and fMRI across multiple cognitive tasks. *Neuroimage, 54* 2808–2821.

Curot, J., Valton, L., Denuelle, M., Vignal, J.-P., Maillard, L., & al., e. (2018). Déjà-rêvé: Prior dreams induced by direct electrical brain stimulation. *Brain Stimulation, 11*, 875-885.

D'Argembeau, A., Jedidi, H., Balteau, E., Bahri, M., Phillips, C., & Salmon, E. (2012). Valuing one's self: Medial prefrontal involvement in epistemic and emotive investments in self-views. *Cerebral Cortex, 22*, 659--667.

Di, X., & Biswal, B. (2014). Identifying the default mode network structure using dynamic causal modeling on resting-state functional magnetic resonance imaging. *Neuroimage, 86*, 53-59.

Domhoff, G. W. (1996). *Finding meaning in dreams: A quantitative approach*. New York: Plenum.

Domhoff, G. W. (2003). Dreaming: An introduction to the science of sleep. *Science, 299*(5615), 1987-1988. Retrieved from <Go to ISI>://0001818 34200025.

Domhoff, G. W. (2011a). Dreams are embodied simulations that dramatize conceptions and concerns: The continuity hypothesis in empirical, theoretical, and historical context. *International Journal of Dream Research, 4*, 50-62.

Domhoff, G. W. (2011b). The neural substrate for dreaming: Is it a subsystem of the default network? *Consciousness and Cognition, 20*, 1163-1174.

Domhoff, G. W. (2015). Friends and friendliness: Could they be the clue in psychiatric patients' dreams? In M. Kramer & M. Glucksman (Eds.), *Dream research: Contributions to clinical practice* (pp. 67-79). New York: Routledge.

Domhoff, G. W. (2018). *The emergence of dreaming: Mind-wandering, embodied simulation, and the default network*. New York: Oxford University Press.

Domhoff, G. W. (2019). The Neurocognitive Theory of Dreams at Age 20: An Assessment and a Comparison With Four Other Theories of Dreaming. *Dreaming, 29*(4), 265-302.

Domhoff, G. W., & Fox, K. (2015). Dreaming and the default network: A review, synthesis, and counterintuitive research proposal. *Consciousness and Cognition, 33*, 342-353.

Domhoff, G. W., Nishigawa, N., & Brubaker, L. (2004). A research note on the male/female percentage in the dreams of Japanese women: A failed attempt at replication. *Dreaming, 14*, 50-53.

Domhoff, G. W., & Schneider, A. (2008). Studying dream content using the archive and search engine on DreamBank.net. *Consciousness and Cognition, 17*, 1238-1247.

Domhoff, G. W., & Schneider, A. (2018). Are Dreams Social Simulations? Or Are They Enactments of Conceptions and Personal Concerns? An Empirical and Theoretical Comparison of Two Dream Theories. *Dreaming, 28*, 1-23.

Domhoff, G. W., & Schneider, A. (2020). From Adolescence to Young Adulthood in Two Dream Series: The Consistency and Continuity of Characters and Major Personal Interests. *Dreaming, 30*(1), In Press.

Durantin, G., Dehais, F., & Delorme, A. (2015). Characterization of mind wandering using fNIRS. *Frontiers Systems Neuroscience, 26*. doi: 10. 3389/fnsys.2015.00045.

Ebdlahad, S., Nofzinger, E., James, J. A., Buysse, D., Price, J., C., & Germain, A. (2013). Comparing neural correlates of REM sleep in posttraumatic stress disorder and depression: A neuroimaging study. *Psychiatry Research, 214*, 422-428.

Eichenlaub, J. B., Bertrand, O., & Ruby, P. (2014). Brain reactivity differentiates subjects with high and low dream recall frequencies during both sleep and wakefulness. *Cerebral Cortex, 24*, 1206-1215. doi: 10.1093/cercor/bhs388.

Foulkes, D. (1982). *Children's dreams: Longitudinal Studies*. New York: Wiley.

Foulkes, D. (1985). *Dreaming: A cognitive-psychological analysis*. Hillsdale, NJ: Erlbaum.

Foulkes, D. (1999). *Children's dreaming and the development of consciousness*. Cambridge, MA: Harvard University Press.

Foulkes, D., Hollifield, M., Bradley, L., Terry, R., & Sullivan, B. (1991). Waking self-understanding, REM-dream self representation, and cognitive ability variables at ages 5–8. *Dreaming, 1*, 41-51. doi: 10. 1037/h0094316.

Foulkes, D., Meier, B., Strauch, I., Kerr, N., Bradley, L., & Hollifield, M. (1993). Linguistic phenomena and language selection in the REM dreams of German-English bilinguals. *International Journal of Psychology, 28*(6), 871-891.

Fox, K., Nijeboer, S., Solomonova, E., Domhoff, G. W., & Christoff, K. (2013). Dreaming as mind wandering: Evidence from functional neuroimaging and first-person content reports *Frontiers in Human Neuroscience, 7*(Article 412), 1-18.

Germain, A., Jeffrey, J., Salvatore, I., Herringa, R. J., & Mammen, O. (2013). A window into the invisible wound of war: Functional neuroimaging of REM sleep in returning combat veterans with PTSD. *Psychiatry Research: Neuroimaging, 211*, 176-179.

Gibbs, R. (2006). *Embodiment and cognitive science*. New York: Cambridge University Press.

Grossmann, T. (2013). The role of medial prefrontal cortex in early social cognition. *Frontiers in Human Neuroscience, 5*(7), 340.

Gusnard, D. A., Akbudak, E., Shulman, G. L., & Raichle, M. E. (2001). Medial prefrontal cortex and self-referential mental activity: Relation to a default mode of brain function. *Proceeding of the National Academy of Sciences, 98*, 4259-4264.

Hall, C. (1966). A comparison of the dreams of four groups of hospitalized mental patients with each other and with a normal population. *Journal of Nervous and Mental Diseases, 143*, 135-139.

Hall, C., & Lind, R. (1970). *Dreams, life and literature: A study of Franz Kafka*. Chapel Hill: University of North Carolina Press.

Hall, C., & Van de Castle, R. (1966). *The content analysis of dreams*. New York: Appleton-Century-Crofts.

Hamaker, E., & Wichers, M. (2017). No time like the present: Discovering the hidden dynamics in intensive longitudinal data. *Current Directions in Psychological Science, 26*, 10-15.

Harrivel, A., Weissman, D., Noll, D., & Peltier, S. (2013). Monitoring attentional state with fNIRS. *Frontiers Human Neuroscience, 13;7:861*. doi:10.3389/fnhum.2013.00861.

Heatherton, T. (2010). Building a social brain. In P. Reuter-Lorenz, K. Baynes, G. Mangun, & E. Phelps (Eds.), *The cognitive neuroscience of mind* (pp. 173-188). Cambridge: MIT Press.

Jenkins, A., & Mitchell, J. (2011). Medial prefrontal cortex subserves diverse forms of self-reflection *Social Neuroscience, 6*, 211-218.

Karagianni, M., Papadopoulou, A., Kallini, A., Dadatsi, A., & Abatzoglou, G. (2013). Dream content of Greek children and adolescents. *Dreaming, 23*, 91-96.

Kihlstrom, J. (2002). Demand characteristics in the laboratory and the clinic: Conversations and collaborations with subjects and patients. *Prevention & Treatment, 5*, http://psycnet.apa.org/index.cfmfa=buy.optionToBuy&id=2003-04137-04003.

Kirschner, N. (1999). Medication and dreams: Changes in dream content after drug treatment. *Dreaming, 9*, 195-200.

Kramer, M. (2010). Dream differences in psychiatric patients. In S. Pandi-Perumal & M. Kramer (Eds.), *Sleep and Mental Illness* (pp. 375-328). New York Cambridge University Press.

Lewis, J. (2008). Dream Reports of Animal Rights Activists. *Dreaming, 18*, 181–200.

Malcolm-Smith, S., & Solms, M. (2004). Incidence of threat in dreams: A response to Revonsuo's threat simulation theory. *Dreaming, 14*, 220-229.

Malcolm-Smith, S., Solms, M., Turnbull, O., & Tredoux, C. (2008). Threat in dreams: An adaptation? *Consciouness and Cognition, 17*, 1281-1291.

Maquet, P., Peters, J., Aerts, J., Delfiore, G., Dequerldre, C., Luxen, A., & Franck, G. (1996). Functional neuroanatomy of human rapid-eye-movement sleep and dreaming. *Nature, 383,* 163-166.

Marquis, L. P., Paquette, T., Blanchette-Carrière, C., Dumel, G., & Nielsen, T. (2017). REM sleep theta changes in frequent nightmare recallers. *Sleep, 40*, 1-12.

Mazandarani, A. A., Aguilar-Vafaie, M. E., & Domhoff, G. W. (2013). Content analysis of Iranian college students' dreams: Comparison with American data. *Dreaming, 23*, 163–174.

McKendrick, R., Parasuraman, R., & Ayaz, H. (2015). Wearable functional near infrared spectroscopy (fNIRS) and transcranial direct current stimulation (tDCS): Expanding vistas for neurocognitive augmentation. *Frontiers in Systems Neuroscience, 9*(March 9).

Meier, B. (1993). Speech and thinking in dreams. In C. Cavallero & D. Foulkes (Eds.), *Dreaming as cognition* (pp. 58-76). New York: Harvester Wheatsheaf.

Moulton, S. T., & Kosslyn, S. (2011). Imagining predictions: Mental imagery as mental emulation. In M. Bar (Ed.), *Predictions in the brain: Using our past to generate a future* (pp. 95-106). New York: Oxford University Press.

Mysliwiec, V., O'Reilly, B., Polchinski, J., Kwon, H., Germain, A., & Roth, B. J. (2014). Trauma associated sleep disorder: A proposed parasomnia encompassing disruptive nocturnal behaviors, nightmares, and REM without atonia in trauma survivors. *Journal of Clinical Sleep Medicine, 10*, 1143-1148.

Nguyen, T., Babawale, O., Kim, T., Jo, H., Liu, H., & Kim, J. (2018). Exploring brain functional connectivity in rest and sleep states: a fNIRS study. *Science Reports, 8*, 16144-xx? doi: 10.1038/s41598-018-33439-2.

Nofzinger, E., Mintun, M., Wiseman, M., Kupfer, D., & Moore, R. (1997). Forebrain activation in REM sleep: An FDG PET study. *Brain Research, 770*, 192-201.

Oberst, U., Charles, C., & Chamarro, A. (2005). Influence of gender and age in aggressive dream content in Spanish children and adolescents. *Dreaming, 15*, 170-177.

Ochsner, K., & Kosslyn, S. (2014). Cognitive neuroscience: Where are we now? In K. Ochsner & S. Kosslyn (Eds.), *The Oxford Handbook of Cognitive Neuroscience* (Vol. 2, pp. 1-7). New York: Oxford University Press.

Pace-Schott, E. (2011). The neurobiology of dreaming. In M. Kryger, T. Roth, & W. Dement (Eds.), *Principles and Practices of Sleep Medicine* (5th ed., pp. 563-575). Philadelphia: Elsevier.

Pagel, J. F. (2003). Non-dreamers. *Sleep Medicine, 4*, 235-241.

Pan, Y., Borragán, G., & Peigneux, P. (2019). Applications of Functional Near-Infrared Spectroscopy in Fatigue, Sleep Deprivation, and Social Cognition. *Brain Topography, Published Online October 29, 2019*. doi: https://doi.org/10.1007/s10548-019-00740-w.

Philippi, C. L., Duff, M. C., Denburg, N. L., Tranel, D., & Rudrauf, D. S. (2012). Medial PFC damage abolishes the self-reference effect. *Journal of Cognitive Neuroscience, 24*, 475-481.

Poza, J., & Marti-Masso, J. (2006). Total dream loss secondary to left temporo-occipital brain injury. *Neurologia, 21*, 152-154.

Raichle, M. E. (2009). A paradigm shift in functional brain imaging. *The Journal of Neuroscience, 29*, 12729-12734.

Raichle, M. E., MacLeod, A. M., Snyder, A. Z., Powers, W. J., Gusnard, D. A., & Shulman, G. L. (2001). A default mode of brain function. *Proceeding of the National Academy of Sciences, 98*, 676–682.

Rosenthal, R., & Ambady, N. (1995). Experimenter effects. In A. Manstead & M. Hewstone (Eds.), *Encyclopedia of Social Psychology* (pp. 230-235). Oxford: Blackwell.

Roth, T., Kramer, M., & Salis, P. (1979). Drugs, REM sleep, and dreams. In B. Wolman (Ed.), *Handbook of dreams* (pp. 203-225). New York: Van Norstrand Reinhold.

Saline, S. (1999). The most recent dreams of children ages 8-11. *Dreaming, 9*, 173-181.

Sasai, S., Homae, F., Watanabe, H., Sasaki, A., Tanabe, H., Sadato, N., & Taga, G. (2012). A NIRS-fMRI study of resting state network. *Neuroimage*, 179-193. doi: 10.1016/j.neuroimage.2012.06.011.

Schacter, D., Addis, D., & Buckner, R. (2008). Episodic simulation of future events: Concepts, data, and applications. *Annals of the New York Academy of Sciences, 1124*, 39-60. doi:10.1196/annals.1440.001.

Simor, P., Horváth, K., Ujma, P. P., Gombos, F., & Bódizs, R. (2013). Fluctuations between sleep and wakefulness: Wake-like features indicated by increased EEG alpha power during different sleep stages in nightmare disorder. *Biological Psychology, 94*, 592-600.

Skancke, J., Holsen, I., & Schredl, M. (2014). Continuity between waking life and dreams of psychiatric patients: A review and discussion of the implications for dream research. *International Journal of Dream Research, 7*, 39-53.

Smith, M., & Hall, C. S. (1964). An investigation of regression in a long dream series. *Journal of Gerontology, 19*, 66-71.

Snyder, F. (1970). The phenomenology of dreaming. In L. Madow & L. Snow (Eds.), *The psychodynamic implications of the physiological studies on dreams* (pp. 124-151). Springfield, IL: Thomas.

Solms, M. (1997). *The neuropsychology of dreams: A clinico-anatomical study*. Hillsdale, NJ: Erlbaum.

Strauch, I. (2003). Träume im Alter [Dreams in old age]. In B. Boothe & B. Ugolini (Eds.), *Lebenshorizont Alter* (pp. 171-187). Zurich: Hochschulverlag AG an der ETH Zürich.

Tartz, R., & Krippner, S. (2017). Cognitive differences in dream content between Japanese males and females using quantitative content analysis. *Dreaming, 27*, 193-205. doi: 10.1037/drm0000054.

In: Parasomnia Dreaming
Editor: James Frederic Pagel

ISBN: 978-1-53617-815-9

Chapter 10

Parasomnias in English Literature: Sleep and Dream Phenomena in Macbeth, A Christmas Carol, Alice in Wonderland, and Harry Potter

Kelly Bulkeley*[*], *PhD
Director, The Sleep and Dream Database

Abstract

This chapter shows how literature can be a valuable ally in scientific investigations of sleep and dreaming. Works of literary art can represent aspects of nocturnal experience in ways that enhance and deepen scientific understanding. The chapter suggests four questions to ask of any work of literature in relation to parasomnias: Biographical; what do we know about the life and background of the author? Literary; what precise aspects of parasomnias are described? Historical/scientific; how do the parasomnias in the story relate to the science of the author's time and culture? And philosophical; in what ways can the whole work be considered a parasomnia? The application of these questions to four popular works of

[*] Corresponding Author's Email: bulkeleyk@gmail.com.

English literature—*Macbeth* (1606) by William Shakespeare, *A Christmas Carol* (1843) by Charles Dickens, *Alice in Wonderland* (1865) and *Through the Looking Glass* (1871) by Lewis Carroll, and the seven *Harry Potter* novels (1997-2007) by J.K. Rowling—will provide illuminating examples of the methodological challenges involved in this kind of study, and the rewards that come from expanding the scope of scientific research into the phenomenology of sleep and dreaming.

INTRODUCTION

Novels, plays, and other works of literature might not seem to be a promising source of knowledge about the nature of sleep and dreaming. These texts portray characters, settings, and events that are explicitly framed as fictional, as untrue. They are fanciful, made up, unreal. They stand at the opposite end of the spectrum from scientific research, which deals with objective truth and external physical reality. For serious scientists, a reference to parasomnias in literature might be useful as a rhetorical flourish, as an epigraph perhaps, but nothing more than that.

This is how it appears from a purely rationalist perspective. The present chapter will argue that such a perspective is based on false assumptions that limit the scope and range of valid empirical inquiry. The chapter will show how literature can be a valuable ally in scientific investigations of sleep and dreaming. The same basic argument can be applied to other art forms, too. Poetry, film, theater, sculpture, music, dance, video games, and graphic novels all have similar potentials as collaborative resources in sleep and dream research. The focus here will be on one specific cultural tradition of literature, a selection of English fantasy writings from the 17th century to the present. This tradition has been chosen not because it is better or more advanced than other traditions, but because it is likely to be familiar and accessible to the English-speaking readers of this book. Many of the same theoretical points made in this chapter could also be made using a study of other literary traditions from other places and times. The goal here is to lay out the fundamental questions that tend to arise in *any* encounter between literature and the science of sleep and dreaming. The application of these questions to four popular works of English literature—*Macbeth* (1606) by William Shakespeare, *A Christmas Carol* (1843) by Charles Dickens, *Alice in Wonderland* (1865) and *Through the Looking Glass* (1871) by Lewis Carroll, and the seven *Harry Potter* novels (1997-2007) by J.K. Rowling—will provide illuminating examples of the methodological challenges involved in this kind of study, and the rewards that come from expanding the scope of scientific research into the phenomenology of sleep and dreaming.

PARASOMNIAS

What are parasomnias? In *The Limits of Dream* (2010), Pagel defines parasomnias as follows:

> "[U]ndesirable physical events or experiences that occur during entry into sleep, within sleep, or during arousals from sleep. These experiences can include sleep-related movements, autonomic motor system functioning, behaviors, perceptions, emotions, and dreaming. Parasomnias become clinical diagnoses when they result in sleep disruption, nocturnal injuries, waking psychosocial effects, or adverse health effects." (120)

This definition covers a wide range of phenomena involving complex interactions of mind, brain, and body. The key feature that unites these phenomena is their "undesirability." They have some kind of negative, disturbing, or potentially harmful impact on the individual. The worse the impact, the more likely they become the subject of clinical diagnosis and treatment.

Researchers in this area may not always appreciate how strange and eerie the world of parasomnias appears to non-specialists. It can be awful to experience or witness a night terror, a REM behavior disorder episode, or a PTSD nightmare. The fact that scientists have explained some of the underlying features of various parasomnias makes them no less a source of wonder and mystery. If anything, it makes them more so. We now know these phenomena are real, not just bizarre legends. They are rooted in the natural somatic processes of the sleep cycle, and they are shaped at a fundamental level by psychological and cultural forces. Some parasomnias are actively dangerous and symptomatic of illness, but others are fairly benign in their effects, despite the short-term emotional distress they cause. This makes them a curious and perhaps unique feature of human life—a spontaneous, universally accessible portal to highly altered states of consciousness. It should come as no surprise that literary artists take great interest in such phenomena.

How to Approach Parasomnias in Literature

The following outline is meant to guide readers interested in the specific topic of parasomnias. It is not meant as a general guide to literature, nor as a statement in favor of, or against, any particular theory of literary criticism. Hopefully the discussion here will be relevant and helpful to anyone interested in literary arts, but it is most directly addressed to readers who already have some familiarity with the science of sleep and dreaming, such as the material presented in other chapters of this book.

Below are four kinds of questions that can be asked of a work of literature in relation to parasomnias.

Biographical

What do we know about the life and personal background of the author? Has the author experienced any parasomnias? If yes, did those experiences influence the writing process? In the discussions below, this question does not play a big role. But with some authors (e.g., Robert Louis Stevenson, Philip K. Dick), it can be very important to know how personal parasomnias interact with literary parasomnias.

Literary

Which precise aspects of the parasomnia are described? What details are and are not included in the description? How does the parasomnia fit into the plot? In what ways does it serve as a literary device contributing to the story as a whole? In what genre has this work been written, and how are parasomnias typically presented in other works in that genre? These are vital questions of textual analysis that cannot be passed over lightly. Readers should resist the temptation to extract short passages from a literary text and use it as evidence in support of a contemporary scientific argument or theory, without taking into account how that passage makes sense in its original narrative context.

Historical/Scientific

How do the parasomnias in the story relate to the science of the author's time and culture? What was known in that historical era about sleep and dreams, and how familiar was the author and audience likely to be with that knowledge? How do the parasomnias in the text relate to the science of our time, in the present? There is no reason to avoid using contemporary scientific language and terminology to discuss parasomnias in an older literary text, as long as the analysis remains a suggested interpretation rather than a definitive diagnosis. The deeper questions here revolve around literature as a potential harbinger of new insights in the science of sleep and dreaming. Where do we find literary descriptions that accurately anticipated later scientific discoveries? Have works of literature portrayed types of parasomnia that contemporary researchers have not yet validated, but perhaps will one day in the future? Is it possible that some literary descriptions are ahead of their scientific time?

Philosophical

In what ways can the whole work be considered a parasomnia? (*Alice in Wonderland* is one example, as discussed below.) Are there parasomnia-like effects of the work on the reader? What does the work tell us about the difference between real dreams and fictional dreams? Between dreaming and waking? Some readers may assume these questions are too speculative to have scientific relevance, but for many works of literature, including several discussed here, such questions are intrinsic to the nature of dreaming itself. Literary artists recognize, perhaps more than scientific researchers do, how parasomnia experiences can have a dramatic impact on people, potentially changing their attitude, behavior, and outlook on life. This can include changes in their ultimate existential beliefs and spiritual views. Writers have long known and appreciated this role of parasomnias in human life, and in their stories they help us understand how these strange experiences of sleep and dreaming can become surprisingly powerful agents of transformation.

These questions will now be used to guide the following discussions of four specific literary texts, as examples of how this model can be applied in practice.

MACBETH

Most likely performed first in 1606, *Macbeth* was written towards the end of the literary career of William Shakespeare (1564-1616). It is considered one of his greatest tragedies, on par with *Hamlet, Othello,* and *King Lear,* and it continues to be performed regularly in theaters around the world. The tremendous popularity of this play over more than four centuries makes it all the more significant that *Macbeth* is virtually a case study in sleep disorders. Or rather, *two* case studies—both Macbeth and his wife Lady Macbeth suffer terrible distress in the realm of sleep and dreaming, and Shakespeare renders their experiences with uncanny poetic power and horrifying grace.

Biographical: Very little is known about the details of Shakespeare's life and background. There is no information about his experiences with sleep and dreaming, nor about any other aspect of his personal beliefs and private behaviors. A few biographical scraps do exist and can be helpful in exploring his works, as noted below.

Literary: Macbeth's parasomnias are interwoven with his uncanny encounters with the "weird sisters," the three witches whose prophecies about his becoming King light the spark of his violent ambition. When he kills the current King in his sleep, Macbeth laments,

> "Methought I heard a voice cry 'Sleep no more, Macbeth does murder sleep'—the innocent sleep, sleep that knits up the raveled sleave of care, the death of each day's life, sore labor's bath, balm of hurt minds, great nature's second course, chief nourisher in life's feast."

This speech highlights the unnatural horror of the act of killing the King while he sleeps as a guest in Macbeth's castle. From this moment forward, Macbeth no longer has restful sleep himself, and when he does sleep, he is tormented by nightmares. Worse still, his bad dreams seem to spill into his waking mind, as he experiences guilt-ridden auditory and visual hallucinations he can neither control nor distinguish from actual waking life.

The worst sufferer of nocturnal distress, however, is Macbeth's wife. In a famous scene (act five, scene one), Lady Macbeth goes on an extended sojourn of sleepwalking, with bouts of guilty sleep-talking that reflect her unbearable suffering over their regicidal crime. The scene is narrated by a servant woman and a doctor, which allows Shakespeare to present a highly detailed account of Lady Macbeth's behavior. The ensuing conversation follows the lead of a medical authority seeking to diagnose a patient. The doctor's conclusion is grim. He sees Lady Macbeth's condition as symptomatic of a "great perturbation of nature," and he worries the ill it bodes is beyond his scope of healing power:

> "Foul whisperings are abroad. Unnatural deeds do breed unnatural troubles. Infected minds to their deaf pillows will discharge their secrets. More needs she the divine than the physician."

After watching her mimic the actions of repeatedly washing her hands, anxiously talking about oceans of blood, the doctor tells the servant woman to remove any objects Lady Macbeth might use to harm herself. He leaves in shock and dismay: "So good night. My mind she has mated [bewildered], and amazed my sight."

Historical/Scientific: The fact that a doctor is present to observe Lady Macbeth's sleepwalking indicates that Shakespeare was familiar with the idea that such behavior would be considered a worrisome sign of pathology, requiring expert consultation. The doctor in the scene represents the current state of medical science, and he asks all the right questions as he watches Lady Macbeth engage in complex actions and monologues, with her eyes wide open yet deeply asleep. By the end of the scene the doctor admits he is incapable of relieving her distress. There is no medical cure for her suffering. Her parasomnias reflect the agony of a soul that has violated its own nature.

Shakespeare's early education suggests he had a deep familiarity with Greek and Roman mythology, primarily through the rendering of the first-century CE Roman poet Ovid in his poem *Metamorphosis*. He was also familiar with the Bible and the Christian faith, although it remains unclear whether he was a Protestant, a secret Catholic, or not religious at all. In both classical and Christian traditions, dreams are believed to have the power of revealing moral sins that people try to hide in waking life. In several of Shakespeare's history plays (e.g., *Henry V*, *Richard III*), Kings have trouble sleeping because of the relentlessly brutal challenges of being a ruler, a phenomenon best summed up in the line, "uneasy lies the head that wears the crown," from *Henry IV part 2*. In the context of Shakespeare's other plays, Macbeth appears as the most violent King suffering the most violent sleep problems.

Seen in the light of contemporary sleep and dream science, *Macbeth* appears as an accurate portrayal of many of the most agonizing forms of parasomnia. Macbeth descends into a dark vortex of insomnia, nightmares, hallucinations, and, in the end, delusions of grandeur and magical invulnerability that mark his complete loss of moral judgment and mental acuity. His wistful comments about the loss of "innocent sleep" ring as true today as they did four hundred years ago. As for Lady Macbeth, her episode of sleepwalking and sleep talking is much longer and more coherent than what present-day clinicians typically see in their patients. These qualities enable her parasomnia to serve as a very effective literary device. Yet her experience is not beyond the pale of possibility in the real world. Indeed, it reflects the emotional extremities of someone utterly consumed by guilt, fear, and despair. Waking-life stress is widely acknowledged as a key factor in many parasomnias, and Lady Macbeth and her husband have brought upon themselves an overwhelming black wave of emotional negativity.

Philosophical: The parasomnias in *Macbeth* are interwoven with themes of fate, guilt, divine retribution, and moral punishment. The play was written during the reign (1603-1625) of King James, who took a keen and punitive interest in witchcraft, demonology, and the occult. Macbeth's dark journey begins with a chance encounter with three witches, and their prophecies about his future lure him into his fateful course of action. In that sense, the story is a cautionary tale about the dangers of demonic sources of knowledge, a moral that King James would surely find praise-worthy. But the philosophical questions go deeper than that, to the nature and scope of free will. If prophecies and parasomnias reveal visions of the future, are they visions of inevitable fate or do they point to possibilities that humans can influence, either avoiding a negative future or acting to gain a positive future? At the end of the play, Macbeth realizes he has fatally misinterpreted the witches' prophecies, and the audience realizes his misinterpretations were motivated by his overweening ambition and myopic selfishness. When Macbeth enters into realms of altered states of consciousness, he does not open himself to the autonomous energies he encounters. Instead, he tries to conquer them, to force them to serve his will. The violent ethos of a warrior might succeed on the battlefield, but it fails miserably as an approach to exploring and understanding the psyche.

A CHRISTMAS CAROL

First published on December 19, 1843, *A Christmas Carol* was one of several short stories that Charles Dickens (1812-1870) wrote during his life about the Christmas holiday. His major novels (*A Tale of Two Cities, Hard Times, Great Expectations, David Copperfield*) have naturally received the most critical attention, but *A Christmas Carol* has a claim to be one of Dickens' most psychologically profound works.

Biographical: Much is known about Dickens' early life and personal development, more than can be covered here. Suffice it to say that when he was twelve years old he had to give up school and take an exhausting job at a factory to support his family, after his father was put in debtor's prison and the family fell into poverty and destitution. The conditions Dickens suffered and saw others suffering during this time made a lasting impression on him. He channeled these searing memories and feelings into his writings. Dickens was just 31 years old when he wrote *A Christmas Carol*, and his own childhood experiences clearly shaped his portrayal of the scenes of poor families trying to survive the harsh cold of winter without enough food, clothing, or warmth.

Literary: Compared to Dickens' other great novels, *A Christmas Carol* is by far his dreamiest and most surreal story, and the only one with supernatural elements at the center of the plot. Many types of parasomnia elements appear during the course of the story. After an unsettling encounter with a door knocker and the ghost of his former partner, miserly old Ebenezer Scrooge encounters three nocturnal spirits. He has almost no volitional control in these encounters; he is startled, scared, and deeply disoriented. The spirits force him to revisit painful memories of childhood trauma and long-buried pain. They compel him to witness the suffering of others in circumstances much worse than his own. They make him contemplate the prospect of his own pitiful death, and the lack of care or concern that others would show at his passing. The spirits bring out feelings and thoughts that he did not know he had—compassion, humility, remorse, charity.

During this long nightmarish adventure, Scrooge finds that time and space do not obey their usual laws of behavior. Everything becomes fluid and porous. Once he finally awakens and realizes it is Christmas morning, Scrooge bursts out with ecstatic joy and good cheer, a newborn spirit dancing in the street, with a kind word for everyone he passes. The central narrative event in this story is a night-long parasomnia, which serves as the agent for transforming Scrooge into a more empathetic and morally self-reflective person. Although often treated as a children's tale, *A Christmas Carol* has more in common with a classic horror story. The disturbing sleep and dream elements help create an atmosphere of inescapable eeriness and dread. During his terrifying nightmare journey, the "old" Scrooge dies; the end of the parasomnia is the beginning of Scrooge's new life.

Historical/scientific: If Dickens was familiar with the early 19th century scientific literature on sleep and dreams, he shows little sign of it. The primary context for understanding the parasomnias in *A Christmas Carol* is the Bible, which contains several references to dreams of prophecy and moral warning. Given the Christian holiday setting of Dickens' story, this would likely be the most familiar framework for the audience to make sense of Scrooge's experiences. There is one significant moment, however, when Scrooge tries to reject the reality of his former partner Jacob Marley's ghost by invoking a materialist, scientific-type of explanation. The ghost asks why he doubts his own senses, and Scrooge replies:

> "Because a little thing affects them. A slight disorder of the stomach makes them cheats. You may be an undigested bit of beef, a blot of mustard, a crumb of cheese, a fragment of an underdone potato. There's more of gravy than of the grave about you, whatever you are!"

This is indeed a common type of explanation among scientists, both in the 19th century and in the present: parasomnias can be explained as purely physiological phenomena, caused by specific physical processes in the body. Any meanings we attach to such phenomena are merely projections of our own fears and wishes. Marley's ghost forces Scrooge to realize that is wrong, and that his mind has become as woefully closed as his purse. The psychological reality of what Scrooge is experiencing cannot be denied, as much as he wants to dismiss it as foolish nonsense. What emerges here is the idea that materialist explanations of parasomnias can have an emotionally defensive quality, protecting the ego against unsettling feelings and disturbing insights.

As for correspondences with the modern science of sleep and dreams, many elements of *A Christmas Carol* can be cited as having plausible realism. The emotional qualities of his experience—fear, confusion, vulnerability, helplessness—are common features of nightmares, sleep paralysis, and several other parasomnias. The distortions of time and space are also prevalent in disturbing dreams. The primary themes in the content of Scrooge's experience also fit well with current knowledge about dreams in relation to traumatic memories, social simulations, future anticipations, and concerns about death. Although fiction, Dickens' story accurately describes first-person details from the kinds of parasomnias that people actually do experience.

Philosophical: As already noted, the central section of the story is one long parasomnia. In the first part of the story, Scrooge appears as a cold, monstrously selfish skinflint. At the end, he has become cheerful, forgiving, generous, and warm-hearted. This remarkable transformation of personality (the religious term "conversion" would not be out of place) is attributed to the parasomnia he experiences on Christmas Eve, seven years to the night after his partner Marley died. At the level of literary analysis, this is a clear instance of a parasomnia as a plot device to move the narrative along. A philosophical analysis pushes the question further, asking what exactly happens within the parasomnia to account for the transformation. Setting aside the supernatural elements, the distinguishing feature of Scrooge's experience in the parasomnia is the *uselessness* of all his waking skills and strengths. He must develop and rely on *other* aspects of his personality to navigate his way through this surreal realm. In that process he is radically changed; moral and spiritual sentiments long dormant within him suddenly spring to life. Once he returns home and awakens, those newly developed capacities carry over into his waking world. The transformative impulse that began in his dream is now fulfilled in waking reality. Scrooge's conversion is not just a rejection of dehumanizing capitalist economics, but also of dehumanizing Enlightenment models that divide mind, heart, and spirit.

ALICE IN WONDERLAND AND THROUGH THE LOOKING GLASS

These two stories, published in 1865 and 1871, have become beloved classics of children's literature. In print, stage, and film versions, the adventures of young Alice have delighted the young and old for many generations. This is all the more remarkable since at the core of both stories are dark, violent, deeply unsettling parasomnias.

Biographical: "Lewis Carroll" was the pseudonym of Charles Dodgson, a brilliant but awkward professor of mathematics at Christ College in Oxford. Legend has it that he first composed the story on a lazy summer day while rowing on a river with a group of friends and children, including 10-year old Alice Liddell, for whom he named the heroine. However, all biographical accounts of the origins of these stories are necessarily speculative, because after Dodgson's death his family removed and presumably destroyed all of his diary entries for the years prior to his composition of the first Alice story. This creates a central mystery in his life, apparently never to be solved and yet tantalizing in its potential significance (why else would his family have destroyed the diary entries, if they did not contain material at odds with his public persona?).

A few rays of insight may be gleaned from another aspect of his biography that also includes missing diary pages. After a fairly idyllic childhood in the English countryside, playing games and making up stories with his ten siblings, Charles was sent at the age of 14 to an all-male boarding school. He stayed there for three years and apparently hated every minute of it, though his diary entries from that time were later inexplicably removed from his personal papers. In letters written years later, he recounted that nights at the school were the worst, because that was when the older boys would sexually terrorize the younger boys, with the awareness and tacit approval of the teachers. His experiences at this school left Charles with a lifelong sense of the cruel brutality hidden beneath the civilized veneer of British society—brutality that came out most clearly at night, in the shadows, during sleep.

Literary: Alice falls asleep, has a long dream, then wakes up. That's the basic plot of both stories, which thus invite readers to regard everything that happens within them as aspects of dreaming, though it is not always clear who is dreaming and who is being dreamed (more on that below). In the first story, Alice enters a world governed by animate playing cards; in the second, she traverses a world arranged like a giant chessboard. Dodgson was an endlessly playful and inventive person who enjoyed creating puzzles, riddles, and intricate paradoxes of mathematical and logical reasoning, and he wove many of these intellectual amusements into the Alice stories, along with surprisingly radical ideas about language, morality, politics, and the monarchy. Dense, multi-faceted layers of literary context await anyone who makes a serious scholarly exploration of these stories. An argument can be made, however, that these multiple layers of significance *enhance* the essential dreaminess of the stories, rather than diluting or diminishing it. Whereas in other texts a dream might serve as a literary device, here in the Alice stories the literary devices seem to serve the overarching shape and texture of the dream.

Historical/Scientific: Charles Dodgson possessed numerous books on current science and medical practice in his library, so he was likely familiar with ideas about the physiological causation of dreams and the various pathologies of sleep. He was also raised by a Protestant minister father, meaning that Christian and Biblical teachings about dreams were likely part of his background, too. Yet almost none of this enters into the two stories. Wonderland exists in a place very far from Christendom. The teachings of Jesus have no purchase here, and the endless array of extremely talkative, magically endowed animals make it clear this is not a realm governed by Enlightenment rationality, either. Dodgson/Carroll guides his young readers into a world free from the normal constraints of reason, morality, and civilized behavior—in other words, into the world of dreaming. Like Alice, readers must learn how to navigate through this strangely delightful yet unpredictably threatening reality. As noted above, any explanations about authorial intentions with this story must remain speculative. But it certainly seems clear from the literary and historical context that Dodgson/Carroll meant his two Alice stories to be read and appreciated *as dreams*, as extended sojourns into the free creativity of the sleeping imagination.

To the extent that notion is true, the stories offer a wealth of opportunities for contemporary researchers to compare features of Alice's adventures to what is currently known about the phenomenology of sleep and dreaming. Examples include the bizarre, surreal, otherworldly atmosphere of Wonderland; the recurrent themes of falling, flying, and bodily transformation; the plethora of aggressive magical animals; the endless puns, slippages of language, and wordplay; the sudden shifts in setting and emotional tonality; numerous elements of "day residue" from her waking life; the loss of declarative memories combined with a heightened metacognitive self-awareness; the frightening vulnerability and loss of agency; the prominence of confusion, uncertainty, and imbalance. Unfortunately, most of the scholarship on the Alice stories ignores these features of dreaming experience, treating them as literary devices only, with no further psychological significance. There is a strand of scholarship that has applied Freudian psychoanalysis to the stories, but this approach sees the dream elements chiefly as symptoms of the unconscious conflicts that presumably motivated Dodgson/Carroll's creative efforts. For present-day readers who are interested in parasomnias, the traditional treatments of the Alice stories leave much of interest unexplored.

Philosophical: Throughout both stories, Alice struggles with the uncertainty of her existential state of being. Indeed, this seems to be the theme that encompasses all the others. The farther into Wonderland she descends, the more she realizes she is no longer in the realm of her ordinary waking life. Where *is* she, then? The challenge becomes most acute when she encounters the Red King, leaning against a tree, sound asleep and snoring loudly. Tweedledee and Tweedledum see that she has noticed him, and they lure her into a profound metaphysical conundrum:

> "He's dreaming now," said Tweedledee; "and what do you think he's dreaming about?"
>
> Alice said, "Nobody can guess that."
>
> "Why, about *you*!" Tweedledee exclaimed, clapping his hands triumphantly. "And if he left off dreaming about you, where do you suppose you would be?"
>
> "Where I am now, of course," said Alice.
>
> "Not you!" Tweedledee retorted, contemptuously. "You'd be nowhere. Why you're only a sort of thing in his dream!"
>
> "If that there King was to wake," added Tweedledum, "you'd go out—bang!—just like a candle!"

The brothers Tweedle have brought Alice to the startling realization that, just as she dreams other characters, perhaps those other characters are dreaming about her. To make it clear that this is the key theme in the stories, the very last paragraph of the second and concluding tale return to this scene, as Alice sits in her easy chair with her cat and reflects on what just happened:

> "Now, Kitty, let's consider who it was that dreamed it all. This is a serious question, my dear, and you should *not* go on licking your paws like that… You see, Kitty, it *must* have been either me or the Red King. He was part of my dream, of course—but then I was part of his dream, too! *Was* it the Red King, Kitty?"

The story ends here, not with a tidy moral lesson as was typical with Victorian-era children's literature, but with a question addressed directly to the reader: "Which do *you* think it was?" There could not be a more explicit invitation to all readers, children and adults, to consider the personal significance of these philosophical issues in their own waking and dreaming lives.

The Harry Potter Novels

The seven novels in the *Harry Potter* series by J. K. Rowling were published between 1997 and 2007. Public excitement and fan mania grew with the arrival of each new title, as a vast audience of young readers followed the exploits of Harry and his friends at the Hogwarts School for Witchcraft and Wizardry. The stories are delightfully inventive, amusing, witty, full of intrigue and adventure—perfect for the children and early teens who are the primary audience for the books. But the passionate response of so many readers, including increasingly large numbers of adults, derived from the profound psychological journey Harry takes through the course of the seven books. Rowling presents Harry's life and development with remarkably insightful attention to the long, painful process of forging one's own identity and character. As with the other three stories considered so far, the *Harry Potter* novels take Harry and his friends (and the readers) on a harrowing journey filled with evil, violence, betrayal, and death. Significant dreams and sleep phenomena play key roles at various points in the stories, in ways both familiar and surprising.

Biographical: Joanne Rowling was born in 1965 in very modest circumstances, and for several years after college she struggled with finances, the end of a marriage, and single-parenting. She only managed to find a publisher for her first novel, *Harry Potter and the Philosopher's Stone*, after many years of effort and multiple rejections. Her rise to literary success derives from the power of her stories to connect with the creative imaginations of her young readers. No specific elements of her biographical information seem directly relevant to the parasomnias presented in the stories, but future research and interviews with Rowling might lead to new insights.

Literary: The first time we meet eleven-year old Harry, he is waking up in the tiny bed under the stairway in the home of his mean-spirited aunt and uncle, the Dursleys. Harry "rolled onto his back and tried to remember the dream he had been having. It had been a good one. There had been a flying motorcycle in it. He had a funny feeling he'd had the same dream before." When Harry mentions the dream during a family car ride, his uncle Dursley, who "didn't approve of imagination," angrily tells him that motorcycles don't fly. Harry ashamedly says he knows they don't, and he admits "it was only a dream." Yet Harry and his readers soon learn that his recurrent dream is *not* just a dream, but rather a true, meaningful, and reassuring reminder of his true origins. Ten years earlier a flying motorcycle did in fact bring the orphan Harry to the Dursley's house. The tension between Harry's uncanny dreams and the stubborn ignorance of the Dursleys foreshadows a major theme across all seven books: the challenge of developing Harry's magical powers despite the anxious determination of non-magical people, or "Muggles," to deny that such powers exist.

Significant dreams occur to Harry at the beginning of the first, second, and fourth novels in the series. In each case the dream anticipates the difficulties and obstacles Harry must overcome in that story. The dreams thus serve quite effectively as literary devices. Yet each dream also conveys meaningful psychological insight into Harry's character, especially as he comes to the realization that his life's mission is to battle and defeat the evil Lord Voldemort. The farther he descends into the darkness of this challenge, the more nightmarish his dreams become. The fifth novel, *Harry Potter and the Order of the Phoenix*, the longest and most psychologically complex in the whole series, opens with Harry suffering terrible nightmares following the murder of a fellow student by Voldemort at the end of the previous novel. Throughout this novel Harry has a vivid, profoundly alluring recurrent dream of a long, dark corridor. He doesn't tell other people about these dreams out of fear they will think he is crazy, but that turns out to be a terrible mistake: his dreaming mind has become a direct portal with the mind of Voldemort himself, and the dark Lord is using that secret dream connection to manipulate Harry into a fatal trap.

Historical/Scientific: As contemporary works, these novels were written during the present era of scientific knowledge of sleep and dreaming. They incorporate many elements of this knowledge into the stories. Rowling's portrayals of Harry's dream experiences can be read and interpreted at many levels of psychological significance. References abound to symbolism in Freudian (broomsticks, snakes, magic wands) and Jungian (tricksters, shadows, heroic journeys) registers. The stories also include detailed accounts of how trauma impacts dreaming. The increasing number of graphic, repetitive nightmares that plague Harry through the books offer a realistic reflection of the horrible experiences of violence and death he has suffered at every turn. The term "post-traumatic stress disorder" is never used in the novels, but it certainly applies to Harry's mental state in the latter half of the series.

Several of his dreams are more mundane and relate to ordinary concerns about school, friends, dating, and Quidditch (a wizarding sport). But other dreams offer shockingly accurate clairvoyant visions of events happening at a distance, either in the present or in the future. In this regard, dreams appear in the novels not simply as psychological mirrors of the self, but as potentially objective means of perception, knowledge, and even communication. Harry's mentor, Professor Dumbledore, tries to warn him against letting Voldemort into his dreams, but Harry is intensely curious to explore this aspect of his mind, which seems to hold so much magical possibility. The association of dreaming with imagination, made so clearly in the first scene of the first book, supports the basic premise of many contemporary psychological models of dreaming: dreams can open paths of insight and understanding beyond the limits of ordinary reason and waking reality.

Philosophical: The dreams in the Harry Potter novels appear within the conventional narrative framework of Harry's time as a student at Hogwarts. Thus, the overall story is not presented as a parasomnia, as for example in *Alice in Wonderland.* Still, readers of Rowling's works are left with profound metaphysical questions about the nature and significance of dreaming. Dreams in these tales can be trivial, wishful, confusing, and forgettable—yet they can also reveal vital truths and insights that go far beyond what the rational mind is capable of grasping. So what *are* dreams, then? The Harry Potter novels do not give a definitive answer, beyond a strong affirmation of their seemingly infinite variability, what the psychologist Harry Hunt memorably called "the multiplicity of dreams." In this fantastic realm of dueling wizards and witches, dreaming emerges as perhaps the best way for Muggles to get a glimpse of what it is like to be magical.

Conclusion

Readers interested in further study of the relationship between dreaming and literature will benefit from a familiarity with several important scholarly works on the topic. These include Carol Schreier Rupprecht's *The Dream and the Text: Essays on Language and Literature* (1993), which offers chapters looking at the sleep and dream references in Shakespeare, Russian literature, and writings from the Italian renaissance; Jane White-Lewis' fascinating essay on her experiences teaching dreams and literature to the underprivileged students of an inner-city high school (1994); Bert States' *Dreams and Storytelling* (1993), which delves deeply into dreaming and literary theory; Patricia Kilroe's thought-provoking article "The Dream as Text, the Dream as Narrative" (2000), in which she probes the nature of figurative language in dreams and literature; and Bernard Welt's brilliant chapter in *Dreaming in the Classroom* (2011) on teaching dreams in courses on writing and comparative literature.

The parasomnias in each of the literary works considered here could be explored, analyzed, and interpreted in much more detail than has been possible given the limited space of this chapter. The goal, however, has not been to conduct an exhaustive exegesis of these texts, but rather to show readers interested in sleep and dream phenomena how to approach a literary work, and how to address the key issues and questions—biographical, literary, historical/scientific, and philosophical—that inevitably arise in such an encounter between art and science. The possibilities for creative interdisciplinary research in this area are endless, limited only by the imaginations of the investigators.

References

Bulkeley, Kelly. 2017. *An Introduction to the Psychology of Dreaming*. Second edition. Westport: Praeger.

Carroll, Lewis. 2009. *Alice in Wonderland* and *Through the Looking Glass.* New York: Macmillan Collector's Library.

Dickens, Charles. 1984. *A Christmas Carol.* New York: Signet Classics.

Kilroe, Patricia. 2000. "The Dream as Text, the Dream as Narrative." *Dreaming* 10 (3): 125-137.

King, Philip, Kelly Bulkeley, and Bernard Welt. 2011. *Dreaming in the Classroom.* Albany: State University of New York Press.

Rowling, J. K. 1997-2007. *Harry Potter*. New York: Scholastic Books.

Rupprecht, Carol Schreier, ed. 1994. *The Dream and the Text: Essays on Language and Literature.* Albany: State University of New York Press.

Shakespeare, William. 1959. *Macbeth.* New York: Washington Square Press.

States, Bert. 1993. *Dreaming and Storytelling.* Ithaca: Cornell University Press.

White-Lewis, Jane. 1994. "Dreams and Social Responsibility: Teaching a Dream Course in the Inner-City." In *Among All These Dreamers: Essays on Dreaming and Modern Society*, edited by Kelly Bulkeley, 3-12. Albany: State University of New York Press.

In: Parasomnia Dreaming
Editor: James Frederic Pagel
ISBN: 978-1-53617-815-9

Chapter 11

USING PARASOMNIAS: THE CREATIVE UTILIZATION OF FRIGHTENING DREAMS BY SUCCESSFUL ARTISTS

J. F. Pagel*, MS, MD
Associate Clinical Professor, University of Colorado School of Medicine, Director Rocky Mt. Sleep Disorders Center, Pueblo, Colorado, US

ABSTRACT

Many artists use frightening dreams (most often described as nightmares) in their work. This small study (N = 12) is based on a series of in-depth interviews with such artists, including their report of any experience of major trauma, an assessment as to any past and current symptoms of PTSD, as well as reports of frightening dream content and phenomenology.

A majority (8/12) of these successful artists had histories of significant trauma. One created nightmare inspired art during a period of grieving and had no ongoing symptoms consistent with PTSD. Seven artists (58% of the

* Corresponding Author's E-mail: pueo34@earrthlink.net.

total sample) reported past experiences of major trauma associated with the onset of their nightmares, as well as symptoms consistent with the diagnosis of PTSD.

Several reported that their best creative work took place after their experiences of trauma. Of the four individuals reporting no major past experience of trauma, two reported life-long recurrent classic nightmares, one reported nightmare content with sleep paralysis, and one reported dream-associated experiences of sleep walking that had occurred during childhood.

USING PARASOMNIAS - THE CREATIVE UTILIZATION OF FRIGHTENING DREAMS BY SUCCESSFUL ARTISTS

Many artists use mentation and images experienced during sleep (dreams) in their creative process. Artists ranging from Goya, to Fuseli, Picasso and Salvador Dali have reported using dreams with negative emotion and frightening content in their work (Van de Castle 1994). This tendency to utilize such dream content persists among a subset of modern artists.

In Sundance Lab studies addressing dream use by filmmakers, significantly elevated levels of nightmare recall and use were reported when compared to that reported from clinical sleep and medical practice populations.

Among this population, the creative grouping of actors, directors, and screenwriters had a significantly higher frequency of reported nightmares than those in the professional and crew filmmaking categories (Pagel Kwiatkowski and Broyles 1999). In another study, sleep laboratory subjects who reported having a creative outlet reported a higher incidence of nightmares than those reporting no creative interest (Pagel and Kwiatkowski 2003).

These studies suggest that for some individuals, the report of experiencing nightmares could be a signifier or a marker for interest and involvement in the process of creativity.

STUDY DESIGN

This small study includes a series of twelve in-depth interviews with successful artists who have produced work from their frightening dreams. Interviews were directed in order to obtain information as to frightening dream content, age of onset, any description of associated trauma, and whether the individual reported any symptoms or functional incapacity consistent with the diagnosis of posttraumatic stress disorder (PTSD). Due to the small group size, this is primarily a descriptive study. This study attempts to determine whether or not these successful artists producing nightmare inspired art have histories of significant trauma and/or symptoms of PTSD. The attempt is also made to determine whether the reported frightening dreams have the typical phenomenology of parasomnas known to be associated with frightening dreams: the nightmares of nightmare disorders, nightmares associated with significant trauma (PTSD), night terrors, hypnogogic hallucinations, and sleep paralysis.

NIGHTMARES AS A SYMPTOM OF POST-TRAUMATIC STRESS DISORDER

Seventy percent of the sleep lab population report having a nightmare more than once a month (Pagel and Kwiatkowski 2003; Pagel and Shocknasse 2007; Pagel 2008; Pagel 2010) (Table 1).

The classic nightmare is a dream that features agonizing dread, a sense of oppression, and the conviction of helpless paralysis (Jones, 1923). Today, this classic definition for the nightmare has been both expanded and contracted, to indicate that a nightmare is a disturbing mental experience that often results in awakening (Zadra and Donderi 2000; Pagel and Neilsen 2005).

Classic nightmares almost always occurs in REM sleep (patients with PTSD in exception) (Ross et al. 1989).

Table 1. Reported recall frequency by questionnaire from three sleep laboratory studies (1999-2003)

Reported Recall	Never	Monthly	Weekly	3 X/Week	Nightly
Dreams	9.6%	27.7%	42.2%	16.5%	3.48%
Nightmares	29.4%	34.7%	26.8%	7.57%	1.34%

N = 1150, Mean Age 49.6, Gender 1.56 (F = 1:M = 2).
Pagel and Kwiatkowski 2003: Pagel and Shocknasse 2007; Pagel 2010.

The most consistent characteristic of REM sleep dreams such as nightmares is their greater report length than the dreams of other sleep stages (Pagel 2014). Other typical nightmare characteristics include the presence of intense negative emotions, bizarreness, contextual imagery, and salience.

There are also frightening dreams reported from other stages of sleep with differing characteristics than those associated with the typical nightmare: the hypnogogic hallucinations of sleep onset (stage 1), the panic attacks of stage 2, and the night terrors, confusional arousals and somambulism (sleep-walking) of deep sleep (stage 3) (Pagel 2014). The typical sleep-stage phenomenology of these different forms of frightening dreams is summarized in Table 2.

Distressing and recurrent nightmares are the most commonly reported symptom of PTSD (Singareddy and Balon 2002). PTSD symptoms are reported by up to 25% of individuals experiencing major physical or psychological trauma (Santiago et al. 2013). Many individuals who experience irreconcilable physical or psychological trauma will have symptoms of PTSD that persist throughout their lives, with many never returning to the level of functioning that they achieved before their experience of trauma (Chopra et al. 2015). It has been suggested that due to diminished functional capacity, individuals with PTSD may be less likely than others to function well in their creative pursuits (Frish et al. 1992).

Table 2. The typical phenomenology of frightening dreams classified as to sleep stage associaiton

	Recall Frequency	Report Length	Narrative structure	Emotion	Memory	Thought	Lucidity	Salience	Bizarreness Hobson (Ho) Hunt (Hu) *	Awakening Disturbance
Sleep Onset Stage 1	> 80%	short	coherent stories	Intense anxiety (PTSD, Paralysis, sleep starts)	Visual and Day Reflective	(High) Perceptual, Memory and time awareness, Affective, Subjective (Low) Analytical, Executive, Intuitive, Operational	at least 17% of lucid dreams	High Hypnogogic Hallucinations and Sleep Paralysis	Ho (low except for discontinuities) Hu (high) Hallucinations Confusion emotion and bizarre personifications	Apparently real - hallucinations , sleep paralysis
Stage 2	40 - 60%, variable across night	short		associated anxiety and panic, autonomic behaviors	Day reflective	Low rationality except for: Perceptual and Operational thought	Possible	Low	Ho (Low) Hu (Low)	Sometimes cannot be distinguished from waking thoughts

Table 2. The typical phenomenology of frightening dreams classified as to sleep stage association (Continued)

	Recall Frequency	Report Length	Narrative structure	Emotion	Memory	Thought	Lucidity	Salience	Bizarreness Hobson (Ho) Hunt (Hu) *	Awakening Disturbance
Deep Sleep Stage 3	40 - 60%	short	fragmented	night terrors: frantic attempts to escape a perceived threat	associated confusion and disorientation	Low rationality except for: affective thought	not described	High	Ho (low except for: Uncertainties (explicit vagueness) Hu (Low) Hallucinations (High) clouding, confusion emotion and bizarre personification	automatic behaviors, extreme confusion, autonomic discharge, scream
REM Sleep	> 80%	long	coherent with detailed plot,	disturbing and intense (nightmare s)	Reality based (PTSD)	Rational: Analytic, Perceptual	classic association	High (Nightmares)	Ho (High) discontinuities, incongruities uncertainties	Agitated awakening,

	Recall Frequency	Report Length	Narrative structure	Emotion	Memory	Thought	Lucidity	Salience	Bizarreness Hobson (Ho) Hunt (Hu) *	Awakening Disturbance
REM Sleep			character or actions			Time awareness Affective, Executive, Subjective and Intuitive.			Hu (Low) hallucinations confusion (High) uncanny emotions	Apodictic (nightmare), sleep paralysis

* Bizarreness: (Hu) denotes H. Hunt Classification (Hunt 1989): (Ho) denotes A. Hobson Narrative Classification (Hobson et. al. 1987).

Sigmund Freud in his seminal work suggested that the psychological origin of creativity was based in trauma, often the commonly experienced trauma of separation and/or family relationship transition (Freud 1916). Freud indicated that the creative process could be therapeutic. The artist; "possesses the mysterious power of shaping some particular material until it becomes a faithful image of his phantasy". So that the traumatized artist can potentially achieve; "...honour, power, and the love of women" (Freud 1908).

In the sleep laboratory population, seventy percent of individuals report at least one nightmare per month (Pagel and Kwiatkowski, 2003: Pagel and Shocknasse 2007; Pagel 2010) (Table 1). A much lower percentage (7.5%) report a nightmare frequency of more than three times a week, but even among these individuals reporting frequent and recurrent nightmares there are many who have not had a previous experience of major trauma. If these nightmares produce sleep disruption, and waking symptoms such as fatigue and difficulty focusing, non-traumatized nightmare sufferers will fit in to the diagnostic classification of nightmare disorder (Pagel and Neilsen 2005). Many of these individuals have the personality characteristic of "thin-borders" as developed by Ernest Hartmann. These individuals, while more likely than others to have some psychiatric disorders such as schizotypical personality, are also more likely to invest themselves in the creative experience, to produce creative products, and to lead creative roles in society. Individuals reporting frequent nightmares often test as unusually imaginative (Hartmann 1984; 2011).

In creative writing classes, students with the most bizarre and fantastic dreams are those most likely to produce superior work (Schecter, Schmeidler and Staal 1965).

METHODOLOGY

The term "successful" as applied to the artists included in this study indicates that these individuals have, at some point in their careers, derived primary income based on their artistic role. The older subjects included in this study had been acknowledged with multiple honors for their creative output (one filmmaker at the academy award level). All were currently working in their creative field at the time of interview (2010 - 2015). The interviews conducted by the author, were transcribed and returned to each interviewee for correction and signed release. Three of the interviewees chose to submit their interview under a pseudonym. Art and or written work based on frightening dreams were submitted by ten of the participants.

Examples of that work have been published in a previous article (Pagel 2015).

Each subject was asked to describe a frightening dream that they had incorporated into their work, noting their age at the onset of that experience, and report whether they had experienced a significant trauma that they associated with their nightmare. If a major experience of trauma was reported, that trauma was described. The individual was then queried as to the severity of that experience and associated symptoms: whether the trauma had produced significant impairment in their ability to function, distressing re-experiencing the traumatic event, avoidance of reminders of the event, negative mood, depression or associated suicidal ideation, and symptoms of arousal and hyper-vigilance (APA -DSM-V 2013). Demographics: Gender - 9 female, 3 males. Mean age at the time of interview - 60.

Eight of the interviewees are visual artists (2 photographers, 7 painters), two are screenwriters, one a playwright, and one a curator. At the time of interview, all but one, continued to use their frightening dream experiences in their work.

TRAUMA

A majority of this grouping (8/12; 66.6%) reported a history of the experience major trauma. Most had experienced this trauma during adolescence. Three of the four women experienced the loss of a parent or sibling. Two of the three men had experienced their trauma during war-associated combat. The mean age of trauma experience was younger for the women respondents (20.8 years of age) than for the men (31.2 years of age).

ASD (Acute Stress Disorder)

One individual had created nightmare inspired art during the six month long period of grieving after caring for her husband during his difficult death from cancer. At the time of interview, her nightmares were occurring at a lower frequency, and she reported no ongoing symptoms consistent with PTSD. Nightmare description:

> I have these nightmares about being in the sea. I wake and my nails have cut grooves into my hands. Something bad is coming! The wave is standing still and huge above me –the tsunami dream - It's a ghastly presence. It's a Hokusai wave, dark and sometimes transparent. At its crest, the curl ruffles and sometimes trails tendrils of fabric. When it's bad, it's dark blue or black. Darkness is the threat, because you can't see. When you're not here, the day will go on, but you won't see. It's like a still life that changes as the light changes through the day. But you can't paint it at night. There's no color. Artists need light – the intimacy of life. Light always transforms the object of your regard. Light is your symbol. Darkness is your extinction. A year ago, I had this dream. The wave was huge and black and curled, standing above us. This wave was a threat, and I knew it was the recurrence of my husband's cancer. This grotesque death stood there shaking the very air around us. Very grim – and I had to stand there with it. I had to live through it (Jan Haworth).

PTSD (Posttraumatic Stress Disorder)

Seven of the twelve subjects with a history of major trauma (58% of the total sample) had symptoms, as based on interviews, consistent with the diagnosis PTSD. All reported trauma associated functional impairment, as well as symptoms in all four of the PTSD diagnostic clusters. All had at some point been treated with antidepressant medication, however none were using such medications at the time of interview. Three did report regularly requiring hypnotic medications to assist them in getting to sleep. Most had used techniques (primarily imagery therapy and lucid dream control) designed to suppress their nightmares. Some now viewed both medication and behavioral therapies as interfering with achieving the emotional intensity required for their work. All had periodically stopped PTSD medication; reasons provided: cost, and drug side effects interfering with clarity and rapidity of thought or diminishing their levels of enthusiasm, emotion and energy.

This group had experienced significant trauma associated stress. Six volunteered descriptions of suicidal ideation and two reported previous suicide attempts. Three experienced nightmares that clearly included re-experience of their trauma.

While these nightmares were not an exact replay of the experience (the dreams include metaphoric memories and associated emotional intrusions), each closely parodied the actual experience of trauma. Example re-experiencing nightmare:

> Every time I try to sleep, the event plays over and over in my mind. I have this dream that started the next night after the crash, a dream that I had every night for the next six months. I'm in the cockpit of an airplane. There are all these wires. I see the wings out the window and they drop away. The pilot leaves the cockpit. He's just no longer there, leaving me to take over flying the plane. There's a flash of light that startles me awake, and I wake, very frightened (Doug Smoot).

Four of these individuals had nightmares that while not re-experiences of the traumatic experience, clearly included trauma associated metaphors. The most severely traumatized (2/7), were able to describe only minimal nightmare content during interviews beyond versions of their nightmare that they had been able to present in their art. Example metaphor of trauma nightmare:

> I'm a toothpick in the dream, moving through a tube – a tunnel, coated with sticky lack, tar-like goo. It doesn't' stick to me. There's a noise – it's a drone, nah – nah, that comes in a wave, up and down. I'm moving towards a light at the end of the tunnel where there's a little pink flower (something like a periwinkle with 4 petals and a hole in the middle). I look down and

> my feet swell. They get big and they get small. And when I get to the flower, I wake up. It's horrible, frightening. And the whole next day the noise is there, droning on and on. Eventually I began to have a different nightmare. In the dream, everyone in my family is on a roller coaster (something that I hate in waking life). My dad and I are in the first car, and as we go up and down, everyone is screaming and yelling. And each time when we come to the level part, the platform at the start, there would be one more car of my family dead. By the end of the dream, everyone in my family is dead except for me and my dad (Anonymous).

Six of these artists reported that their most notable work was accomplished after their experience of trauma, with that work either based on or in reaction to the experience of trauma or the associated nightmares. Despite the intensity of their traumatic experience, for 4 of 7 (56%) it was after their trauma that they went on to achieve preeminence in their chosen artistic field.

Frightening Dreams in Successful Artists without a History of Major Trauma

Four of the twelve subjects (33% of the total sample) reported no history of trauma beyond childhood social conflicts, separations, and expected bereavements.

The nightmares reported by this group were phenomenologically more diverse than in the trauma grouping. Two individuals reported classic nightmares, one a nightmare associated with sleep paralysis, and one a recurrent frightening dream (limited to childhood) associated with awakening confusion and somnambulism.

Nightmare Disorder

For two of these individuals without a history of major trauma, frightening dreams began in childhood and continued into adult life.

Reported dreams were long and involved with multiple storylines. They were often repetitive, and included nightmare typical content including experiences of being trapped, being engulfed by a tidal wave, and experiencing a crucifixion (Schredl and Goritz 2018).

Example nightmare:

> There is a huge metal ship in the harbor – somewhat like the Titanic in scale. I'm with my father, a small child somewhere between the age of three and five, and we are meant to board, but running late. He takes my hand and we run to catch the ship that is already pulling away from the dock heading out to sea. The passage through which the ship is moving is lined with iron rods, narrow like rebar, protruding out from the concrete lining the passage. My father holds my hand as we leap from narrow rod to narrow rod, trying to reach the ship that is moving away. The surf is crashing below, rising higher and higher. I leap for a rod and miss, and lose the grip of my father's hand. I fall – terribly frightened, and I wake (Joan Tewkesbury).

Sleep Paralysis

One artist reported recurrently experiencing an intense and frightening experience of paralysis on awakening from her frightening dream. She reported a life-long history of nightmares, however, the nightmare association with paralysis did not begin until in her thirties after the birth of her daughter. While sleep paralysis like nightmares is most commonly associated with REM sleep, her nightmare report is short in length, has an intense visual focus, and extreme emotion set within a contextual image - a pattern of phenomenology fitting better with the other sleep stage (sleep onset or stage 1) in which sleep paralysis sometimes occurs (Pagel 2014). Described frightening dream:

> The usual nightmare is the image of a tidal wave – an immense wall that moves forward behind buildings. The scene is very realistic, colored in the various shades of gray of Paris where she lives. She describes beauty in the gigantic size of the vision that extends, "my soul beyond human

being." She wakes from this dream bewildered, paralyzed, and unable to physically move (Valerie Eguchi).

Night Terror/Somnambulism

One individual reported a frightening dream that fit best into the phenomenology of a deep sleep parasomnia (Table 2). This frightening dream, not associated with significant trauma, occurred repetitively, if somewhat changed, throughout early childhood. The subject, a playwright, reported no history of any such a dream experience occurring after adolescence. The dream content is limited, and the associated emotion profound. The associated sleep behaviors include confusion on waking, autonomic discharge (nausea and loss of balance), and sleepwalking (waking in a different room on a different floor of the house). The dream itself was of great importance to the subject, serving as a basis for his award-winning playhouse version of Kafka's *Trial*, in which the audience, divided into small groups, enters and interacts with a rotating and constantly changing set.

Described frightening dream experience:

> I am standing on a street in Chicago – across the street from the Audubon Hotel where a revolving door leads from the street into the building. I cross the street with difficulty, pulling behind me a ball and chain that's attached to my ankle. There is a doorman and he bows to me as I enter the door, that spins and rotates and dumps me back out on the street again. The scenario repeats over and over again, until I wake in a fright.
>
> This dream happened every night, and sometimes when I woke from the dream I would find that I was in a different room, once even on a different floor from where I had gone to bed. I started to try and make plans before bed of how I wanted to change the dream. I started to realize that if I stopped in the dream, the dream stopped at that point. My motion was driving the dream. I decided to look down the street to my left.

At the edge of the street was a circular void that stuck out into the street. In the wall of that void were a series of compartments. A different version of me in different situations at different times in my life was in each compartment and as I watched the scenes in each compartment were moving from one into another (Bob Benedetti).

Discussion

Despite being small and descriptive, this study is sufficient to answer several questions as to the use of the imagery from frightening dreams by successful artists.

(1) Even among this small grouping, the nightmare/ frightening dream experience utilized in their works was diverse, fitting into different diagnostic and parasomnia categories.

(2) Eight of twelve subjects had experienced major and significant trauma, with seven of these individuals having symptoms consistent with PTSD. Most of these subjects (6/8; 75%) reported that their best and most acclaimed work occurred after their trauma. This indicates that significant trauma and symptoms of PTSD do not preclude, and may actually contribute to the artistic process for some successful artists.

(3) One third of this group (4/12) had no history of significant trauma. Two had typical recurrent nightmares consistent with the diagnosis of nightmare disorder, one described a nightmare associated with the parasomnia - sleep paralysis, while the other described frightening dreams consistent with a the experience of a childhood arousal parasomnia (somnambulism). These phenomenologically very different experiences were usefully incorporated by these successful artists into their creative output.

(4) All of these successful artists used frightening dream content based on parasomnias in their art. As based on typical phenomenology, two artists incorporated frightening dreams from parasomnias other

> than nightmares (sleep paralysis, and sleep walking) into their work, indicating that dream-associated content from multiple types of parasomnia can be creatively utilized by some artists.

During this series of interviews, several of the artists who had experienced symptoms of PTSD indicated that they felt that their creation of nightmare-based art had been quite useful in helping them in their attempts to address what was apparently irreconcilable trauma. Some of those interviewed suggested that their approach of creative immersion allowed them to project their experiences into an artistic rendering outside of themselves. They noted that such creation could sometimes be transformative, incorporating the experience of trauma and suffering into transcendent art. This art was sometimes of extraordinary quality, sometimes at a level even higher than possible before the experience of trauma. For these artists creative endeavor provided an alternative potential for realignment after trauma, a realignment that improved their level of functioning as artists in measurable terms, both in terms of quality of life, as well as in the quality of their artistic product.

CONCLUSION

In this small exploratory study, 2/3 of the successful artists who incorporated frightening dreams into their work reported a previous experience of significant trauma. Seven of those traumatized (58% of the total grouping) had symptoms and functional impairment consistent with the diagnosis of PTSD. This finding strongly suggests that neither significant trauma nor PTSD preclude the capacity for an individual to become a successful artist. It seems clear that some artists can utilize the frightening dreams associated with traumatic experience to create high quality art.

Other artists who have frightening dreams without a history of significant trauma also successfully utilize such dreams in their work. Some incorporate the frightening dreams of nightmares into their work. In one case that nightmare also included the experience of sleep paralysis - a parasomnia associated with REM sleep and sleep onset (Stage 1) sleep. One individual incorporated a "nightmare" into his work that had the characteristic phenomenology of somnambulism or sleep-walking (a deep sleep [stage 3] parasomnia). This small study indicates that artists experiencing a diverse spectrum of parasomnias from different stages of sleep can successfully incorporate the associated mentation into the creation of some of their best work.

Parasomnias are generally described as unwanted behaviors occurring during sleep (APA -DSM V 2013). However, the artists included in this study are successfully utilizing cognitive aspects of these parasomnia experiencces in their work.

It is a medical and therapeutic tendency to treat a diagnosis such as a parasomnia. This is particularly true when a therapy or medication is available that can be used to eliminate the "illness" (in this case, frightening dream experiences). A variety of effective therapeutic options can be used to suppress nightmares and the frightening dreams associated with other parasomnias. In such a case, treatment may be recommended even when there is no clear indication that the symptom (e.g., the experience of the parasomnia) leads to a decline in performance or waking function.

These individuals, this group of successful artists, have chosen to avoid the suppression of their nightmares, attempting to integrate their parasomnia experiences into their lives and work. They have chosen to use their parasomnia-based dreams: the frightening dreams of nightmare, the overwhelming powerlessness of sleep paralysis, and even the strange and inordinate discombobulation of an arousal disorder. These artists have discovered their own paths to creative and social / financial success by using this approach.

REFERENCES

American Psychiatric Association: Diagnostic and Statistical Manual of Mental Disorders, Fifth Edition. (2013) Arlington, VA, American Psychiatric Association.

Chopra, M., Zhang, H., Pless-Kaiser, A., Moye, J., Llorerte, M., Oslin, D., Spiro, A. (2015). PTSD is a chronic, fluctuating disorder affecting the mental quality of life in older adults. *Am. J. Geriatr. Psychiatry,* 22: 86 - 97.

Freud, S. (1908). Creative Writers and Daydreaming, in *Introductory Lectures on Psychoanalysis,* The Standard Edition (Lecture XXIII – The Paths to the Formation of Symptoms. (1966). J. Strachey (Ed. and Trans.). New York, W. W. Norton and Co. p. 375.

Freud, S. (1916/1951). *Beyond the Pleasure Principle*, Vol. 18, The Standard Edition. (Trans. and Ed.) J. Strachey, London, Hogarth.

Frisch, M. B., Cornell, J., Villanueva, M., Retzlaff, P. J. Clinical validation of the Quality of Life Inventory. A measure of life satisfaction for use in treatment planning and outcome assessment. *Psychol. Assess.,* 1992; 4(1):92 - 101. http://dx.doi.org/10.1037/1040-3590.4.1.92.

Hartmann, E. (1984). *The Nightmare*, New York. Basic Books.

Hartmann, E. (2011). *The Nature and Functions of Dreaming*, Oxford UK, Oxford University Press, p. 7.

Hobson, J., Hoffman, S., Hefland, R.. Kostner, D. (1987). Dream bizarreness and the activation-synthesis hypothesis. *Human Neurobiology,* 6:157 - 164.

Hunt, H. (1989). *The Multiplicity of Dreams*. New Haven, CT. Yale University Press pp. 223 - 4.

Jones, E. (1923). *On the Nightmare*, London, The International Psycho-analytic Library, p. 20.

Pagel, J. F., Kwiatkowski, C., Broyles, K. (1999). Dream Use in Film Making, *Dreaming,* 9(4);247 - 296.

Pagel, J. F., Kwiatkowski, C. F. (2003). Creativity and dreaming: Correlation of reported dream incorporation into awake behavior with level and type of creative interest. *Creativity Research Journal,* 15(2 and 3); 199 - 205.

Pagel, J. F., Nielsen, T. (2005). Parasomnias: Recurrent Nightmares – *The International Classification of Sleep Disorders – Diagnostic and Coding Manual* (ICD-10): American Academy of Sleep Medicine.

Pagel, J. F., Shocknasse, S. (2007). Dreaming and insomnia: Polysomnographic correlates of reported dream recall frequency. *Dreaming,* 17(3):140 - 151.

Pagel, J. F. (2008). *The Limits of Dream – A Scientific Exploration of the Mind/Brain Interface*, Academic Press (Elsiever) Oxford.

Pagel, J. F. (2010), The Nightmares of Sleep Apnea: Nightmare Frequency Declines with Increasing Apnea Hypopnea Index (AHI), *Journal of Clinical Sleep Medicine,* 6(1);69 - 74.

Pagel, J. (2014). *Dream Science*: *Exploring the Forms of Consciousness*, Oxford UK, Academic Press.

Pagel, J. (2015). *The Creative Nightmare*, Dreamtime Spring, 2015.

Ross, R., Ball, W., Sullivan, K., Caroff, S. (1989). Dream disturbance as the hallmark of post-traumatic stress disorder, *Am. J. Psychiatry,* 146:697 - 707.

Santiago, P., Ursano, R., Gray, C., Pynoos, R. S., Spiegel, D., Lewis-Fernandez, R., Friedman, M. J., Fullerton, C. S. (2013). A systematic review of PTSD prevalence and trajectories in DSM-5 defined trauma exposed populations: intentional and non-intentional traumatic events. *PLOS One,* 8; e59236.

Schecter, N., Schmeidler, G., Staal, M. (1965). Dream reports and creative tendencies in students in the arts, sciences and engineering. *J. Consulting Psychology,* 29:415 - 21.

Schredl, M., Goriz, A. Z. (2018). Nightmare themes: An online study of most recent nightmares and childhood nightmares. *JCSM,* 15(9).

Singareddy, R. K., Balon, R. Sleep in posttraumatic stress disorder. *Ann. Clin. Psychiatry,* 2002; 14(3):183 - 90.

Van de Castle, B. (1994). *Our Dreaming Mind.* New York. Ballantine Books.

Zadra, A., Donderi, D. (2000). Nightmares and bad dreams: Their prevalence and relationship to well-being. *J. Abnormal Psychology,* 109:273 - 81.

ABOUT THE EDITOR

J. F. Pagel MS, MD
Associate Clinical Professor
University of Colorado School of Medicine

Dr. Pagel is an Associate Clinical Professor at the University of Colorado School of Medicine and Director of Rocky Mt. Sleep. His books include The Limits of Dream – A Scientific Exploration of the Mind / Brain Interface (2008), Dream Science: Exploring the Forms of Consciousness (2014), and Machine Dreaming and Consciousness (2017) co-written with Philip Kirshtein. He edited Dreaming and Nightmares (2010) and is the co-editor of the major Sleep Medicine textbook: Primary Care Sleep Disorders A Practical Guide (2007, 2014). He has chaired the Education committee for the American Academy of Sleep Medicine and served on the Sleep Medicine Test and Policy Committee for the American Board of Internal Medicine. He has authored 40 scientific papers, 26 general and review articles, and 45 book chapters on different aspects of sleep and dreams.

INDEX

A

B

C

F

G

H

I

J

K

L

M

N

O

P

R

S

T

U

V

W

Z